Studies in Universal Logic

This series is devoted to the universal approach to logic and the development of a general theory of logics. It covers topics such as global set-ups for fundamental theorems of logic and frameworks for the study of logics, in particular logical matrices, Kripke structures, combination of logics, categorical logic, abstract proof theory, consequence operators, and algebraic logic. It includes also books with historical and philosophical discussions about the nature and scope of logic. Three types of books will appear in the series: graduate textbooks, research monographs, and volumes with contributed papers.

More information about this series at http://www.springer.com/series/7391

Francine F. Abeles • Mark E. Fuller
Editors

Modern Logic 1850-1950, East and West

 Birkhäuser

Editors
Francine F. Abeles
Departments of Mathematics and Computer
  Science
Kean University
Union, NJ
USA

Mark E. Fuller
University of Wisconsin
Janesville, WI
USA

ISSN 2297-0282   ISSN 2297-0290   (electronic)
Studies in Universal Logic
ISBN 978-3-319-24754-0   ISBN 978-3-319-24756-4   (eBook)
DOI 10.1007/978-3-319-24756-4

Library of Congress Control Number: 2016939218

Mathematics Subject Classification (2010): 01A55, 01A60, 03A05, 03F03, 03G25

This book is published under the trade name Birkhäuser
The registered company is Springer International Publishing AG Switzerland
(www.birkhauser-science.com)

Contents

Contributors

Francine F. Abeles is Professor Emerita at Kean University in Union, NJ, USA. For the ten years beforehand, she was Distinguished Professor of Mathematics and Computer Science and head of the graduate programs (master's level) in mathematics, computing, statistics, and mathematics education. She has co-edited a proceedings of the Canadian Society for The History and Philosophy of Mathematics, and edited three volumes in the pamphlets of Lewis Carroll series for the University Press of Virginia. Dr. Abeles wrote a chapter on Charles L. Dodgson's work in voting theory in *Lewis Carroll Observed*, a chapter on his mathematical work in *yours very sincerely C. L. Dodgson (alias "Lewis Carroll")*, a chapter on the popularizers of mathematics Martin Gardner, Richard Proctor, and Charles Dodgson in *A Bouquet for The Gardner*, and a chapter on Dodgson's work in logic in volume 6 of *Logica Universalis*, "Toward a Visual Proof System: Lewis Carroll's Method of Trees". She is the author of nearly one hundred papers in journals on topics in geometry, number theory, voting theory, linear algebra, logic, and their history, most recently: "Nineteenth Century British Logic on Hypotheticals, Conditionals, and Implication" (*History and Philosophy of Logic* **35** (1), 2014, 24-37) and "Chiò's and Dodgson's Determinantal Identities" (*Linear Algebra and Its Applications* **454**, 2014, 130-137). Dr. Abeles was a member of the editorial board and Reviews Editor of the international journal, *Modern Logic* and its successor, *The Review of Modern Logic*. Currently she is a member of the Publications Committee of The Lewis Carroll Society of North America and managing trustee of the Morton N. Cohen Publications Trust. A recipient of a grant from the American Philosophical Society and numerous faculty research grants from the State of New Jersey, she received an award from Kean University for distinguished research and a leadership award from Stevens Institute of Technology.

Valentin A. Bazhanov is Chairperson in Philosophy at Ulyanovsk State University, a leading researcher at Tomsk State University, and member of the Russian Basic Research Foundation and the Russian Foundation for Humanities expertise committees. His major works include a biography of the logician, *N.A.Vasiliev (1880–1940)*, Nauka, 1988, as well as the following: *The Completeness of Quantum Theory*, Kazan Univ. Press, 1983; *Science as Self-Reflexive System*, Kazan. Univ. Press, 1991; *The Interrupted Flight. The History of University Philosophy and Logic in Russia (XIX - mid XX centuries)*, Moscow Univ. Press, 1995; *Essays on the Social History of Logic in Russia*, Mid Volga Research Center, 2002; *British Philosophical and Social Sciences Ideas Reception in Russia*

(XIX - early XX centuries), Ulyanovsk Univ. Press, 2005; *History of Logic in Russia and the USSR*, Canon+, 2007; and *Nicolai A. Vasiliev and his Imaginary Logic. The Revival of One Forgotten Idea*, Canon+, 2009. In addition, Dr. Bazhanov has authored 480 published papers. Volumes which he has edited include: *Mathematisation of Natural Sciences*, Kazan Univ. Press, 1984; *Vistas of Sciences*, Kazan Univ. Press, 1989; *The Scientists and Universities of Russia in the Transitional Period*, Ulyanovsk, 1998; and *Logic and Epistemology in Russian Philosophy of First Half of XX century*, Rosspen, 2012. He also compiled A.V. Vasiliev's book *N.I. Lobachevsky (1792-1856)*, Nauka, 1992. Dr. Bazhanov is an Associate Editor of *The Review of Modern Logic* (formerly *Modern Logic*) since its founding in 1989, and an Associate Editor of *Epistemology and Philosophy of Science*; *Reflexive Processes and Governing*; *News of Russia's Universities.Volga Region (Humanities)*;*Kant Studies;* and *Herald of Nizhny Novgorod University*. He is the recipient of several research grants (British Academy, RBRF, RFH, Soros Foundation), including most recently a Lakatos Research Grant (2008).

Philippe de Rouilhan is Directeur de Recherche émérite at the Centre National de la Recherche Scientifique (CNRS), and a member of the Institut d'histoire et de philosophie des sciences et des techniques (IHPST, of which he was the head for a long time). His works belong to logic *lato sensu*, or, more specifically, to formal ontology, formal semantics, philosophy of logic, philosophy of mathematics, philosophy of language. He is currently working on two topics: 1°) truth and logical consequence, and 2°) hyperintensional logic. Dr. de Rouilhan is the editor of *Russell en héritage. Le centenaire des Principles* (special issue of *Revue Internationale de Philosophie* **58** (3), 2004). He is the author of the books *Frege - Les paradoxes de la representation*, Editions de Minuit, 1988; *Russell et le cercle des paradoxes*, P.U.F., 1996; *Logique épistémique et philosophie des mathématiques* (with Paul Gochet), Vuibert, 2007. He is also the author of numerous publications, including more recently: "On What There Are" (*Proceedings of the Aristotelian Society* **102**, 2002, 183-200); "Carnap on Logical Consequence for Languages I and II" (in *Carnap's Logical Syntax of Language*, Palgrave Macmillan, 2009, 121-146); "Carnap and the Semantical Explication of Analyticity" (in *Carnap's Ideal of Explication and Naturalism*, Palgrave Macmillan, 2012, 144-158); "In Defense of Logical Universalism: Taking Issue with Jean van Heijenoort" (*Logica Universalis* **6**, 2012, 553-586); "Putting Davidson's Semantic to Work to Solve Frege's Paradox on Concept and Object" (in the forthcoming *Unifying the Philosophy of Truth*, Springer); "Logic and Content. One Way to the Philosophy of Logic" (in the forthcoming *Handbook of Philosophy of Science*, Oxford UP).

David DeVidi is Professor of Philosophy and Department Chair at the University of Waterloo in Waterloo, Ontario, Canada. He is author (with John L. Bell and Graham Solomon) of *Logical Options: An Introduction to Classical and Alternative Logics*, Broadview Press, 2001, and editor of *A Logical Approach to Philosophy: Essays in Honour of Graham Solomon* (with Tim Kenyon), Springer, 2006, as well as of *Logic, Mathematics, Philosophy: Vintage Enthusiasms: Essays in Honour of John L. Bell* (with Michael Hallett and Peter Clark), Springer, 2011. He is Assistant Editor for Philosophy of Mathematics for the Western Ontario Series in Philosophy of Science. His research has largely focused on issues at the boundary of metaphysics and logic, including Carnap's views on metaphysics, the knowability paradox, realism and antirealism, empirical negation in constructive logic, operators like Hilbert's "logical choice function" in constructive logics, and varieties of mathematical constructivism. In the face of a series of heavy administrative roles in his university, his research output has been slowed for the past few years, but a series of papers arising out of his most recent Social Sciences and Humanities Research Council of Canada grant (2009-2013) "Pluralisms, Logical and Mathematical," should appear soon. Some of his recent research focuses on philosophical issues involving people living with disabilities, an interest arising out of volunteer work he has been doing for two decades. An example is "Advocacy, Autism and Autonomy" (in *The Philosophy of Autism*, Rowman and Littlefield, 2012). He was the inaugural winner of the University of Waterloo's Equity and Inclusivity Award, in addition to several other awards in recognition of his work on behalf of a more just and equitable academy and society.

Mark E. Fuller is Professor of Mathematics at the University of Wisconsin—Rock County. His interests are in logic, philosophy of logic, philosophy of mathematics, and pedagogical issues in teaching mathematics and logic. He was managing editor of *Modern Logic* and the *The Review of Modern Logic* from 1998 until 2009. His publications include "The Normal Depth of Filters on an Infinite Cardinal" (with C.A. Di Prisco and J.M. Henle) (*Z. Math. Logik Grundlag. Math.* **36**, 1990, 293-296); "Normality of a Filter over a Space of Partitions" (*J. Symbolic Logic* **59**, 1994, 529-533); a review of *Understanding the Infinite*, Shaughan Lavine (*Modern Logic* **7**, 1999); and a review of *Engines of Logic*, Martin Davis (*The Review of Modern Logic* **9** (3&4), 2004).

Richard E. Hodel is Associate Professor Emeritus at Duke University. His research interests include set-theoretic topology (metrization theory and cardinal functions), set theory and logic. He co-authored, with D. W. Loveland and S.G. Sterrett, *Three Views of Logic: Mathematics, Philosophy, and Computer* Science, Princeton University Press, 2014 and is the author of *An Introduction to Mathematical Logic*, Dover, 2013. Published papers which Dr. Hodel has authored include "Cardinal Functions I" (in the *Handbook of Set-Theoretic Topology*, North Holland, 1984, 1-61); "Combinatorial Set Theory and Cardinal Function Inequalities" (*Proceedings of the American Mathematical Society* **111**, 1991, 567-575); "Restricted Versions of the Tukey-Teichmüller Theorem that are Equivalent to the Boolean Prime Ideal Theorem" (*Archive for Mathematical Logic* **44**, 2005, 459-472); "Arhangel'skiï's Solution to Alexandroff's Problem: a Survey" (*Topology and Its Applications* **153**, 2006, 2199-2217); "A Theory of Convergence and Cluster Points Based on κ-nets" (*Topology Proceedings* **35**, 2010, 1-40); and "Nagata's Research in Dimension Theory" (*Topology and Its Applications* **159**, 2012, 1545-1558). He has been a frequent reviewer for *The Review of Modern Logic* (formerly *Modern Logic*).

Herbert Korté is Adjunct Professor of Philosophy at the University of British Columbia and Professor Emeritus of Philosophy at the University of Regina. His research has mainly focused on the philosophical and mathematical foundations of space-time theories, especially Einstein's General Theory of Relativity. He is currently working on biological intentional agency and ethics, and a naturalistic foundation for animal rights. Dr. Korté was an invited research fellow at the Zentrum Für Interdisziplinäre Forschung (ZiF) at the University of Bielefeld, Germany, with a bi-national team of leading researchers on the foundations of space-time theories. The latter fruitful collaboration resulted in Dr. Korté being invited by the editor of Erkenntnis to put together a related double-issue volume of *Erkenntnis*. For over twenty years, he has been the recipient of research grants from the Research Council of Canada (NSERC and SSHRC). Dr. Korté's publications include "Why Fundamental Structures are of First or Second Order" (with R. A. Coleman) (*Journal of Mathematical Physics* **35(4)**, 1803-1818, 1993); "A New Semantics for the Epistemology of Geometry I, Modeling Spacetime Structure" (*Erkenntnis* **42**, 141-160, 1995a); "A New Semantics for the Epistemology of Geometry II, Epistemological Completeness of Newton-Galilei and Einstein-Maxwell Theory" (*Erkenntnis* **42**, 161-189, 1995b); "Hermann Weyl: Mathematician, Physicist, Philosopher" (with R. A. Coleman) (in *Hermann Weyl's Raum – Zeit – Materie and a General Introduction to His Scientific Work*, volume 30 of *Deutsche Mathematiker-Vereinigung Seminar*, 161-386, Birkhäuser, 2001); "Einstein's Hole Argument and Weyl's Field-Body Relationalism" (in *Physical Theory and its Interpretation: Essays in Honor of Jeffrey Bub*, chapter 9, 183-211, Springer Verlag, 2006); and "Hermann Weyl" (with John Bell) (in the online *Stanford Encyclopedia of Philosophy*, 2009).

Vladik Kreinovich received his MS in Mathematics and Computer Science from St. Petersburg University, Russia, in 1974, and his PhD from the Institute of Mathematics, Soviet Academy of Sciences, Novosibirsk, in 1979. From 1975 to 1980, he worked with the Soviet Academy of Sciences; during this time, he worked with the Special Astrophysical Observatory (focusing on the representation and processing of uncertainty in radio astronomy). For most of the 1980s, he worked on error estimation and intelligent information processing for the National Institute for Electrical Measuring Instruments, Russia. In 1989, he was a visiting scholar at Stanford University. Since 1990, he has worked in the Department of Computer Science at the University of Texas at El Paso. In addition, he has served as an invited professor in Paris (University of Paris VI), France; Hong Kong; St. Petersburg, Russia; and Brazil. His main interests are in the representation and processing of uncertainty, especially interval computations and intelligent control. Dr. Kreinovich has published six books, edited eleven books, and published more than 1,000 papers. He is a member of the editorial board of the international journal *Reliable Computing* (formerly *Interval Computations*) as well as several other journals. In addition, he is the co-maintainer of the international website Interval Computations (www.cs.utep.edu/interval-comp). Dr. Kreinovich served as President of the North American Fuzzy Information Processing Society from 2012 to 2014; is a foreign member of the Russian Academy of Metrological Sciences; was the recipient of the 2003 El Paso Energy Foundation Faculty Achievement Award for Research awarded by the University of Texas at El Paso; and was a co-recipient of the 2005 Star Award from the University of Texas System.

Roman Murawski is Full Professor and Chairman of Mathematical Logic at the Faculty of Mathematics and Comp. Sci. of Adam Mickiewicz University in Poznań, Poland, former vice-dean of this faculty, and former president of the Polish Association for Logic and Philosophy of Science. His interests are in mathematical logic and foundations of mathematics, philosophy and history of logic and mathematics. He has published the following books: *Philosophy of Mathematics. Anthology of Classical Texts* (in Polish), Scientific Press of Adam Mickiewicz University, 2003; *Philosophy of Mathematics. An Outline of History* (in Polish), Scientific Press of Adam Mickiewicz University,

2013; *Contemporary Philosophy of Mathematics* (in Polish), Scientific Press PWN, 2002; *Mechanization of Reasoning in a Historical Perspective* (with W. Marciszewski), Editions Rodopi, 1995; *Euphony and Logos* (with J. Pogonowski), Editions Rodopi, 1997; *Recursive Functions and Metamathematics. Problems of Completeness and Decidability, Gödel's Theorems*, Kluwer, 1999; *Zählen: Grundlage der elementaren Arithmetik* (with Th. Bedürftig), Verlag Franzbecker, 2001; *Philosophie der Mathematik* (with Th. Bedürftig), Walter de Gruyter, 2012; *Essays in the Philosophy and History of Logic and Mathematics*, Editions Rodopi, 2010; *Logos and Máthēma. Studies in the Philosophy of Mathematics and History of Logic*, Peter Lang Internationaler Verlag der Wissenschaften, 2011; and *The Philosophy of Mathematics and Logic in the 1920s and 1930s in Poland*, Birkhauser Verlag, Basel 2014. He is a member of the editorial boards of *The Review of Modern Logic*, *Studies in Logic, Grammar and Rhetoric*, *Antiquitates Mathematicae*, *Diametros*, *Studia Humana*, and *Filosofija Nauki*.

Alberto Peruzzi is Full Professor of Theoretical Philosophy at the University of Florence, Italy. One of the last students of Giulio Preti, he started his research in the National Research Council of Italy (1976-1981), working on the theory of definition and applied logic (to semantics of natural language), before focusing on the category-theoretic architecture of mathematics and its meaning for epistemology. He co-directed the "Pianeta Galileo" project in Tuscany from 2004 to 2011 (for which he received a medal from the President of the Italian Republic in 2010) and founded the journal *Humana Mente*. Dr. Peruzzi's present research is on the theory of precategories (a notion he introduced in 1994). Among his books, he has authored *Definizioni. La cartografia dei concetti* (1983), *Noema. Mente e logica attraverso Husserl* (1988), *Il significato inesistente. Lezioni sulla semantica* (2004), *Modelli della spiegazione scientifica* (2009), *Dialoghi della ragione impura*, 3 vols (2009-2011), *Scienza per la democrazia* (2009), and *La treccia di Putnam* (2013) and edited *Mind and causality* (2004), and *Pianeta Galileo Atti* (2004-2011). Articles authored by Dr. Peruzzi include: "The Theory of Descriptions Revisited" (*Notre Dame Journal of Formal Logic* **30**, 1988, 91-104); "Prolegomena to the Theory of Kinds" (in *Logical Foundations of Cognition*, Oxford University Press, 1994, 176-211); "The Geometric Roots of Semantics" (in *Meaning and Cognition*, John Benjamins, 2000, 169-201); "ILGE-interference Patterns in Semantics and Epistemology" (*Axiomathes* **13**, 2002, 39-64); "Hartmann's Stratified Reality" (*Axiomathes* **12**, 2001, 227-260); "The Meaning of Category Theory for 21st Century's Philosophy" (*Axiomathes* **16**, 2006, 425-460); "Au-delà d'un clivage: Giulio Preti et la querelle entre analytiques et continentaux" (*Diogène* **216**, 2006, 56-74); and "Logic in Category Theory" (in *Logic, Mathematics, Philosophy,* Springer, 2011, 287-326).

Jonathan P. Seldin is Professor Emeritus in the Department of Mathematics and Computer Science at the University of Lethbridge, Alberta, Canada. His main areas of interest are combinatory logic and lambda-calculus, both untyped and typed. He is also interested in proof theory, in the philosophy of mathematics (and related philosophical issues), in the history of mathematics in general, and in applying these ideas to the teaching of mathematics. Books which Dr. Seldin has co-authored include the following: *Lambda-Calculus and Combinators: an Introduction* (with J. R. Hindley), Cambridge University Press, 2008; *Introduction to Combinators and λ-Calculus* (with J. R. Hindley), Cambridge University Press, 1986; *Introduction to Combinatory Logic* (with J. R. Hindley and B. Lercher), Cambridge University Press, 1972; *Combinatory Logic*, Vol. II (with H. B. Curry and J. R. Hindley), North-Holland, 1972, and *Studies in Illative Combinatory Logic*, Ph.D. Dissertation (University of Amsterdam, 1968). Dr. Seldin co-edited (with J. R. Hindley) the volume *To H. B. Curry: Essays on Combinatory Logic, Lambda Calculus and Formalism*, Academic Press, 1980. Some of his more recent published articles include "The search for a reduction in combinatory logic equivalent to $\lambda\beta$-reduction" (*Theoretical Computer Science* **412**, 2011, 4905-4918); "Curry's formalism as structuralism" (*Logica Universalis* **5**, 2011, 91-100); "Variants of the Basic Calculus of Constructions" (with Martin Bunder) (*Journal of Applied Logic* **2**, 2004, 173-189); "Interpreting HOL in the calculus of constructions" (*Journal of Applied Logic* **2**, 2004, 173-189; "Extensional Set Equality in the Calculus of Constructions" (*Journal of Logic and Computation* **11**, 2001, 483-493); and "A Gentzen-style Sequent Calculus for the Calculus of Constructions with Expansion Rules" (*Theoretical Computer Science* **243**, 2000, 199-215). Dr. Seldin has been the recipient of numerous research grants from NSERC and FACR (in Quebec).

Jean Paul Van Bendegem is at present Professor at the Vrije Universiteit Brussel, where he teaches courses in logic and philosophy of science. He studied mathematics and philosophy at the Rijksuniversiteit Ghent, and he wrote a doctoral thesis on the problem of strict finitism, *i.e.*, the possibility to eliminate the notion of infinity from mathematics and still retain a workable mathematics. In 1985 he spent a semester as post-doctoral researcher at the University of Pittsburgh, for which he received a Fulbright-Hays grant. While strict finitism is still one of his main research projects, the study of mathematical practice has become equally important, *viz.*, to try to understand what it is mathematicians do when they do mathematics. Some of his relevant publications include: "A defense of strict finitism" (*Constructivist Foundations* **7**(2), 2012, 141-149); "Significs and mathematics: creative and other subjects" (*Semiotica* **196**(1/4), 2013, 307-323); and "The inconsistency of mathematics and the mathematics of inconsistency" (*Synthese*, **191** (13), 2014, 3063-3078). With co-editor Bart Van Kerkhove, he edited *Perspectives on Mathematical Practices. Bringing together Philosophy of Mathematics, Sociology of Mathematics, and Mathematics Education*, Springer, 2007. He is furthermore director of the Center for Logic and Philosophy of Science (www.vub.ac.be/CLWF/) officially created in 1995, where currently nearly twenty researchers are working. He is, since 1988, the editor of the logic journal *Logique et Analyse* (www.vub.ac.be/CLWF/L&A/). He is member of several editorial boards, among them, *Philosophia Mathematica* and *Philosophica*, and continues to do refereeing work for a number of journals, ranging from the *Journal of Symbolic Logic, Studia Logica*, to the *British Journal for the Philosophy of Science* and *Synthese*.

Jan Woleński is Professor Emeritus at the Jagiellonian University, Krakow, Poland, Professor at the Academy of Information, Technology and Management, Rzeszów, Poland, and a member of the Polish Academy of Sciences, the Polish Academy of Arts and Sciences, and the Institute International of Philosophy. He was awarded the Prize of the Foundation for Polish Science in The Humanities and Social Sciences in 2013. Professor Woleński's current interests are in logic and its history, in applications of logic to philosophy, and in epistemology. He is the author of the books *Logic and Philosophy in the Lvov-Warsaw School,* Kluwer, 1989; *Essays in the History of Logic and Logical Philosophy*, Jagiellonian University Press, 1999; *Essays on Logic and Its Applications in*

Philosophy, Lang, 2011; and *Historico-Philosophical Essays*, Copernicus Press, 2013. A selection of volumes and articles of which Professor Woleński is a co-editor include *Alfred Tarski and the Vienna Circle* (with E. Köhler), Kluwer, 1999; *Handbook of Epistemology* (with I. Niiniluoto and M. Sintonen), Kluwer, 2004; *Provinces of Logic Determined. Essays in the Memory of Alfred Tarski*. Parts I, II and III (with Z. Adamowicz, S. Artemov, D. Niwiński, E. Orłowska and A. Romanowska) (*Annals of Pure and Applied Logic* **126**, 2004); *Provinces of Logic Determined. Essays in the Memory of Alfred Tarski*. Parts IV, V and VI (with Z. Adamowicz, S. Artemov, D. Niwiński, E. Orłowska and A. Romanowska) (*Annals of Pure and Applied Logic* **127**, 2004); *Church's Thesis After 70 Years* (with by A. Olszewski and R. Janusz), Ontos, 2006; *The Golden Age of Polish Philosophy. Kazimierz Twardowski's Philosophical Legacy* (with S. Lapointe, M. Marion and W. Miśkiewicz), Springer, 2009; and *Jewish and Polish Philosophy* (with Y. Senderowicz and J. Bremer), Austeria, 2013.

Introduction

Francine F. Abeles and Mark E. Fuller

This volume of *Logica Universalis* honors the memory of Irving H. Anellis (1946–2013). Irving's interests in logic, mathematics, and their history were both wide and deep. In Mathematical Logic and Set Theory he specialized in meta-mathematics, proof theory, set-theoretic foundations of analysis and number theory, universal algebra (especially Boolean algebra and algebraic logic), and foundations of group theory.

In the History of Logic his particular interests were in proof theory, algebraic logic, the work of Charles Sanders Peirce, and that of Bertrand Russell, mathematical logic in Russia and the Soviet Union, universal algebra (especially Boolean algebra, algebraic logic and lattice theory), set theory and foundations.

In the History of Mathematics he was especially concerned with universal algebra, group theory, Russian mathematics from the tenth to the seventeenth centuries, and the history of mathematics education in Russia.

In these three areas and a few others he produced more than 430 publications, including four books, 36 edited works, 103 articles on the history of logic, 78 pieces on a variety of topics, and 211 reviews, abstracts, and notes. His final complete publication in 2012 was as the guest editor (and author of four articles) of a volume in the series, *Logica Universalis*, whose chief editor is Jean-Yves Béziau, titled: *Perspectives on the History and Philosophy of Modern Logic: van Heijenoort Centenary.* Irving's academic degrees were all in philosophy. He earned his Ph.D. from Brandeis University under the supervision of Jean van Heijenoort with a thesis on "Ontological Commitments in Ideal Languages: Semantic Interpretations for Logical Positivism". Irving became involved in the history of logic as a doctoral student.

At his death on 18 July 2013, he was a contributing editor and a research associate with the Peirce Edition at the Institute of American Thought based at Indiana University—Purdue University Indianapolis, where he was in charge of writing and revising annotations for Peirce's texts on logic and mathematics. In April 2012, Irving was given the principal responsibility for the annotations of volume 11 in the series (containing all the chapters of Pierce's unpublished 1894 manuscript "How to Reason: A Critick of Arguments"). The announcement of this auspicious assignment stated that the "principal responsibility for that volume's annotations has been assigned to Dr. Irving Anellis, whose competence as a historian of logic and mathematics is simply prodigious". The Minutes

© Springer International Publishing Switzerland 2016
F.F. Abeles, M.E. Fuller (eds.), *Modern Logic 1850-1950, East and West*, Studies in Universal Logic, DOI 10.1007/978-3-319-24756-4_1

1

of the Charles Sanders Peirce Society for 5 April 2012, added that "Dr. Irving Anellis . . . has long built a strong reputation as a historian of logic and mathematics". Irving also was a contributing editor for logic and mathematics for volume 7 of the Peirce Edition Project at the Université de Québec à Montréal. And he was on the advisory board for the Hilbert—Bernays Project at the Universität des Saarlandes.

Irving was the owner, editor and publisher of Peirce Publishing from 2005 to 2007, and of Modern Logic Publishing from 1989 to 1998. As the editor of the journal, *Modern Logic: International Journal for the History of Mathematical Logic, Set Theory, and Foundations of Mathematics* from 1990 to 1998, a unique journal at that time, Irving published articles and essay reviews of books on the history of modern mathematical logic and its foundations, thereby making accessible to a wide academic audience the work of writers in the United States, Canada, Europe, and the former Soviet Union on topics not usually published in standard journals. Under the leadership of a new editor, Mark E. Fuller, the journal was renamed *The Review of Modern Logic* which continued to be published through 2009.

Although he had been invited to speak at many conferences, most recently at the Fourth World Congress and School on Universal Logic held in Rio de Janeiro in March and April, 2013, and at Terceiro Simpósio Internacional de Filosofia da Linguagem na UFF in Campus de Gragoatá, Brazil in April 2013, the last conference Irving attended (where he presented a paper) was the Association for Symbolic Logic's Annual Meeting in March and April 2012 held at the University of Wisconsin in Madison where Thomas Drucker, a special advisor for this volume, organized a special session on "The History of Logic on the Centenary of the Birth of Jean van Heijenoort".

During the time when Irving owned **Modern Logic Publishing** and edited *Modern Logic*, he lived in Ames, Iowa. In 1989 when he became a contributing editor at the Peirce Edition Project (PEP), an invitation encouraged by Nathan Houser and given to him by the then director of the PEP, Christian J. W. Kloesel, he moved to Indianapolis, Indiana, where he lived until his death. The logic community is indebted to Houser who first introduced Peirce as a logician to Irving.

Irving's large presence in this volume takes the form of two papers. The first is an unpublished piece titled "Preface and Prospectus" that he intended as the introduction to a book on historiography, a new interest he was developing. Appearing on his web page and dated 9 May 2013, about two months before his death, it wove together several closely related historiographical questions: Was there a "Fregean" revolution in logic? What became of the logic of the nineteenth century? What is the relationship between the algebraic logic of the nineteenth century and the mathematical logic of the twentieth century? Irving intended the book to be a social and intellectual history of logic, as well as an historiographical and philosophical study of the history of logic between 1850 and 1950.

The second is a joint paper, "The Historical Sources of Tree Graphs and the Tree Method in the Work of Peirce and Gentzen", a collaborative project Irving and Francine Abeles had been working on informally for more than 20 years beginning with their discussions of the relevant work of A. Cayley, A. Kempe, W. Clifford, and J. Sylvester, and which finally began to be realized in 2012. Anellis and Abeles wanted to provide the historical background for Peirce's visual proof methods and for his classification of logical arguments. The tree form of deduction which first arose in the work of E.Beth,

J. Hintikka, and R. Smullyan culminated in the work of P. Hertz and G. Gentzen. Peirce's development of his entitative and existential graphs arose from his work on truth-functional logic and in his experimentation with diagrammatic methods to analyze proofs. Along the way, A. MacFarlane, A. Marquand, C. Ladd-Franklin, and C. Dodgson developed their own more restrictive diagrammatic methods.

The other papers included in this volume reflect Irving's interests in myriad ways. In "Logic and Argumentation in Belgium: The Role of Leo Apostel", Jean Paul Van Bendegem deals with the Signific Movement, a little known philosophical project that began in the Netherlands in the early part of the twentieth century and migrated to Belgium in the second half of the century. Recognizing that the work of Leo Apostel (1925–1995) was central to the development of that project and ultimately to logic and argumentation theory in Belgium, the author first presents an overview of Apostel's work and the intellectual influence of C. Perelman, R. Carnap, and J. Piaget, as well as the impact of the Erlangen School and that of the strict finitist, D. Van Dantzig, on his thinking.

Philippe de Rouilhan in his paper, "Tarski's Recantation: Reading the Postscript to 'Wahrheitsbegriff'", examines the postscript to the second (German) edition of Tarski's book, originally written in Polish and first published in 1933: "The concept of truth in the languages of deductive sciences", to argue that Tarski abandoned not only (1) the basic principles of the theory of semantical categories, but additionally (2) logical universalism itself, and that the path Tarski took from (1) to (2) lies outside the realm of mathematical proof and therefore is essentially part of philosophy.

In "The Paradox of Analyticity and Related Issues", Jan Woleński formulates the paradox of analyticity in a way similar to that of the Liar Paradox. A sentence A is an analytical if and only if C(A), where C refers to a condition that is satisfied by an analytical, e.g. "is true in all possible worlds", with many other possible meanings. T-equivalences are used to generate both puzzles. (A T-equivalence compresses the left part of a formula $\mathbf{T}$ (A) $\leftrightarrow$ (A) to its right part.) Woleński goes on to show that retaining T-equivalences requires using resources in a suitably selected meta-language **ML** of the object language **L**.

David DeVidi and Herbert Korté present a modified version of natural deduction for classical predicate logic in their paper, "Naturalizing Natural Deduction". In their system, the rules of universal introduction and existential elimination are given a new proof theoretic role in the form of *commonizing quantifiers* (but universal elimination and existential introduction are retained). The authors argue that their system is easier to use and teach, and from a philosophical viewpoint, that their system restores "naturalness" as reflected in unformalized rigorous reasoning. In addition, viewing natural deduction in an intuitionistic setting, they show that embedding arbitrary object semantics can be avoided.

Alberto Peruzzi's paper, "Category Theory and the Search for Universals: A Very Short Guide for Philosophers", provides an introduction to the fundamental elements of category theory as they apply to analytic philosophy. He uses the categorical concept of adjunction to capture the important philosophical notion of universality. The motivations, topics of interest, and historical developments leading to the categorical reformulation of logic beginning in the late twentieth century are presented in the first half of the paper. In the second half, the author gives his reasons for considering adjoint functors as having a central role in mathematics and particularly in logic.

In his paper, "On the Way to Modern Logic—The Case of Polish Logic", Roman Murawski presents the work of two Polish logicians who, although not well known in the West, represent the state of logic that existed in Poland prior to the establishment of the Warsaw School of Logic whose members S. Leśniewski, J. Łukasiewicz, and A. Tarski developed mathematical logic in the twentieth century. These early logicians, Henryk Struve and Władysław Biegański, were active in the period 1863–1916.

In his paper, "Russia's Origins of Non-Classical Logics", Valentin A. Bazhanov addresses three questions: What were the conditions that aided the development of non-classical logic in Russia? Who were the principal Russian participants? What were the socio-cultural factors that contributed to its development? The author principally discusses the work of two relatively unknown logicians: N.A. Vasiliev in connection with the rejection of the laws of excluded middle and non-contradiction, and Ivan E. Orlov who, in the 1920s, sought to develop an appropriate modal logic for natural science.

Vladik Kreinovich recounts events in the 1970s and 1980s when he was a graduate student in the mathematics department at St. Petersburg University in his paper, "Constructive Mathematics in St. Petersburg, Russia: A (Somewhat Subjective) View from Within". After providing a brief history of constructive mathematics in the Soviet Union up to the 1960s, he surveys the state of constructive mathematics during the time when he was studying it: its principal participants, how it progressed, the challenges it dealt with, and its successes and failures. All of this is portrayed against the background political climate of the time.

In his paper, "On Normalizing Disjunctive Intermediate Logics", Jonathan P. Seldin examines the natural deduction versions of the ten disjunctive intermediate logics given by T. Umezawa in 1959, one by López-Escobar in 1982, and a suitably formulated classical logic to prove a normalization result for them. These intermediate logics are obtained from intuitionistic logic by adding a disjunction. The results also hold if the axioms in question are added to minimal logic instead of intuitionistic logic. Seldin shows that the normalization procedure is not as complete as it is for intuitionistic and minimal logic.

Using the algebraic structure $(B, \wedge, ', 0)$, Richard E. Hodel in his paper, "A Natural Axiom System for Boolean Algebras with Applications", proves Stone's Representation Theorem for Boolean algebras (stating that every Boolean algebra is isomorphic to a field of sets). This theorem was first proved by O. Frink, Jr. in 1941. Hodel claims his proof is more advantageous. One of his applications yields a simple proof of the Deduction Theorem, an important result in Proof Theory.

F.F. Abeles (✉)
Departments of Mathematics and Computer Science, Kean University, Union, NJ 07083, USA
e-mail: fabeles@kean.edu

M.E. Fuller
Department of Mathematics, University of Wisconsin – Rock County, Janesville, WI, USA

Preface and Prospectus to a planned "History vs. Philosophy of Logic" Text

Irving H. Anellis

Editors' Note This piece was an introduction to a text that Irving Anellis was working on at the time of his death. The working title was "History vs. Philosophy of Logic". No bibliography was found, but we felt the reader would find it nonetheless both sufficiently interesting as well as revelatory of his outlook and plans.

Charles F. Breslin's (b. 1928) lengthy essay "Idea and Process in the Historiography of Logic" (Breslin 1973) may be summarized as averring that the concept of *formal logic* has evolved over the extent of its history "from Plato's dialectic to Gentzen's system of rules for natural deduction..." and that the proliferation of varieties of logic make it unfeasible to arrive at a definition of the subject that would satisfy either logicians or historians of logic. John MacFarlane (2000), for example, attempted a classification of formality, which, however, Catarina Dutilh Novaes (2011) found too limited, insofar as MacFarlane's purpose was to distinguish logic from other disciplines and to ascertain a line of demarcation between formal logic and other fields, in particular, in order to characterize what shall count as logic. Dutilh Novaes identifies two categories of formality in attempting to determine what is the formalality of formal logic, each of which contains three species. Despite the differences, what they all seem to have in common is that terms or objects are replaced by letters, that each is neutral with respect to content and meaning, and that propositions may likewise be replaced by letters, the schematic connection between which are determined by the nature and definition of the copula holding between them, regardless of content or meaning of the subject-matter. Breslin also pointed out the—unremarkable—truism that doing logic is not the same as writing its history. More pertinently, he argued (Breslin 1973, 654) that, although Frege "very subtly adumbrated" his "methodological advantage" over his predecessors in distinguishing formal from informal or intuitive considerations and formal from natural language concepts, nevertheless, "[e]ver since Frege's epoch making work it has not been possible to formulate a convenient analytic definition of logic and an historical description of what logic is hardly yields a precise and universally acceptable statement of the general principles." That being the case, how does one assess and explain the nature of the evolution of, or developments in the history of formal logic over the entire course of its history, and still more, in its more recent history, as Aristotle's syllogistic gave way to first-order functional logic?

© Springer International Publishing Switzerland 2016
F.F. Abeles, M.E. Fuller (eds.), *Modern Logic 1850-1950, East and West*, Studies in Universal Logic, DOI 10.1007/978-3-319-24756-4_2

This "Preface" may serve as an abstract for the present study, which asks, and seeks to provide answers to, several closely related historiographical questions.

Was there a "Fregean revolution" in logic? What became of the algebraic logic of the nineteenth century pioneered by Boole and De Morgan and brought to its [presumed] apogée by Peirce and Schröder before being submerged by the mathematical or "symbolic" logic of the twentieth century typified by the *Principia Mathematica*? (I say "presumed" insofar as the announcement of the demise of algebraic logic, even if it now is regarded as merely a subspecialty of the new "mathematical" logic stemming from the work of Frege and Russell and their heirs, is belied by the work of Tarski and his heirs beginning in the 1940s.) What relationship, if any, holds between the algebraic logic of the nineteenth century and the mathematical logic of the twentieth? Was the Boolean tradition really nothing more than an interesting but minor sidelight to the history of mathematical logic; and if so, how did this "new" mathematical logic replace the "old" algebraic logic in the mainstream of the development of formal logic? Had Kant's famous—now infamous—remark, that "as regards the history of Logic, we will only mention [that] ... it is derived from Aristotle's *Analytic*. [...] Since Aristotle's time Logic has not gained much in *extent*, as indeed its nature forbids that it should. [... It] admits of no further alteration [...] and indeed we do not require any new discoveries in Logic ..." appeared in, say 1870 or 1880, rather than in 1800 (in Kant 1800, 18; see, e.g. Kant 1963, 10–11), how might we today, as historians of logic, have reacted?; how might we, as contemporaries of Boole, De Morgan, Jevons, Peirce, or Schröder, have reacted? Could we legitimately or calmly accept the charge of the inconsequence of the work of Boole, De Morgan and their immediate successors? If there was indeed what some historians and philosophers termed a "Fregean revolution" in logic, did that entail the demise of the traditional [Aristotelian] logic?; of algebraic logic?

These are the questions that this investigation attempts to answer. As such, this multi-layered investigation is essentially historiographical (and sociological and philosophical in some aspects) rather than historical or technical. It deals with the perception which logicians and historians of logic have held during the crucial period when the "Fregean revolution" is alleged to have been carried out and with the nature of their work and that of their colleagues; and at the same time it attempts to assess how closely this perception matched the actual research carried out during that period. It also deals with the retrospective perception which more recent and contemporary historians have had regarding the "Fregean revolution" and with the corollary perception which recent and contemporary scholars hold regarding the nature of the research carried out by the logicians of the era of the "Fregean revolution". And it assesses the impact which these perceptions have had on the course of the development of history of logic as a science. The present study consequently goes well beyond the typical chronological account of the development of logic, and seeks to ascertain the attitudes of those who contributed to that development, as well as the attitudes of those who, while working in logic, have since come to be ignored as irrelevant to, or marginalized by, the developments of logic towards its current condition. We wish to establish, insofar as possible, what logic was like for those both who contributed to the attainment of logic as we know it and those who are seen not to advance logic toward its current state. Hence we investigate how logic was understood by those working in logic who formulated its contemporary status in the past, whether or not they participated in developing it towards our own understanding of logic, and what they knew of logic while they were working.

Certainly the announcement of the "death" of a theory, whether of Aristotelian logic, or of algebraic logic, or of invariant theory, or any other that suggests itself, unlike the premature announcement in the Japanese press, and noted in the *New York Times* (see New York Times 1921) of the death in China of Bertrand Russell on 28 March 1921, in the midst of his lecture tour during his bout with pneumonia, indeed seem to require, not merely a factual verification, but also historiographical judgments and demand plausible explanations. As philosopher Emil Ludwig Fackenheim (1916–2003) remarked (1994, xiv): "Establishing the facts, the historian must also explain them." One is here reminded also of the response to London-based reporter Frank Marshall White of the *New York Journal* by Mark Twain (Samuel Langhorne Clemens; 1835–1910) in 1897 to false rumors circulating of Twain's death, that it was an "exaggeration" (for the factual circumstances of the origin of the rumors and an account of how the exact quote morphed into "greatly exaggerated" (see White 1897; see also Clemens 1945). In the difference between reportage and verification lies one of the more significant distinctions, so deeply troubling to historians and philosophers of history, between myth and history (see, e.g. Anellis 1992d for a discussion on the application of this issue to history of logic and mathematics, and on the use of rumor and other unpublished or unverified sources in history). Herodotus of Halicarnassus (484–425? BCE) in his *Histories* attempted both, not always successfully, and occasionally quite fitfully. But it was his *insistence* upon *both* factual reporting *and* rational explanation that rendered his history precisely that, a research (ιστορία; *historia*), and earned him the title of "father of history".

It has been recognized by late nineteenth-century logicians that logic was not a monolithic enterprise, and various schools of logic competed with one another. It was the fate of some (the Aristotelian-neo-scholastic, Kantian, Hegelian, empiricist or "Millian", and psychologistic, even the "Boolean" or algebraic, types of logic, for example) that they largely receded into comparative insignificance, if not into total anonymity and oblivion, after the appearance of *Principia Mathematica*, although they each held some prominence through much of the nineteenth century and into the early years of the twentieth. It is our contention that we cannot fully appreciate or comprehend the rise of the mathematical logic or logistic of Frege and Whitehead and Russell if we do not also recognize the role of these other logics in the historical condition of logic in the second half of the nineteenth century and first decades of the twentieth century, and if we do not attempt to explore the relationships which these logics and their advocates had during that period with the work of the "logicistians" Frege and Russell. Much of the present investigation, therefore, is devoted to an exploration of how the logicians of all of the various competing nineteenth-century schools reacted to one another and to their various conceptions of logic, and to the question of what these researchers understood both about their own work in logic and about the work of those of their colleagues pursuing other conceptions of or directions in logic. This is, then, both a social and an intellectual history of logic, *and* an historiographical and philosophical study of the history of logic.

In attempting to account for the twin phenomena of the strength of the Russello-Fregean approach to mathematical logic in the twentieth century and the decline of the Boolean for one, and of the influence of Russell and the lack of influence of Peirce in contemporary histories of logic for the other, it was simply not enough to explain them away merely by noting that at the end of the nineteenth and beginning of the twentieth century, Peirce's (and Schröder's) career was coming to its end just as Russell's was entering its beginning. A serious account demanded a detailed exploration of the attitudes of those who were working during these years and whose own careers as logicians overlapped both the sunset of Peirce's and the sunrise of Russell's as well as an examination of both the professional and the "extra-logical" influences which Peirce and Russell exerted on both their colleagues and their logical and philosophical heirs.

Along the way, it becomes clear that there were numerous efforts, many stemming directly from the influence of Kant and his *Critik der reinen Vernunft*, that those who investigated the question of the nature and scope of logic were not in uniform agreement either in their choice between the Frege-Russellian attitude towards mathematical logic or the Boolean attitude or in their preference of the classical Aristotelian formal logic and the new mathematical logic. Most philosophers of the period, in particular prior to the publication of the work of Russell, with his 1903 *Principles of Mathematics*, and often as late as the early 1920s with the publication of the second edition of Whitehead and Russell's *Principia Mathematica*, were as likely to prefer syllogistic as their formal logic over either the Boolean or Russello-Fregean competitors (see, e.g. Anellis 2011). The adherents of the Frege-Russellian camp in particular were also commonly as likely to stress a difference, indeed a dichotomy, between their *mathematical* logic and the Boolean *algebraic* logic. As (Pulkinnen 2005, 22–23) wrote, "no agreement regarding the essence of the new mathematical logic existed. [. . .] The different attempts to reform logic can be divided into two main groups," namely the "Booleans" and the "Russello-Fregeans" or 'logisticians". We thus treat these two unavoidable themes of rivalry: (1) between Aristotelians and "mathematicians"; and (2) within the "mathematicians'" camp, between the "Booleans" and the "Russello-Fregeans". Without examining both combats, we cannot fully understand or appreciate how the Boolean, or algebraic, tradition came (to use a Hegelian turn of phrase), to be subrepted by, the Russello-Fregean, or logistic tradition. And not only that, but to subrept *in situ* the syllogistic logic as well. What we shall find, embedded in the discussions emerging from Kant's views on logic and the competing views emergent with Hegel, is what we might describe as a turf battle or a borderline dispute, as philosophers, psychologists, linguists, and their allies argued whether logic belongs more properly to philology, to rhetoric, or psychology, or, if to philosophy, whether to metaphysics or to epistemology, or is it a distinct field of philosophy unto itself. Few, if any of those entering the lists from the mid-nineteenth century into the earliest years of the twentieth century—other than the mathematicians themselves—, were prepared to concede logic to the mathematicians. Yet even among those who were crucial in the development of mathematical logic (in this case the "algebraists"), there were those who raised questions about the provenance of the logic that they helped to create. Thus, for example, for Charles Sanders Peirce, who was both a mathematician and a philosopher as well as, primarily (in his own account) a logician, logic belongs to mathematics. But it had other aspects as well, outside of mathematics. He defined logic to include semiotics, or theory of signs, as well, and he likewise divided logic, taken in the stricter sense as a *method* of critical reasoning, into deduction, induction, and abduction (see, e.g. Hilpinen 2004, 644–653). The latter term—abduction—has an ambiguity about it, and can perhaps best be understood as encompassing all those tools and techniques of critical reasoning which are neither exclusively and precisely deductive nor exclusively and imprecisely inductive (see, e.g. Kapitan 1997 on abduction). Thus, a major theme in our account of the shift from algebraic logic to logistic will have to include an examination of the debates which occurred in the latter part of the nineteenth century and into the early years of the twentieth that German philosophers formulated as: *Was ist Logik?*

   The investigation which is carried out in this book shows that, during the crucial period from around 1880 to around 1910 or 1920, the majority of logicians and historians of logic—with but few notable exceptions, the most conspicuous exceptions being Frege and Russell—perceived themselves and their colleagues to be *building upon and expanding* the work of the algebraic logicians from Boole and De Morgan to Peirce and Schröder, rather than *creating an altogether new* endeavor. This investigation also suggests that the early work of Hilbert, of Peano and his school, and of the American postulate theorists in the years from the 1890s to the 1910s and into the 1930s forms a natural "bridge" between the "Booleans" and the "Fregeans", both chronologically and intellectually. By the nature of this investigation, what is proffered is not a chronological history of logic, but rather a thematic discussion of the transformation of logic, wrapped around the one central theme that not a "Fregean revolution" took place, but a natural evolution, from Aristotelian logic to modern mathematical logic, in which algebraic logic played a significant, preparatory, role. It began with the recognition (in Anellis and Houser 1988, 1991; Anellis 1995b) that there was not only a shift away from the algebraic logic of the nineteenth century, typically dated to the publication in 1879 of Frege's *Begriffsschrift* and attributed in large measure to Bertrand Russell's influence in advocating and disseminating the Russello-Fregean conception of logic, but that the work of the algebraic logicians of the nineteenth century was being either neglected or even denigrated by defenders of the Russello-Fregean conception of logic. It has been shown, for example by Volker Peckhaus 2004, that many aspects of the presumed distinction, formulated by Jean van Heijenoort (1912–1986) (1967a) between algebraic logic as a calculus and modern mathematical logic as a language as well as a calculus, were present in the work of Schröder, and that Schröder can, in many respects, even be rightfully considered a logicist (see Peckhaus 1990/1991, 1993). Beyond that, it has been shown (in Anellis 2012) that all of the criteria enumerated by van Heijenoort as distinguishing Frege in the *Begriffsschrift* as the founder of mathematical logic, were already present to some extent in the work of Peirce and his colleagues, although not necessarily developed all at once and in a unified, systematic, work.) The present work is in large measure an attempt to understand why and how this state of affairs came about.

   In order to attain an accurate perspective on the developments of the past and their significance, it is necessary to recognize not only what is original to the period being considered, but what of the past is embedded in the present, and what the past embeds of its own past; or, in the words of social historian Beryl Satter (2012, 5), "the past has a profound impact on the present." That is, what does the period under investigation carry from its past, and how is the work of the past reflected in the present conception of the field and of the present conception of the past? Is there an accumulation? And, if so, what of the past survives into the present? Not only that, but is our understanding of the developments of the past the same or different from the understanding that those who contributed to the new developments and of the developments that preceded their work? What this really means is that the historian must attempt, insofar as possible, to see the work under investigation at three levels, and in an important sense see each level both simultaneously and separately and distinctly. The attempt to view past intellectual achievements on their own level, as those who created it viewed it, may be the most difficult task. But it is a task necessary for an accurate comprehension of the value and significance of those developments for those who were responsible for them. In her "Preface" to Burt C. Hopkins' (2011) *The Origin of the Logic of Symbolic Mathematics, Edmund Husserl and Jacob Klein*, Eva Brann (b. 1929), translator of Jacob Klein's (1899–1978) (1934–1936, 1968) history of the Greek origins of algebra, wrote of Husserl's "intentional history" by saying (Brann 2011, xxiv) that it 'enjoins on the historian one chief task, that of "desedimentation" and . . . "reactivation." These terms refer to a scraping away of the accumulated strata of tradition and the reanimating recollection of the thinking that had been skewed, superseded, and "ruptured" at crucial moments, but had remained embalmed—one might say, semi-consciously preserved—within those modern concepts that are so effective precisely because the burden of their origin is ignored.' It is in this sense that Hopkins (2011, 6) understands Edmund Husserl (1859–1938) as conceiving of epistemology as an historical science in his efforts to uncover the meaning and historical significance of the achievements in geometry of Descartes in "Die Frage nach dem Ursprung der Geometrie als intentional-historisches Problem" (Husserl 1939, composed in 1936) and in physics of Galileo in "Die Krisis der europäischen Wissenschaften und die transzendentale Phänomenologie. Eine Einleitung in die phänomenologische Philosophie" (Husserl 1936), and of Klein's (1934–1936, 1968) similar effort to uncover the meaning and significance of the achievements of François Viète (1540–1603) in his (1591) *In artem analyticam isagoge sue Algebra nova* towards establishing algebra as an "analytic art", that is, as an abstract formal or symbolic, rather than a merely concrete, numerical, science. I see it as my task, in attempting to answer the question of whether there indeed was a Fregean revolution in logic or not, and, if so, in what it really consisted, to attempt these twin efforts to "desediment" and . . . "reactivate" those aspects of the history of logic that (1) were the cause and contents of the Fregean revolution, if indeed such a revolution occurred, and (2) to explain how, why, and whether, the historiographical conception of a Fregean revolution arose. The twin processes of de-sedimentation and reactivation, howsoever flawed or incomplete, are likely the best that the historian can hope to attempt in the effort to approach "*wie es eigentlich gewesen.*"

As we explore this historiography, it becomes increasing evident that—as Martin Davis said (1995, 277) in another context, one which is, however equally applicable in the current case—

> Studies like this one can have the effect of bolstering a sense of our superiority to our logical forbears. But this would be a serious mistake. The lessons to be learned are rather that the development of the outlook on our subject that today we take for granted was attained only with great difficulty.

If we wish to further justify our approach of immersing ourselves in the fads and fallacies, streams and rivers, both continually flowing and long since evaporated, we can do little better than remember that Jean van Heijenoort, among the foremost historians of logic of the mid-twentieth century and compiler of the seminal and highly influential *From Frege to Gödel: A Source Book in Mathematical Logic, 1879–1931* (van Heijenoort 1967), with all its shortcomings, nevertheless was convinced that the best way to learn a subject was to study its history, along with logician Leon Henkin's remark, in an article on mathematics education (Henkin 1995, 3), that:

> Waves of history wash over our nation, stirring up our society and our institutions. Soon we see changes in the way that all of us do things, including our mathematics and our teaching. These changes form themselves into rivulets and streams that merge at various angles with those arising in parts of our society quite different from education, mathematics, or science. Rivers are formed, contributing powerful currents that will produce future waves of history.

Beyond Acknowledgments I became interested in and involved with the study of the history of logic as a doctoral student of the late Jean van Heijenoort, the editor of the well-known and influential anthology *From Frege to Gödel* (van Heijenoort 1967), and it is from him that I first began learning the history of logic. Van Heijenoort (1967, vi) dated the "rebirth" of logic to the middle of the nineteenth century and saw Boole, De Morgan, and Jevons as the "initiators of modern logic"; but he dated the *true* origin of modern mathematical logic from the publication of Frege's *Begriffsschrift* (Frege 1879) and he held the previous period, the period during which the Boolean tradition of Boole, De Morgan, and Jevons (and, we must of course add, of Peirce and Schröder), flourished to be one which "would not count as a great epoch" (1967, vi). Indeed, in a manuscript of 1974 that was first published in 1992, van Heijenoort (1992, 242) began with the unequivocal and unqualified declaration that: "Modern Logic began in 1879, the year in which Gottlob Frege (1848–1925) published his *Begriffsschrift*. In less than ninety pages this booklet presented a number of discoveries that changed the face of logic." Most historians and philosophers of logic, not only van Heijenoort, counted either the virtually simultaneous publications of Boole's *The Mathematical Analysis of Logic* (Boole 1847) and De Morgan's *Formal Logic* (De Morgan 1847), or the publication of Boole's *An Investigation of the Laws of Thought* (Boole 1854) and De Morgan's *Formal Logic* as heralding the birth of modern mathematical logic. But at the same time they likewise regarded it as relatively trivial for the development of logic as we know it today in comparison with the publication of Frege's *Begriffsschrift* in 1879. Thus, for example, Rudolf Carnap (1954, 1; 1958, 1) wrote that "Symbolic logic (also called mathematical logic or logistic) is the modern form of logic developed in the last hundred years." He also wrote (Carnap 1954, 1958, 2–3): "Symbolic logic was founded around the middle of the last century . . . "; which, however, he regarded, as he continued his remark, that logic in that guise in which it existed as developed in the mid-nineteenth century is "essentially of historical interest only" (Carnap 1954, 1958, 2–3). An obvious question must be: If the *birth* of modern logic is to be traced to the work of Boole (and/or De Morgan), why is it that their work and the work of their followers, such as Peirce and Schröder, is properly regarded as only of passing historical interest, and, more importantly, properly regarded merely a sidelight in the history of modern logic?

I first seriously studied the work of Bertrand Russell as a graduate student, when I took a reading course with Morris Weitz (1916–1981) on Russell in 1974, who, while working on his University of Michigan doctorate, took time to study with Russell while Russell was visiting the University of Chicago. The upshot of Weitz's study with Russell was the doctoral thesis, *Method of Analysis in the Philosophy of Bertrand Russell* (Weitz 1943) which became the basis of Weitz's (1944) contribution on "Analysis and the Unity of Russell's Philosophy" to the Russell volume of Paul Arthur Schilpp's (1897–1993) "Library of Living Philosophers". The plan was for us to go through Russell's most important work in their chronological order. We began, if I recall correctly, with Russell's (1897) *Essay of the Foundations of Geometry*, and from there to his (1900) book on Leibniz, to some of the essays gathered in *Logic and Knowledge: Essays 1901–1950* (Russell 1956a), and concluded with *The Principles of Mathematics* (Russell 1903). Peirce's work in logic first came to my serious attention only much later, when I heard historian of logic Nathan Houser of the Peirce Edition Project [PEP] at Indiana University–Purdue University at Indianapolis, Indiana, USA, speak on Peirce's work on the law of distributivity in the logic of relatives during an American Mathematical Society history of logic session in Chicago in March 1985 at which I spoke on the history of Russell's discovery of his paradox. Until I heard Houser's talk and had an opportunity to compare notes with him, Peirce was for me just a minor, almost invisible, figure in the history of logic, notable only for the few scant, and largely indirect, mentions in van Heijenoort's *From Frege to Gödel*, and in particular and primarily for little more than Schröder's adaptation of Peirce's notation for quantified formulas that were employed by Löwenheim and Skolem in their establishment of the Löwenheim-Skolem Theorem, and to Russell's scattered handful of criticisms of Peirce in the *Principles*. My understanding of Schröder's significance was that it was limited essentially to the role of his formulation of quantifiers in the work of Löwenheim and Skolem.

I gradually became interested in the specific historiographical questions being posed in this study through my work as a research associate at the Bertrand Russell Editorial Project [BREP] at McMaster University in Hamilton, Ontario, Canada in June 1982–May 1983 and during my visit to PEP on 11 April 1986. During my first visit to PEP at that time, I had the opportunity to meet and talk with Houser, now past director of PEP and editor of PEP's edition of Peirce's writings, and the late Max Harold Fisch (1899–1994), a philosopher who during his lifetime was doyen of Peirce studies. I discussed with them comparisons of the respective contributions of Russell and Peirce to the history of logic, and considerations as well of comparisons between BREP and PEP.

During my tenure at the BREP, I had had the opportunity to examine archival materials, including manuscripts and, unpublished writings of Russell, from the Bertrand Russell Archives located at the McMaster University library. Subsequently, during visits to the PEP and through my appointment as a contributing editor to PEP, I have also been provided access to archival materials and Peirce manuscripts, including unpublished writings. The ability to examine these materials led to questions not only of the relative strengths, merits, and weaknesses of Peirce and Russell as logicians, but to questions, such as this investigation attempts to explore and answer, of when, why, and how, or even *whether*, the "Boolean" tradition in logic gave way to the "Fregean revolution" and the "Russello-Fregean" tradition that arose from this "revolution" in the early years of the twentieth century. Raised as well were questions of the roles which Peirce and Russell played in this historic and presumably monumental shift in the character and direction of technical developments in logic. The impetus for undertaking a full-scale comparison of the contributions to logic of Peirce and Russell arose in large measure from discussions with Houser. It was from Houser that I also first learned that historian of logic Benjamin S. Hawkins, Jr. had also undertaken a comparative study of Peirce and Russell, and it was through Houser that I first established contact with Ben. We all three considered that Peirce's contributions to logic and mathematical acumen were generally underestimated by logicians and historians of logic, whereas Russell's were exaggerated. It was my research with both the Russell and the Peirce materials in subsequent years that enabled me to undertake a comparative survey and evaluation of the respective contributions and influences of Peirce and Russell and which led to the publication of (Anellis 1995b), two years before Ben was able to see his own comparative study (Hawkins 1997) in print. Although the general conclusions drawn by (Anellis 1995b) and (Hawkins 1997) in assessing the comparative logical strengths and weaknesses of Peirce and Russell largely coincided and on the principal points agreed, my study was essentially historiographical and social history, whereas Hawkins's was primarily technical and philosophical. Moreover, although Ben began work on his study a decade before I took up the issue, and although we examined many of the same materials for our respective studies, I was able to work faster, having had more direct and wide-ranging access to both the Peirce and Russell archives and (in some cases) wider latitude in the use of the materials accessed than did he. Such comparative studies have steadily gained in the methodology of general history, as, beginning in the 1950s, specialists in ancient history have applied comparative structures to study the complexity of the political and cultural dynamics, e.g., of ancient Mesoamerica, Peru, Egypt, Mesopotamia, and, most recently, to the Bronze Age Aegean civilizations and the Iron Age (pre-classical) Greek and Italian (Etruscan) civilizations (see, e.g. the studies in Terrenato and Haggis 2011).

The present study, however, goes beyond a comparison between Peirce and Russell as logicians, beyond a comparison of their respective impact upon the development of logic, and even beyond a comparison of the influences which they exerted upon perceptions and perspectives of the history of logic. The principal goal here is to examine the changing conception of the nature of the history of logic and to understand the reasons for the perceived "Fregean revolution" that historians and philosophers of logic detected in the generations that were raised upon Frege's *Begriffsschrift*, Russell's *Principles of Mathematics*, and Whitehead and Russell's *Principia Mathematica*. This historiographical question and the related questions that we listed as those under consideration in this study have been raised only in the last few years of the final decade of the twentieth century, brought to the fore by Donald Angus Gillies's (b. 1944) exploration of the question whether revolutions occur in the history of logic not unlike the "Copernican revolution" in physics and astronomy that Thomas Samuel Kuhn (1922–1996) first brought to the attention of philosophers of science in his *magnum opus* (1962) *The Structure of Scientific Revolutions*. The leader in this new direction in the historiography of logic has been Volker Peckhaus of the University of Erlangen-Nürnberg (and now at the University of Paderborn), and his recent explications and elucidations (e.g. Peckhaus 1997, 1997a) of the so-called "Fregean revolution" in social terms, as amounting to a "professionalization" or "institutionalization" of logic at the hands, beginning in the last years of the nineteenth century, of mathematicians rather than in the hands of the philosophers who had, for most of logic's history, led the researches in logic, has been an important point of departure for the present study. These issues were also being vigorously pursued by historian of logic and mathematics Ivor Grattan-Guinness of the mathematics department at Middlesex University in England (see, e.g. Grattan-Guinness 1999, 2000). The latter, (Grattan-Guinness 2000), is a particularly rich source of historical information and for references to original work of sometimes forgotten workers in mathematics, logic, and history and philosophy of mathematics and logic.

An investigation of changing conceptions of the nature of the history of logic and to understanding the reasons for the perceived "Fregean revolution" became—*ipso facto*—a study as well of the history of logic of the period, as a part of which history the history of the philosophies of logic of the day, of the textbooks and the pedagogy of logic of the period had necessarily to be included as well. In essence, this study has considered the history of logic for the period 1850–1930, with especial concentration on the half-century 1870–1920, examined as the period during which the "Fregean" revolution, if such there was, would have arisen and reached its fruition. James W. Van Evra (2000, 115) has noted that: "Since each of the major logicians before and after the transition" from Aristotelian syllogistic to algebraic logic "had significant dealings with the syllogism, observing it in transit through the change is a convenient vantage point" from which to observe, evaluate, analyze, and comprehend the historical transitions and the complexities of the mechanics, perceptions, technicalities, and background of the transition. And he adds (Van Evra 2000, 115) that this "tracking" can enable the historian to observe and attempt to understand "the forces operating in logical theory throughout the period," and bring it "into clear relief". This is what he undertakes in examining one process in the history of logic, namely the rise of algebraic logic from the seventeenth century to the end of the nineteenth, as logicians undertook to study and modernize, or algebraicize, traditional logic. Van Evra's (2000) study is limited, however, to the work of British workers in logic in the years 1700–1900.

The nature of the historiographical-sociological aspects of these investigations have suggested and encouraged, in some cases even demanded, apparently tangential, and almost always lengthy, discussions of little known, and sometimes ignored, aspects and episodes of the history of logic and of logic education, and I have permitted myself pursuit of these within these pages, since they frequently provide the intellectual background or contextual framework for the historical and historiographical points of the main thesis. Thus, to take but one example: I have gone to considerable lengths to apprise the reader of the developments in the knowledge of logic in Russia from the medieval period through the nineteenth century, including the trends, from interest in Raymond Llully in the seventeenth century, to the Wolffian logic taught by German professors in Saint Petersburg in the early eighteenth century, and thence the various brands of idealistic and empiricist logics which competed for attention with Aristotelian logic in Russia in the nineteenth century in order to allow the reader to gauge the significance of Russian contributions in the late nineteenth century to algebraic logic. Similar efforts to explore and elucidate in these pages the various historical trends in logic from the early modern period through the first part of the twentieth precipitated the question of whether any connection, however tenuous, could be detected between interest in, if not devotion to, Ramist or Cartesian logic in the early modern era and interest in, if not contributions to, algebraic logic in the latter nineteenth century or "mathematical" logic, in the Frego-Russellian spirit at the end of the nineteenth century and early twentieth century.

A sociological approach, such as suggested by Randall Collins (b. 1941) in his (1998) *The Sociology of Philosophies: A Global Theory of Intellectual Change* for conceptualizing the history of philosophy, provides not only an apparatus for exploring the dynamics of history, but opens the path to a comparison of the particularities and peculiarities of various periods of time within the total environment. And it enables a more in-depth comparison of the intellectual currents of specific times and places, as well as of the thought-patterns that not only emerge within a specific environment, but which were the presuppositions carried over from the past and also those which were preserved in other times and places, and which were abandoned. It is a commonplace of Russian political and intellectual history that, given the circumspection imposed by the regime in power upon intellectuals whose energies would otherwise be absorbed in writing about political philosophy and other sensitive areas, literary efforts, in the guise of fiction, whether prose or poetry, carried out in an "aesopian" language, could serve as the carriers of thoughts that could not, without censorship, be expressed explicitly as nonfiction. What this means is that what often passes for *petite histoire* and is generally disregarded by the "serious" historian, can offer distinct insight into the "temper" of the times and illustrate, for those capable of reading the literary works, the products of the plastic and performing arts, the products of material culture, the folkways, customs, and mores, the everyday items, the depth to which patterns of thought and often the hidden thoughts themselves, are embedded within a culture. By *petite histoire*, we do not, then, indicate either the peculiarly local or the minutiae of daily living, but rather those elements in a cultural and temporal setting and explanatory modes of knowledge which are expressions of the particular milieu in which they occur. These illustrate the background against or within which the broad narrative of *grande histoire*, the social, economic, political, military, intellectual, philosophical, religious, cultural, literary, and scientific currents flow, those aspects of a time and place which, more than individual figures or events, express the temper of the time and place, its *Zeitgeist*. This idea can perhaps be traced to Voltaire, who, in such historical works as *Histoire de Charles XII, roi de Suède* (Voltaire 1731–1732), *Le siècle de Louis XIV* (Voltaire 1751), *Essai sur les mœurs et l'esprit des nations et sur les principaux faits de l'histoire, depuis Charlemagne jusqu'a Louis XIII* (Voltaire 1756–1757), and *Histoire de l'Empire de Russie sous Pierre le Grand* (Voltaire 1759–1763), sought to characterize the internal cultural and intellectual essence of the era of late seventeenth and early eighteenth century. And here, our concern is in particular with the intellectual *Zeitgeist* that informs the development of modern mathematical logic in its various stages and processes. In Russian cultural history, to provide just a handful of examples, historian of Russian literature Nicholas Rzhevsky (b. 1943), in his (1983) *Russian Literature and Ideology: Herzen, Dostoevsky, Leontiev, Tolstoy, Fadayev* (see also Anellis 1987g), has elucidated the ideological elements of some of the greatest of Russian writers of fiction. Ethelbert Courtland Barksdale, Jr. (b. 1944), in his (1979) *Daggers of the Mind: Structuralism and Neuropsychology in an Exploration of the Russian Literary Imagination* (see also Anellis 1983), has attempted to plumb the subconscious of Russian writers on the basis of a hermeneutical and psychological and neuropsychological study of their writings. Alexander D. Nakhimovsky and Alice Stone Nakhimovsky in their translation of essays (1985) *The Semiotics of Russian Cultural History* by Yurii Mikhailovich Lotman (1922–1993) and his school of semioticians of history and culture (see also Anellis 1987h, have presented us with efforts to understand the conceptions and mental frames of references behind the symbolic representations that occur. And in "Perun's Revenge: Understanding the *duxovnaja kul'tura*" (Anellis 1984), I have undertaken to elaborate an essential pattern of thought that, through a dynamic interaction of religious conflict between paganism and Christianity, might explain the reversals, *perelomy*, that periodically appear to occur on the surface of Russian history. The ethnohistorians of mathematics, such as Marcia Ascher (b. 1939), in such studies as *Code of the Quipu: A Study in Media, Mathematics, and Culture* (Ascher and Ascher 1981), *Ethno-mathematics: A Multicultural View of Mathematical Ideas* (Ascher 1991), and *Mathematics Elsewhere: An Exploration of Ideas Across Cultures* (Ascher and Ascher 2002), examine the artifacts of daily life of various "primitive" cultures that have no discernible written material and preserve no discernible formal expression of their mathematical knowledge, on the basis of which they attempt to analyze the extent of their inherent, but not expressed, knowledge of mathematics. It allows us to go more deeply into the origins, often lost in pre-historical mists, of the evolution of mathematics and of mathematical concepts, than might be found in the largely speculative reconstructions of Abraham Seidenberg (1916–1988) (see, e.g. Mathews 1985). Similarly, cultural anthropologist Cheryl Ann Silverman, in her (1989) Columbia University doctoral thesis, *Jewish Emigrés and Popular Images of Jews in Japan*, has utilized elements of *petite histoire*—such as popular literature, newspaper accounts and Japanese translations of William Shakespeare's (1564–1616) *The Merchant of Venice* in its depiction of the Jewish moneylender Shylock—to document the Japanese attitudes towards the Jewish community residing in Japan, that help document the absorption, from the Meiji period to mid-1980s, of European ideas and attitudes. (Shakespeare's own attitude towards Shylock was in turn almost certainly framed and formed by the popular perceptions of his era, as depicted in English society and in literary, historical, religious, and socio-political English Elizabethan-era culture and writings. And it is unlikely that he had any actual personal experience with Jews; see, e.g. Weinmann 2008.) What Silverman's work demonstrates is that the flotsam and jetsam of culture can provide a view of the milieu that is inhabited by a culture that the culture may not openly or clearly express on the surface in and through its deliberate intellectual articulations. The depiction by Feodr Mikhailovich Dostoevskii (1821–1881) in his literary works of the venal and ridiculous *zhid Yankel* and his courteous and cordial journalistic exchanges with the assimilated journalist Avraham Uri (Arkadii Grigor'evich) Kovner (1841/1842–1909) in the pages of Dostoevskii's journal *Dnevnik pisatelya* [*Diary of a Writer*] open the way for social historians such as David I. Goldstein (1981) to explore the ambiguity which these contrasting articulations may tell us about the social status of Jews in late imperial Russia and about the complex relationships between the majority of Orthodox Russians and the religious and ethnic minorities in Russia in a personalized, and perhaps even explanatory, manner that confirms or disconfirms, and in any case supplements, the official record of social history.

The wider lesson, gleaned from the work of Ubiratan D'Ambrosio (b. 1932), the founder of ethnomathematics, is that to fully appreciate and obtain a comprehensive conception of a field of knowledge, as applied to mathematics, and consequently, I shall add, as applied to logic, it is essential to examine three strands or streams: history as taught in schools, history as developed through the creation of mathematics, and the history of that mathematics which is used in the street and the workplace. Continuing this fluvial metaphor, a comprehensive understanding of the history of a field of knowledge requires an examination of their respective sources, or headwaters; of the course of their respective passages through the broader landscape; and of their intermingling in the lake or sea where they emerge and merge. Using another metaphor, that of strands, it is the complex interweaving of these strands into a single, complex, unified pattern, that forms the field of knowledge of our study; and the view of the field of knowledge that presents itself at a specific time and place is articulated by the stage of the weaving of these strands into the cloth that emerges in that time period.

What these studies have in common is that they employ the full panoply of written and unwritten material of a time and place in order to attempt to understand and appreciate the inner and often "hidden" workings of that time and place, and apply them towards understanding the background behind the external and visible productions of that time and place. Applied to the present study, this means examining not only the seminal documents that form the technical and historical bases for mathematical logic—such as Jean van Heijenoort asserted that he attempted to do in his (1967) anthology, but taking digressions to examine as well the philosophical disputations regarding the question: 'Was ist Logik?' and exploring the contents of sometimes forgotten texts in logic that help us view in intimate fashion what logicians were teaching to their students and what logic their students were learning, along with refamiliarizing ourselves with the logic that was known before, during, and after the appearance of the pivotal, groundbreaking works. How better else, than to deeply immerse ourselves in the daily practice of logicians and attempt to conceive of logic as they conceived it—to place ourselves, insofar as possible, within the framework of *petite histoire* as lived by those who made it—, at critical moments, to understand whether or not there was a Fregean revolution, and, if so, when it occurred, how it occurred, and how long it took to accomplish? This "archaeological" study in essence amounts, in our case, to a crucial aspect of the process of "de-sedimentation" and "reactivation" that Brann, the founder of historicism, explains as fundamental to understanding the historical evolution of mathematics and of mathematical concepts, namely allowing the primary sources to "speak for themselves" (2011, xxiv).

Van Heijenoort may, in an important sense, be said to have been a pupil of Leopold von Ranke (1795–1886), the founder of historicism or "scientific history", of allowing the primary sources to "speak for themselves. Ranke made it a practice during his vacations to scour the book stalls, rag markets, and paper markets of Italy it search of texts and remnants of texts containing old documents, and it was to a significant extent on the findings that he thusly acquired that became the basis for his (1878) *Die römischen Päpste in den letzten vier Jahrhunderten*, relying to a large extent upon discarded correspondence between popes and Vatican diplomats and the various posts across Italy where they served, and members of the ruling families of the various city-states. Primary sources, speaking "for themselves", served as the basis of the "scientific history" that enabled the authentic original voices to tell the historian in their own words "*wie es eigentlich gewesen*" (Ranke 1824, vii). Ranke's follower Jacob Christoph Burckhardt's (1818–1897) seminal *Die Cultur der Renaissance in Italien* (1860), well known in English as *The Civilization of the Renaissance in Italy* (Burckhardt 1878), meanwhile extracted, wholly integrated *grande histoire* and *petite histoire*, studying the political and social history of the Italian city-states in the Renaissance on the basis of the monuments that were produced, from historical documents and political writings as well as artistic works; for example, the view of the *condottieri* from paintings and statues as well as historical writings; political theory not only from the works of Niccolò Machiavelli (1469–1527), but from the correspondence and treatises as well as the poetry, of Dante (Durante degli Alighieri; 1265–1321) and Petrarch (Francesco Pertrarca; 1304–1374). The epic poet Jacopo Sannazaro (1458–1530) is probably better known thanks to his portrait by Titian (Tiziano Vecelli, or Vecellio; ca. 1488/1490–1576), even in his own day, than for his own literary production. For Burckhardt, the politics and economics of Renaissance Italy was the background for the social, literary, artistic, and intellectual life of the time and place, and served as a two-way mirror in which the culture was reflected in the political and economic existence and the political and economic life reflected in the monuments and documents of the era.

Ranke's view was shared by Charles Peirce, to the extent that Peirce in his manuscript of 1901 "On the Logic of Drawing History from Ancient Documents, Especially from Testimonies" (Peirce 1901b; see Peirce 1998, 75–114) held that the testimony provided in authenticated historical documents ought to be given credence over the objections of those who believe the assertions made by authors seem "unlikely" or "unproven" according to the Humean version of the theory of balancing likelihoods of probability theory. The theory of balancing likelihoods operates under the presupposition that, if a statement sounds unlikely, it is more probably false than true; to which Peirce objected that, when a concordance of independent testimonies tend towards a given assertion, it is most likely true, since such statements were not made in a vacuum, and these testimonies need to be assessed, therefore, not as independent statements, but as part and parcel of the entire body of evidence within which they occur. In history, Peirce (1901b; 1998, 113) wrote, "the facts are to be explained are, in part, of the nature of monuments, among which are to be reckoned the manuscripts; but the greater part of the facts are documentary; that is, they are assertions and virtual assertions which we read either in the manuscripts or upon inscriptions. The latter class of facts is so much in access," he immediately added, that "history may be said to consist in the interpretation of testimonies, occasionally supported or refuted by the indirect evidence of the monuments."

The value of a sociological approach to the history of mathematics was made express by D'Ambrosio when he wrote (1989, 3) that the history of the sciences and of mathematics "tends to minimize and in some cases ignore the cultural atmosphere and motivations" behind advances in the various fields. The aim of the sociological or ethnomathematical approach is to help recover this background and to restore the balance between the achievements which we regard as significant and the broad background against which, and within which, advances were made. This means, in Kuhnian terms, we are interested not only in the results of "revolutions"—whatever we mean by that—in logic or mathematics, but also the contributions in "normal" developments in logic or mathematics, and in how "normal" and "revolutionary" work in logic or mathematics relate in the over-all portrait of logic or mathematics, both in historical perspective and within the context of their own time and place.

The case can be made for these apparent digressions by appealing to the conception of history set forth by Numa Denis Fustel de Coulanges (1830–1889), as articulated in his University of Strasbourg inaugural address for 1862 (see Fustel de Coulanges 1901) and reiterated in the introduction to his history of political institutions of ancient France (Fustel de Coulages 1891, xi–xiv). He made it a point that in order to fully understand the present—or for that matter any historical era with which we are concerned—it is crucial to obtain an over-arching and complete conception of all of history up to that point. The need to trace the entire course of development arises from the dynamics of development through history, inasmuch as the mutability of intellectual powers entails changes in beliefs, trends, and changes of ideas, along with things that are transformed with these ideas, such as laws, institutions, arts, and science. And if life were static, one could examine the present state of affairs, and have no need for history. Any misunderstanding of developments or perceptions of developments of the past on the part of those looking back are due to examining these as abstractions, without reference to the state of mind and beliefs of those in the past holding those beliefs, ideas, perceptions. More concisely, in the words of Frederick Jackson Turner (1861–1932) in "The Significance of History" (1938, 53), to try to understand one time, it is essential to understand what went before. *Mutatis mutandis* for logic. There would be no need to ask whether there was a Fregean revolution if Frege and Russell had presented Aristotle; the question would not even arise.

The case for apparent digressions can also be made in terms of the uniqueness, or non-duplicability, of historical processes. History is at least as much explanation and judgment as narrative. The historical explanation for facts, for events, or processes, resides within the framework of the narrative. That is to say, history is essentially empirical, although the source for the data resides, not in an external "nature", be it biological, chemical, physical, geological, astronomical, but in human activity and thought. As such we can express the concept of history in empiricist terms, as expressed by historian of science Matthew Norton Wise (b. 1940) (2011, 370), writing about physics as natural history, as follows: "of what historians normally do in exploring historical narratives. Historians are constitutionally empiricist. They generally attempt to write a story with a beginning, middle, and end which incorporates as much factual information as possible into a coherent account, where coherence depends above all on continuity of people and processes. The history may be written from a cultural, social, political, economic, or other perspective; the facts may derive from a wide variety of sources; and interpretation and judgment are always in play, but in general, conviction comes with narrative coherence and empirical adequacy. This much will seem unproblematic to most historians. But the analogue in simulated histories, may help to remind us of what historicity means at a deeper level." Wise's point is that whereas laws of nature, as explanations for experimentally or empirically observed scientific phenomena are general, or global, historical phenomena (facts, events, processes) are local. Nevertheless, a narrative account, at the appropriate (global or local) level provides the explanation for the phenomena under consideration. And, the more data accumulated, that is, the more detailed the narrative, the more accurate and feasible is the explanation.

What historian of geometry Friedrich Engel (1861–1941), in his review of Russell's *Principles of Mathematics* wrote (Engel 1904) with regard to the attitudes of the working mathematician towards foundations or logic of mathematics has typically applied with equal force to the attitudes of philosophers of mathematics, philosophers of logic, and working mathematicians and logicians towards the history of logic; that "die meisten produktiven Mathematiker überhaupt nicht viel Neigung haben, sich mit solchen philosophischen Spekulationen über die allerersten Grundlagen ihrer Wissenschaft abzugeben, ebensowenig wie etwa der wirkliche Musiker das Bedürfnis hat, sich darüber Rechenschaft zu geben, worauf die musikalische Logik, die ihn sein Ohr lehrt, eigentlich beruht," that they generally do not give considerable attention to their history. I would go so far as to suggest that the working mathematician, the working logician, is interested in the history of a topic only to the extent that there is a result, a theorem, that they require for the immediate advancement of their current investigation; and that this attitude helps to sustain the acceptance of the popular or folkloric historical conceptions that have been comfortably—even authoritatively—established.

The efforts that have been undertaken since the last decades of the twentieth century to teach mathematics using an historical approach and have grown considerably since pioneered by a handful of logicians and mathematicians such as V. Frederick Rickey (see Rickey 1992, 1995, 1997, 2010) have as their aim something deeper than the endeavor to introduce students to the conception that mathematics is not a cut-and-dried body of immutable and unchanging rules for rote memorization carved in stone from the beginning of time. One of Rickey's practices has been to employ historical texts in place of the potted exercises that so many mathematics textbooks provide for students. Today, these homework problems can appear to be artificial at best. But, seen from the perspective of the authors of ancient textbooks, they elucidate the development of concepts and techniques that were active puzzles and problems for the authors of those early books, rather than canned homework problems to test and torture the student. At the same time, taken within the historical context rather than pulled, seemingly at random, from a contemporary textbook, the material illustrates how and why concepts and techniques were developed, what their intellectual background and parentage were, and why and how we use these concepts and techniques rather than some others. Among the aims that Rickey lists in favor of this approach are: (1) to give life to your knowledge of mathematics; (2) to provide an overview of mathematics—so you can see how your various courses fit together and to see where they come from; (3) to show you that mathematics is part of our culture. This is clearly in full consonance with the notebook comment (as quoted by Ore 1957, 138; Swetz et al. 1995, vii) of Niels Henrik Abel (1802–1829) that "if one wants to make progress in mathematics, one should study the masters and not the pupils." The case for using the classics is also made by (Roy 2011), providing concrete examples to illustrate not only how and why mathematicians developed their ideas, but learned to extend the implications of their own results and those of others. The over-arching benefit, beyond illustrating inspirations, motivations, and exploring potential new avenues for further development, it is concluded, is to display mathematics as "dynamic and evolving" (Roy 2011, 1285). As Morris Klein (1908–1992) has reminded us: Charles-Émile Picard (1856–1941) noted (1905, 5) that "If Newton and Leibniz had known that continuous functions need not necessarily have a derivative, the differential calculus would never have been created."

As the present study developed, it seemed increasingly important and relevant that a history of the historiography of logic be undertaken at some point by someone, although such a task would indubitably be dauntingly difficult, time-consuming, and in all likelihood gargantuan, and even more certainly beyond the capabilities or resources of the present author, or perhaps of any single individual researcher. The body of the present investigation has consumed close to a hundred pages of text per decade if counted in terms of its focus on the period 1850–1930; nor is this surprising: historian of mathematics Florian Cajori (1859–1930) (1918, 282) projected, on the basis of the nearly one thousand pages of the four volumes of Moritz Cantor's (1829–1929) (1880–1908) *Vorlesungen über Geschichte der Mathematik* (vol. 1: Antiquity to 1200 AD; vol. 2: 1200–1668; vol. 3: 1668–1758; vol. 4: 1758–1799) that, at the same scale and depth of coverage, fourteen or fifteen volumes of one thousand pages apiece would be required for the nineteenth century. Joong Fang (1923–2010) (1972a, 43), continuing this projection, estimated that the period 1900–1960 alone, with the same depth of coverage, would require two or three scores, if not more, of volumes of at least 1000 pages each! Perhaps the present study will serve as an invitation to others to undertake such a project. The sheer size of the enterprise undertaken, the amount of material covered, the structure of the arguments and the nature of the problematics, also made it virtually inevitable that there would be some repetitions, and readers are therefore asked for their indulgence in that matter.

Recognition of the role which history of mathematics can play in clarifying issues in the philosophy of mathematics, and the integration of history and philosophy of logic, aiding in the clarification of epistemological questions in history and historiography of mathematics in general, as well as in the history and historiography of logic, given that these connections are a significant factor in the present study, can indubitably be traced to a graduate philosophy of science course which I took with Richard M. Burian in 1976. The focus was on the epistemology of science and the conception, owed in large measure to the work of Thomas Kuhn, Karl Raimund Popper (1902–1994), Imre Lakatos (1922–1974), and Paul Feyerabend (1924–1994), that an historical examination of the development of scientific theories, gives a more accurate representation of the reality of scientific investigation than does a rational reconstruction, such as proposed by the logical positivists. As Burian (1977, 1) wrote in the abstract of his paper "More Than a Marriage of Convenience: On the Inextricability of History and Philosophy of Science": "History of science, it has been argued, has benefited philosophers of science primarily by forcing them into greater contact with "real science."" He argues that "additional major benefits arise from the importance of specifically historical considerations within philosophy of science. Loci for specifically historical investigations include: (1) making and evaluating rational reconstructions of particular theories and explanations, (2) estimating the degree of support earned by particular theories and theoretical claims, and (3) evaluating proposed philosophical norms for the evaluation of the degree of support for theories and the worth of explanations." He also argues that theories develop and change structure with time, that (like biological species) they are historical entities. Accordingly, both the identification and the evaluation of theories are essentially historical in character." He had already raised the question of the degree of incommensurability between apparently competing theories (Burian 1975), and concluded (Burian 1975, 20) that a careful historical investigation suggests that there are constraints the "full-fledged adoption of a revolutionary theory" of which should "enable one to understand the continuities in the development of science and to define the sense in which there is progress in science even in the face of the wildest variation in the fundamental ontology of our theories." Applying to mathematics some of the insights gained in this context, it was easy to locate examples from the history of mathematics that demonstrated that mathematical progress is not always linear (see, e.g. Anellis 1989a). But this raises the question of whether there are also revolutions in mathematics, or in logic, that are at least as definitive for the history of mathematics and logic as the Copernican revolution in astronomy was considered to be by Kuhn.

The other factor in my approach to the history of mathematics and history of logic was formulated by a deep and long-lasting interest in history that I developed long before turning to the study of philosophy and mathematics. This interest in history was encouraged by Norman Rosenblatt (1928–1991) of Northeastern University in Boston, a specialist on the history of medieval Spain who displayed a wicker figure of Don Quixote aboard his horse atop his office file cabinet, and who, while I was still an undergraduate student, nominated me for membership in the Mediaeval Academy of America. My study of historiography in Robert Arnold Feer's (ca. 1926–1970) "Historian's Craft" course at Northeastern in 1967 and David L. Wilmarth's (1924–1996) two-part "History of Science" course the following year, which stressed the interrelations between the three elements of the "STS complex" of science, technology, and society, have, in conjunction with J. Rosson Overcash's three-part "Introduction to Earth Sciences" which included (in 1965) a history of astronomy from the Babylonians through Newton, along with what I learned from Burian, been one of the guides to my approach to the history of mathematics and history of logic. Feer taught not only that history was "great good fun" (perhaps echoing the sentiments of Marc Léopold Benjamin Bloch (1886–1944) in the *Historian's Craft* (1962, 7–8), one of the assigned readings for the historiography course that Feer taught), but the need to be critical, analytic, one assignment intended to teach critical analysis of sources being the writing of a review of Einhard's (ca. 770–840) *Life of Charlemagne* (*Vita Caroli Magni imperatoris*; see Einhard 1966), and the necessity of familiarizing and availing oneself of a wide variety of documentary materials from a wide variety of sources. Feer described the course as: "A discussion of the ways in which the historian studies the past and the nature of historical statements. Problems considered include research techniques, changing conceptions of historical knowledge, and the relationship between the historian and the society in which he works," thus offering not only technical tools, but seeking to ingrain an integrated approach to study not just events, but the entire backdrop against which, and within which, they occurred, in order to develop a more complete understanding of the events, their significance, and their context. Suzanne L. Hamner, in "The Rise of Nation States" (in 1967), covering the period in European history from the thirteenth to seventeenth centuries and included an integrated history of monarchies, economic development of capitalism and international trade, and the conflicts between church and state from the *Investiturstreit* through the Reformation, taught the need to go beyond the glamorous developments, such as the political, military, and diplomatic, but to the often very mundane aspects of history that often make political, diplomatic, and military arrangements possible (for example, the development of mining and the manufacture of saltpeter for gunpowder), as well as a study of the essential ideas formulated during a period, such as the treatises on natural law, domestic and international law and diplomacy, not just of the very familiar, such as Machiavelli, but of the somewhat less familiar, such as Jean Charlier de Gerson (1363–1429), Jean Bodin (1530–1596), and Hugo Grotius (Huig de Groot; 1583–1645), in works such as the 1960 edition of John Neville Figgis's (1866–1919) (1907) *Political Thought from Gerson to Grotius, 1414–1625*. The lesson that I suggest may be extracted from all this, as applied to history of mathematics in general and history of logic in particular, and indeed to any field of intellectual history, is that in order to obtain a better appreciation of the contributions and advances made by "leading" or pioneering figures, it is also useful, even necessary, to pay attention as well to the more pedestrian work of their predecessors and contemporaries. (I have elsewhere (Anellis 1992d) considered more fully the implications and applications for history of mathematics, and especially for history of logic of this approach.)

In *The Historian's Craft* by Marc Bloch, one of the leading experts on medieval feudalism of his day, the author demanded (Bloch 1962, 81) what he called a healthy "criticism of the documents of the archives," and what can be called an historical understanding of the past through reliving that past. This is perhaps the basis behind the conception by Jan Marius Romein (1893–1962), as described by Pieter Catharinus Arie Geyl (1887–1966) (1961), of "bowing to the spirit of the age" as an effort to understand the temper of the times one studies as the means to understanding, if not experiencing, the lived realities that went into the decision-making processes of those aspects of history that are regarded as worthy of description in the wider historical narrative, ... and into the history textbooks.

The conception that a scientific theory could be better understood by reference to its history is not incompatible with Charles Peirce's pragmatic conception of truth as a communal enterprise in which there is an increasingly greater approximation to truth through increasing stores of observational data. Thus, there is an evolutionary process in the formation of scientific theories as data accumulate and lead to increasingly refined and accurate theory formation, and thus a self-correcting process. Again, this raises the possibility of revolutionary change within science, but does not yet answer the question of whether scientific revolutions can or do in fact occur: *mutatis mutandis*, we may say, for revolutions in mathematics, and perforce in logic.

It is still not uncommon to assert without qualification, as William R. Everdell (b. 1941) (1997, 43) did in attempting to understand the concept of "modernity" as arising in the twentieth century and to identify its characteristics in the sciences, humanities, and arts, that, in respect to mathematics generally and logic in particular, "Gottlob Frege has the honor of being the true begetter of mathematical logic," and his belief that this view of

Frege is still shared by "common consent" among mathematicians, historians, and even philosophers. But this view is outdated, as, fully a decade prior, even Quine, who for at least half a century beginning at least as early as the 1930s (see Quine 1934–1935; 1934–1935a), had once held the position, came to admit, in 1985 (see e.g. Quine 1985, 1995, 1995a), that much of the accomplishments which had been attributed to the Frege of the 1897 *Begriffsschrift*, had become common currency and entered the mainstream of logical research through the publications and influences of Peirce in the 1870s and 1880s, long before logicians, beginning with Russell in 1900 began to take cognizance of Frege's work. That this view remains entrenched, despite efforts of historians of logic to redress the imbalance since the early 1980s—starting with the interest of Hilary Putnam (1982) and Quine (1985) to begin to reexamine and reevaluate the work in logic in particular of Peirce, to enhance the handful of scattered efforts beginning in the 1950s (e.g., Quine's student George David W. Berry (ca. 1915/1916–1986) (1952)) to consider various aspects of Peirce's work in logic, the conception of the singular centrality and vital significance of Frege as the founder of modern logic remains sufficiently pervasive that J. Brent Crouch (2011) still thought it to be typical enough in the first decade of the twenty-first century, that he cited (Crouch 2011, 155) the historical note (Quine 1961, i) from the 1961 edition of Quine's *Methods of Logic* as evidence of the status of Peirce and Frege in the contemporary historiography of logic.

If we examine the sociological literature on the theme of intellectual transmission across temporal and cultural (including both territorial and what Hegel would call *zeitgeistige*), as Randall Collins has done for the history of philosophy (Collins 1998), we find a useful terminological framework for our own investigation of the dynamics of paradigm shifts without necessarily accepting the particulars of the political and social forces that operate within and between networks. In *The Sociology of Philosophies: A Global Theory of Intellectual Change*, Collins (1998, xviii) speaks of "intellectual networks", which are the links between communities of thinkers whose shared (sets of) ideas are passed from one generation to the next. But these complexes of ideas do not remain static over time; neither do they necessarily remain static across communities that are temporally contemporaneous. And whether over time or across networks these complexes evolve or revolve in paradigm shifts, there is a dynamic that underlies and explains the change, which occurs either gradually or swiftly and suddenly, and in which, none, a few, some, many, or all elements of the shared complex of ideas that typify an intellectual network are overthrown or replaced. A task of the historian of ideas is to investigate and attempt to understand the dynamics that lead one intellectual network to morph into or replace, or succeed the next and to explore and understand the array of ideas involved in these networks and both those that remain and those that are displaced or altered. Collins (1998, 1) also notes that behind the dynamics of network change, there is a limited focus on a select number of topics, and the conflict that leads to the succession of networks or paradigm shifts is carried out not by individuals but by a small number of "warring camps." Here we have undertaken to examine the dynamics that occurred in and precipitated the putative "Fregean revolution," to attempt to understand, why, how, and, for that matter, whether, there was a Fregean revolution in virtue of which, in Everdell's words, "Gottlob Frege has the honor of being the true begetter of mathematical logic," and to determine the historical as well as intellectual factors that were, or might have be, involved. Thus, for example, we are obliged to ask whether the "Booleans" belonged to the Aristotelian or to the "Fregean" network, whether, and if so, to what extent, they played a role, if any, in the paradigm shift of the "Fregean revolution", and, coincidently, whether the dichotomy between "Booleans" and "Fregeans", to use the terms coined by Hans-Dieter Sluga (b. 1939) (1987) and endorsed, in different terminology by historians of logic such as Jean van Heijenoort (e.g. in his 1967a) and philosophers of logic such as Jaakko Hintikka (b. 1929) (e.g. in his 1997), is historically justified, or is a false dichotomy. In other words, part of what we are after is a full account of the relation between the "warring camps" of "Booleans" and "Fregeans" that led to Frege having the honor of being "the true begetter of mathematical logic."

I am conscious, nevertheless, that however detailed the ensuing discussion may be, it is neither complete nor entirely objective. The erudite scholarship demanded by von Ranke and his followers required to present the past *"wie es eigentlich gewesen"* (Ranke 1824, vii), based entirely on critically examined and verified documents and entirely without the imposition of the historical researcher's presuppositions has been challenged by more recent historians, such as Carl Lotus Becker (1873–1945) (1931, 1955) and Alexander Alexandrovich Goldenweiser (1880–1940) (1936), who acknowledge that even the supposedly simplest hard "fact"—e.g., that "Caesar crossed the Rubicon"—has a multitude of associations and congeries of facts that contributed to the meaning and significance of the crossing in 49 BC of the Rubicon by Gaius Julius Caesar (102/100–44 BC). Moreover, there were clusters of other facts associated with the very act of that river crossing that may or may not have played a peripheral or decisive role attendant to the river crossing that the historian who records the event of Caesar's crossing of the Rubicon cannot overcome his or her own circumstances, beliefs, or historical situation in presenting the fact that Caesar crossed the Rubicon, his purposes in doing so, the activities and motives of others that motivated Caesar's crossing, and the correlation and selection of facts that are recorded as explanations of, or forwarded as background to, Caesar's crossing of the Rubicon. That is to say: the historian cannot, in a fully-fledged Rankean sense, provide an account pure and simple of "wie es eigentlich gewesen", but only of the historian's own best conception of "wie es eigentlich gewesen."

In history, perception can play as much a part as reality. This is a view that, on the surface, runs counter to the historicism of von Ranke. This was brought home in a rather trivial way by Raymond H. Robinson (b. 1927), with whom I took "U.S. History to 1866" at Northeastern University in the Fall of 1970, as he described his first face-to-face meeting with a descendant of George Washington (1732–1799), in connection with research on Washington and his family, and how, after telephonic interviews, neither, upon meeting in person, corresponded with their expectations of the other. He recounted that, when she opened the door, they simultaneously burst out laughing, because neither of them turned out to meet the expectations that they had formulated: he expecting an elderly society matron of the colonial *grande dame*, Daughters of the American Revolution type, she expecting a wizened old academic of the *zerstreute Professor* type. Many decades later, in "The Marketing of an Icon", published in *George Washington: American Symbol* by Barbara J. Mitnick, Robinson (1999, 109) explored the lives of artists who supported themselves with depictions of Washington. Robinson wrote: "In the year of the 1876 centennial and for several decades afterward—a period known today as the "Colonial Revival"—George Washington perfectly symbolized for these Americans both the nation's past and its hopes for the future." I suppose it was at least in part Robinson's example that I had in mind when, in "On the Selection and Use of Sources in the History of Logic", I argued the case that rumors and hearsay, even false or partially false, nevertheless constitute part of our conception, or better, perception, of the past, and so are relevant to an understanding of the intellectual milieu being investigated, even as the false or partially false aspects are to be refuted.

To obtain some idea of the complexities involved in comprehending "wie es eigentlich gewesen", and what "eigentlich gewesen" means, we may turn to the famous account by the Roman historian Suetonius (Gaius Suetonius Tranquillus; ca. 69–after 130 AD), in his *De vitus Caesarum*:

> consecutusque cohortis ad Rubiconem flumen, qui prouinciae eius finis erat, paulum constitit, ac reputans quantum moliretur, conuersus ad proximos: 'etiam nunc' inquit, 'regredi possumus; quod si ponticulum transierimus, omnia armis agenda erunt.'
> [32] Cunctanti ostentum tale factum est. quidam eximia magnitudine et forma in proximo sedens repente apparuit harundine canens; ad quem audiendum cum praeter pastores plurimi etiam ex stationibus milites concurrissent interque eos et aeneatores, rapta ab uno tuba prosiliuit ad flumen et ingenti spiritu classicum exorsus pertendit ad alteram ripam. tunc Caesar: 'eatur,' inquit, 'quo deorum ostenta et inimicorum iniquitas uocat.'
> [33] Iacta alea est,' inquit.

or, in the well-known translation, *Lives of the Twelve Caesars*, by Robert Ranke Graves (1895–1985): (Suetonius 1957):

> Caesar overtook his advanced guard at the river Rubicon, which formed the frontier between Gaul and Italy. Well aware how criticial a decision confronted him, he turned to his staff, remarking: 'We may still draw back: but, once across that little bridge, we shall have to fight it out.' As he stood, in two minds, an apparition of superhuman size and beauty was seen sitting on the river bank playing a reed pipe. A party of shepherds gathered around to listen and, when some of Caesar's men, including some of the trumpeters, broke ranks to do the same, the apparition snatched a trumpet from one of them, ran down to the river, blew a thunderous blast, and crossed over. Caesar exclaimed: 'Let us accept this as a sign from the Gods, and follow where they beckon, in vengeance on our double-dealing enemies. The die is cast.'

To Ranke's "wie es eigentlich gewesen", one must, assuredly therefore, add the words of the Marschellin in "Der Rosenkavalier" of Richard Strauss (1864–1949) and Hugo von Hofmannsthal (1874–1929) (see Strauss and Hofmannsthal 1911, 42): "*und in dem 'Wie', da liegt der ganze Unterschied.*" By itself, Caesar's crossing of the Rubicon is insignificant. What gives it its significance are the surrounding circumstances, the intentions; the direction of the crossing; whether fabricated or real, mythical or mundane, behind it; whether with the army or alone, summoned to Rome or prohibited by law, authority, or tradition from entering Rome with an armed host, in particular in a hostile manner, rather than alone and unarmed; and the resulting consequences, whether actual or symbolical. As historian William Hardy McNeill (b. 1917) noted (McNeill 1986, 1998), bare facts alone, no matter how certain, are not the contents of history; they assume significance only to the extent that they fit within a pattern that provides meaning that constitutes their significance, and pattern recognition is interpretive, for which reason he proposes that the borderline between myth and history fluctuates as historians argue over the patterns, and hence the truth and falsity of their various interpretations, even should they unanimously agree that, e.g., Caesar crossed the Rubicon on 10 January 49 BC. All history, then, is better understood as what McNeill proposes terming "mythistory": "*Quid est veritas?*" This is fully in consonance with McNeill's memoir, which is titled "The Pursuit of Truth" (McNeill 2005), emphasizing the quest rather than the absolute certainty supposedly behind the search for "wie es eigentlich gewesen".

As summarized by Harry Elmer Barnes (1889–1968) in his history of historiography (1962, 271), "historical facts present no significance whatever until they have been selected, sifted, analyzed, and interpreted to show their bearing upon the flow of civilization." To stop with the mere recordation of barren facts is tantamount, he adds, to the scientist who never went beyond laboratory experiments and recording his observations in his notebooks, without attempting to determine to what conclusions the laboratory results lead and how they fit with, expand, and add to or detract from the general body of scientific knowledge. To Ranke's call for the historian to tell *wie es eigentlich gewesen*, then, we must add the equally essential task called for by historian of ideas James Harvey Robinson (1863–1936) (1912, 62), to tell *wie es eigentlich geworden*—how, and why, *wie es eigentlich gewesen* came to be.

Combining the views of Ranke and Robinson with those of Marc Bloch and historian George Malcolm Young (1882–1959), who (as quoted, without citation in Stern 1956, 28) enjoined the historian to "go on reading till you hear the people speaking. The essential matter is not what happened, but what people thought, and said about it . . . ", one may well come to the historical desideratum of Jan Romein, as described by Pieter Geyl (1961), of "bowing to the spirit of the age" one studies, which entails a description of the *Zeitgeist* under consideration. When Romein advocated "bowing to the spirit of the age," he declared that there are two spirits of any age, the *true spirit* and the *false spirit* (the latter encompassing all manifestations that are not true). The proper aim of the historian is to depict the former, "the certainty as I understand it" (see Geyl 1961, 325). A similar sentiment was expressed by Romein's and Geyl's colleague, the social historian of the early modern period Johan Huizinga (1872–1945) (1959, 60) in his essay on "The Task of Cultural History". Although Huizinga's style of history, of "bowing to the spirit of the age, was undertaken as well by George Gordon Coulton (1858–1947) in such books, with self-explanatory titles as *Social Life in Britain from the Conquest to the Reformation* (Coulton 1919) and *Medieval Panorama* (Coulton 1938), social history as an independent discipline within professional history did not become fashionable until the late 1960s and the 1970s, with the advent of the civil rights, human rights, and social consciousness-raising movements. It is still more recent that this sociological approach has entered the historiographical streams in history of philosophy, of history of science, and finally of history of mathematics as well.

The sociological approach to the history of sciences, then, developed later than the economic approach by Marxist historians and philosophers of science such as John Desmond Bernal (1901–1971), is a twentieth-century, or at least a modern, or post-industrial historiographical development, which took its cue primarily from general history. Along similar lines, the STS complex that Wilmarth propounded, is comparatively new, in the sense, for example, that, even while seeking to teach the history of science to humanities students, James Edward McClellan, III (b. 1946) and Harold Dorn (b. 1928), in their *Science and Technology in World History* (McClellan and Dorn 1999, 2006), which examines, from earliest times to the present, the role of science and technology in history, with emphasis on the role of the ruling elite and the intellectual establishment which they supported in directing the development of science and technology, focus their attention on the impact which technology had on science. They argue that prior to the twentieth century (and the establishment of a broad scientific community that could be harnessed for research to enhance and carry out national goals), the direction within the STS complex always flowed in the single direction from technology to science; while in the twentieth century, the flow became bidirectional. What largely distinguishes the first (1999) from the second (2006) edition of their book is the shift away from an internalist, or, as historian of science Herbert Butterfield (1900–1979) would have called it, a "whiggish" interpretation of the history. The sociological perspective of history of the scientific enterprise is developed in such studies as *Never Pure* (Shapin 2010) by historian of ideas Stephen Shapin, in which, as explained in the subtitle, "historical studies of science as if it was produced by people with bodies, situated in time, space, culture, and society, and struggling for credibility and authority" are undertaken. This is, in essence, what I am attempting in the present study with respect to the history of logic.

The distinction between the true spirit and the false spirit that Romein distinguished, is still newer to history of mathematics, and is echoed in Butterfield's (1931, v–vi) concern for a whig interpretation of history and the distinction between internalist and externalist histories of science and histories of mathematics. This distinction, expressed by Ivor Grattan-Guinness (1997a, 7), is between history as concerned with what happened in the past on the one hand and histories which ask merely which mathematics of the past led to the mathematics of today, and how. The latter question is satisfied merely to reformulate or translate the mathematics of the past in the terms of present-day mathematics, however distorting that might appear to the mathematicians of the past. The former question begs us to attempt to understand the history of logic not merely in terms of revolutions or in terms of pivotal or putatively pivotal advances by leading figures or pioneers, whether Aristotle, Boole, or Frege and their presumptive intellectual equals, but in terms as well of the pedestrian plodders, whether obscure or even unknown, if only in the effort to understand (a) what significance the contributions of the Aristotles, Leibnizes, Booles, Freges and their equals of the history of logic have had, and (b) what reactions and conceptions their contemporaries and successors had to their accomplishments. We may in this regard take seriously Newton's acclaimed profession (in his letter of February 5 1676 to Robert Hooke (1635–1703)), that if he had "seen farther" than others, it is because he stood "on the shoulders" of others. Examining more closely the relevant passage, Newton tells Hooke (as quoted in Éspinasse 1956, 11): "You defer too much to my ability in searching into this subject," adding that "What Descartes did was a good step. You have added much several ways . . . . If I have seen further, it is by standing on the shoulders of giants." Taking our cue, then, from Kuhn, we cannot fully understand or appreciate the significance of "revolutionary science" without examining it alongside the development of "normal science". Applied to history of logic, we can argue, as Thomas Drucker (personal communication, 16 May 2010) has, that while certain authors tend to be looked at again and again because everyone feels more comfortable writing about Frege and Hilbert rather than some of the less well-known contributors to the discussion. And if we feel more comfortable with those authors, it is because the secondary literature is large enough that they can find someone else on whom to pin their own interpretation. This may lead to an ingrained sense of the inevitability of the conclusion that Frege and Hilbert are more influential than other, "minor", figures. It certainly is a cyclical argument. Trying to argue against this narrow base involves the claim (1) that there is something interesting in the authors at whom you are looking and (2) that they well may have had some influence on the their own, and perhaps on the next couple of generations of logicians. And if so, we can easily miss the background against which the "major" figures established themselves as original innovators and became dominant influences. This delving into the work of the forgotten legions of obscure logicians, as Drucker [private communication, 17 May 2010] has it, is the means to point to "the undoubted shift in logic over a certain period." To this end it seems desirable to spend as much time with Pedro Teixeira's (d. 1641) (1666) *Compendium logicae* as with Leibniz's (1666) *De Arte Combinatoria*, Joseph Devey's (1825–1897) (1854) *Logic* as with Boole's (1854) *Laws of Thought*, and with Alexander MacFarlane's (1851–1913) (1879) *Principles of the Algebra of Logic* as with Frege's (1879) *Begriffsschrift*, and to examine equally the logical writings of Henry Philip Tappan (1805–1881) as of Adolf Trendelenburg, of Richard Whately (1788–1863) as of Wilhelm Wundt, or of George Leonhard Rabus (1835–1916) or Carveth Read (1848–1941) as of Bertrand Russell.

Thomas Henry Buckle (1821–1862) undertook in his *History of Civilization in England* to develop history as a scientific discipline and to employ the assembled evidence and artifacts of history in its detail to formulate general laws of human action and nature by induction from the collection and organization of historical facts. In writing of the proximate causes of the French revolution, he expressed the notion that, more important than the particulars of change are the intellectual transformations. Thus, he wrote (Buckle 1864, II, 761–762): "That to which attention is usually drawn by the compilers of history is, not the change, but is merely the external result which follows the change. The real history of the human race is the history of tendencies which are perceived by the mind, and not of events which are discerned by the senses," such as won or lost battles, dynastic escapades and the like, that is, "matters which fall entirely within the province of senses, and the moment in which they happen can be recorded by the most ordinary observers." These form the data upon which historical laws can be formulated. But: "those great intellectual revolutions upon which all other revolutions are based, cannot be measured by so simple a standard. To trace the movements of the human mind, it is necessary to contemplate it under several aspects, and then co-ordinate the results . . ." This is what I have endeavored in attempting to understand whether there was, indeed, a "Fregean revolution" in logic, and if so, of what it consisted and how it occurred.

It should also be recognized that those who participate in creating historical change do not always themselves recognize the changes which they help to bring about. It is far easier to see the short term than to forecast the long-term consequences of one's work. An essential aspect, then, of comprehending historical developments, in addition to comparing "revolutionary" science against the background of "normal" science, is to recognize that long-term impacts may not necessarily be discernible in immediate circumstances, however powerful and compelling they may prove themselves to be in retrospect. Thus, wrote Huizinga in "The Idea of History" (Huizinga 1956, 292): "The historian . . . must always maintain towards his subject an indeterminate point of view. He must constantly put himself at a point in the past at which the known factors still seem to present different outcomes." Here is one more reason for going into detail into the work not only of Frege and Russell and their contemporaries, but also those who came before and those who came after, if we are to have a possibility of recognizing and comprehending the significance and implications of their accomplishments, as well as of how they were understood by their contemporaries and successors in the field. What is desired, if not demanded, by an understanding and appreciation of "wie es eigentlich gewesen", in fact of the actuality and essence or nature, of a "Fregean revolution", is, then, a sense of *how* the logicians of that day conceived of their field in their own time. An essentially negative reason for delving with a microscope into a period of transition to examine the thought of largely forgotten figures, those whose work have little intrinsic historical interest from the standpoint of tracing the major developments of a field, was given by Collins (1998, 64), namely, for the sake of what he calls historical "realism". Collins (1998, 89 ff.) also makes the point, using Aristotle as an example, that it is the subsequent reactions of generations of researchers who define one of their number as a significant, even pivotal and path-breaking, figure. It is the effort of multiple generations of scholars who, largely forgotten, toying with various aspects of his thought, modifying it, criticizing it, commenting upon it, teaching it, that lent Aristotle the reputation, *circa* 1250, as "ille Philosophus"—"the Philosopher", requiring, in Thomas Aquinas's (1225–1274) writings, via the influences of Averroës (Abul-Walid Muhammad Ibn Roshd; 1126–1198), no further identification. What this allows us to infer is that it is only through the elevation of some figure's work by successive workers of minor to modest quality and renown that one research's work is understood as significant, central, pivotal, crucial, or revolutionary. And from this, we might well conclude, it is not necessarily the originator of an idea, paradigm, or school that guarantees the significance of a researcher's work, or of his path-breaking contribution, but those who, in hindsight, judge it and deploy it, who render it central to, even revolutionary in, the development of their field. This suggests that, to understand and appreciate the path of, even the very existence of, a Fregean revolution in logic (or a Boolean, or, for that matter, any other putative field-altering development), we again must examine not only Frege's work, but the broad context in which his work took place, before, during, and after, and the attitudes and perspectives of others towards his achievements.

James van Evra's (2000) study of the development of logic, and in particular of the relation between the algebra of logic and the Aristotelian syllogistic logic, is thus an example of the effort to examine in detail the work not only of Boole and De Morgan and their contemporaries, but also those who came before and those who came after, from 1600 to 1900, to have a possibility of recognizing and comprehending the significance and implications of their accomplishments, as well as of how it was understood by their contemporaries and successors in the field. To this end, Van Evra observed the subtle and not so subtle differences in the understanding of the nature of the syllogism in the writings on logic during the three hundred years within which Boole's and De Morgan's contributions, and those of their contemporaries, occurred, starting with Robert Sanderson's (1587–1663) widely popular and wildly influential *Logicae Artis Compendium*, which went through many multiple editions since its appearance in 1615—ten editions alone by 1700, at least ten thousand copies sold by 1678 (see Van Evra 2000, 118), and Henry Aldrich's (1647–1710) even more popular and influential successor, *Logicae Artis Compendium* (1691), which included nineteenth-century editions by John Hill (Aldrich 1821) and Henry Longueville Mansel (1820–1871) (Aldrich 1823) and an English version (Aldrich 1823a); taking into account the early nineteenth-century views on the syllogism of the more influential British philosophers writing on logic, Thomas Reid (1710–1796) and William Stirling Hamilton (1788–1856); before taking up consideration of the work in logic in the mid-nineteenth-century of George Boole and Augustus De Morgan, who developed algebraic logic and the logic of relations; before finally examining the work in the later nineteenth century of their most prominent continuator, Charles Sanders Peirce—all of whom had something to say about the syllogism and undertook to improve or go beyond what they inherited from Aristotle by way of the medieval logicians. The development that Van Evra considers is a segment, albeit a crucial aspect, of the whole development that I seek here to explore. And therefore

I trace a wider historical swath than did Van Evra. But the methodology is essentially the same: to attempt to place what is pivotal in the work of the "Aristoteleans", "Booleans", and "Fregeans" in order to assess what the so-called Fregean revolution in logic, if any, may be, and what role the "Booleans" and the "Fregean's" played in the process of that development.

Nathan Houser read several drafts of this work and provided encouragement for its publication. Benjamin Hawkins read an early version and likewise proved to be an important source of encouragement for this undertaking, as well as for suggestions for improvements and corrections.

I have also benefited from reading the writings of many other historians of logic, and there are indubitably instances where I have been unable to provide a proper attribution; in such cases, only a quote from Francis Herbert Bradley (1846–1924) may come close to making the necessary and proper acknowledgments which might otherwise be missed; he said (as quoted at Keen 1971, 10–11): "I don't claim originality ever for anything because I have read various writers of various schools and my memory is so bad that I may recall something without remembering its source or that it had a source." The opinions expressed herein nevertheless—it may be obligatory, but perhaps obviously and necessarily redundant, to attest—remain those of the author. Along similar lines, chemist and novelist Carl Djerassi (b. 1923) (1991, 59) wrote in his novel *Cantor's Dilemma*: "You can't help but remember what you read, and after a while, say a few months or even weeks later, you forget where you first saw it and gradually you think it's your own idea." Pursued to its ultimate extremity, people such as these should perhaps be identified either with those who literary critic Vissarion Grigor'evich Belinskii (1811–1848) described (see Belinskii 1976, 31) as "people who lack the ability to have their own opinion on things, and who accept another's opinion completely as something they no longer need to think about," or, slightly less unflatteringly, with those who

> while constantly living on another's opinion, have the ability to make it their own, to develop it, to extract new corollaries from it, to discover other ideas through it. This ability so deceives people of this type that they are very sincerely convinced of their own ability to think; with their lively and receptive natures, they themselves do not know and do not understand who transmitted a certain idea to them because everything from without adheres to them almost unconsciously, instinctively. They have only to speak with an intelligent person or to read a good book and immediately a whole series of new ideas that they cannot help accepting as their own rises within them. . . .

If I have absorbed the leading findings of the leading current workers in history of logic, from Grattan-Guinness, Hawkins, Houser, Peckhaus and others, and combined them in a significant and meaningful way and with what I learned from van Heijenoort, to formulate from all these a new unified and coherent synthesis, I shall be gratified. If along the way I have managed to also contribute to a better understanding of, appreciation for, and convincing perspective on a crucial period in the history of mathematical logic, I shall deem myself most fortunate.

Last but not least, I am also significantly obliged to my host since 2008, the Peirce Edition Project, and the Institute for American Thought, both at Indiana University-Purdue University at Indianapolis, and the IAT and PEP staff and IUPUI library, for providing a most congenial setting for my work and for the intellectual stimulation provided by my IUPUI colleagues Cornelius de Waal, André De Tienne, Jonathan Eller, Nathan Houser, David Pfeiffer, John Tilley. I must also gladly note that Valentin A. Bazhanov of the philosophy department of Ulyanovsk State University, Thomas Drucker of the mathematics department of the University at the Wisconsin at Whitewater, and Nathan Houser have been long-term enthusiastic supporters of this undertaking.

And, finally, an *apologia*.

Fustel de Coulanges, again in his inaugural lecture of 1862, also expressed the view, common to many historians, that the best guarantee for understanding the thoughts of those living in the past is to study their writing first-hand, rather than by attempting to glean their frame of mind second-hand, and rather than to seek to rewrite their thoughts in terms of the ideas and languages of our present.

For this reason, I have also at times undertaken to provide quotes, sometimes fairly lengthy, from original sources, or to provide an extended summary of writings during a crucial period. Ivor Grattan-Guinness (see, e.g. Grattan-Guinness 2004, 2009; see also Anellis 2011a) and many historians of mathematics have, in recent years, argued in favor of history rather than heritage, to ask what happened in the past rather than how did we get to where we are today, in an attempt to understand the mathematicians not only on our terms but on their own terms as well, rather than simply on our present terms. This echoes the advice of Turner (1938, 55) on the obligation of the historian who wishes to be accurate in depicting the ideas of our predecessors, to "not judge the past by the canons of the present, nor read into it the ideas of the present." Thus we have the advice of many of the most profound mathematicians through history recommending that the best way to learn mathematics is to study the classics of the past, the works of the foremost and most highly regarded mathematicians of the past. It is for this reason that we agree with Abel that: "It appears to me that if one wants to make progress in mathematics, one should study the masters and not the pupils."

I can, I fear, do no better, therefore, than to quote, *in extenso*, from the preface of the three-volume *History of the Crusades* by the British medievalist and Byzantinist historian Steven Runciman (1903–2000) in an effort to justify, if not defend, the present extremely lengthy exercise. He wrote (1951, I. xiii):

> A single author cannot speak with the high authority of a panel of experts, but he may succeed in giving to his work an integrated and even an epical quality that no composite volume can achieve. Homer as well as Herodotus was a Father of History, as Gibbon, the greatest of our historians, was aware; and it is difficult, in spite of certain critics, to believe that Homer was a panel. History-writing to-day has passed into an Alexandrine age, where criticism has overpowered creation. Faced by the mountainous heap of minutiae of knowledge and awed by the watchful severity of his colleagues, the modern historian too often takes refuge in learned articles or narrowly specialized dissertations, small fortresses that are easy to defend from attack. His work can be of the highest value; but is not an end in itself. I believe that the supreme duty of the historian is to write history, that is to say, to attempt to record in one sweeping sequence the greater events and movements that have swayed the destinies of man. The writer rash enough to make the attempt should not be criticized for his ambition, however much he may deserve censure for the inadequacy of his equipment or the inanity of his results.

We may, then bring this lengthy disquisition to a close with the words of Fustel de Coulanges (1877, 1):

> Il est difficile à une homme de notre temps d'entrer dans le courant des idées et des faits que leur ont donné naissance. Si l'on peut espérer d'y réussir, ce n'est que par une étude patiente des écrits et des documents que chaque siècle a laissés de lui. Il n'existe pas d'autre moyen qui permettre à notre esprit de se détacher assez des préoccupations présents et d'échapper assez à toute espèce de parti pris pour qu'il puisse se représenter avec quelque exactitude … d'autrefois.

Or (in the translation of Stern 1956, 188):

It is difficult for a man of our own time to penetrate the mainstream of ideas and of conditions which brought them forth. The only hope of success lies in the patient study of the literature and documents which each century has left behind. There is no other way which allows our mind to free itself sufficiently from our immediate concerns and from every kind of partisanship so that it can depict with exactitude . . . of times gone by.

But Huizinga would argue that no apology is required for going into great depth on a minute fragment of history. In doing so, he formulates an apology nevertheless, writing (Huizinga 1960, 23–24) that "it is not necessary for the researcher in details to justify the scholarly importance of his work with an appeal to its preparatory character. His true justification lies much deeper. He meets a vital need . . . . Whether his work yields tangible fruits for later research is, relatively, of secondary importance. In polishing one facet out of a billion . . . he achieves living contact of the mind with the old that was genuine and full of significance." He asks, in doing so, not *merely* how did we get here, but the far broader question of what happened in the past, a past that contains, as a small but significant subset, the answer to the question of how we got to where we now are. The important, and difficult, task in assembling and recording the data, is to keep in mind the underlying question for which the data are being accumulated. An examination of the historical examples that Collins (1998) recites demonstrates pointedly that, for the history at least of philosophy, we recover insights into how and why the main lines of the history of philosophy unfolded, and gain some deeper explanation of the tissues that link the various strands of the main lines of development, and of the figures and ideas that contributed to that development.

I am also conscious of the potentiality, in examining historical background, that there is a danger of providing either a surfeit of historical information for historians or the opposite danger of providing insufficient information for mathematicians. In his review of Martin Davis's *Engines of Logic*, Richard Wallace wrote (http://www.alicebot.org/articles/wallace/mathematicians.html), using as an example Davis's treatment of the origins of World War I that: "When the mathematician approaches the subject of history, however, the results can be somewhat curious. [Davis] states the bare facts of August 1914, but one wonders about the intended audience. Surely not historians interested in the history of mathematics, but if it is mathematicians interested in history, one has to wonder why they would be ignorant of such basic facts." Wallace fails to consider the alternative, that this is intended to provide mathematicians who require it a minimal amount of information sufficient to explain the historical and political circumstances that reflected and surrounded Frege's emotional situation at a particular time in his life, and was intended neither as a detailed account of the origins of World War I nor as a hint sufficient for political historians to fill in the details of the war's origins. With this discussion in mind, I have assumed that both logicians and historians of logic have the technical knowledge of the logic that is under discussion, such that I shall not have to provide a logic textbook, and that logicians and historians of logic know enough about the history of logic such that I shall not have to consider more history than is required to present my thesis and ground its arguments, nor be obliged to reconstruct in detail for either those aspects of the history that seem to me to be irrelevant to my thesis or to the arguments raised regarding that thesis. When in doubt, I hope that I shall err on the side of more, rather than less, detail.

To understand, then, whether, and if, there was a Fregean revolution in logic, and in what it may consist, I urge that we need to understand as well the life and times of logic as it was, and was perceived by logicians, before, during, after, and as a result of, that Fregean revolution, whether real or merely putative, and how and why such an event or concatenation of events came to be formulated by the historiography of logic. That is what I undertake to attempt here.

The result will perhaps, leave logicians dissatisfied, that there is insufficient technical apparatus to fully compare Aristotelian syllogistic, the algebraic logic of Boole, Peirce, and Schröder, the calculi of Frege, Russell and White, and Hilbert, while the philosophers and historians may be dissatisfied that there is either too much technicality, or insufficient historical or philosophical analysis, or both. This will be more likely if one accepts as a truism Grattan-Guinness's (2000a, 159) supposition, that: "Symbolic Logics tend to be too mathematical for the philosophers and too philosophical for the mathematicians; and their history is too historical for most mathematicians, philosophers, and logicians." In response, I may appeal to two masters, nearly twenty-five centuries apart: Aristotle, the "father" of formal logic, wrote that "If you would understand anything, observe its beginning and its development"; Jean van Heijenoort held that the very best way to learn logic was to study its history, and his anthology was built upon and guided by that principle. In the words of Anita Feferman (1993, 274), speaking of van Heijenoort's *Source Book*: "Not everyone who works in a subject thinks it necessary to know all the work that has been done in that subject, but van Heijenoort's approach was total immersion."

If you would understand anything, observe its beginning and its development.—Aristotle

One's special knowledge of logic can be a painful cross to bear but duty demands that you fulfill your calling.—Charles Sanders Peirce to Rev. John Wesley Brown (1892)

History has a way of crucifying its most creative children, but at the same time history has a way of settling scores. Those who forget the mistakes of history do indeed repeat them.—Richard S. Wallace (ca. 2001)

I.H. Anellis
Peirce Edition Project, Institute for American Thought, Indiana University-Purdue University at Indianapolis, Indianapolis, IN 46202, USA

The Historical Sources of Tree Graphs and the Tree Method in the Work of Peirce and Gentzen

Irving H. Anellis and Francine F. Abeles

Abstract Charles Peirce's development of his diagrammatic logic, his entitative and existential graphs, was significantly influenced by his affinity for tree graphs that were being used in chemistry. In his development of systems of natural deduction and sequent calculi, Gerhard Gentzen made use of the tree (tableau) method. In presenting the historical sources of both these tools, we draw on unpublished manuscripts from the Peirce Edition Project at the University of Indianapolis, where for many years the first author was a member of the research staff.

Keywords Diagrammatic logic · Dodgson · Gentzen · Graphs · Hertz · History of logic · Mathematics in the nineteenth century · Mathematics in the twentieth century · Peirce · Tree method

Mathematics Subject Classification Primary 03-03 · 01A55 · 01A60 · Secondary 03F03 · 03B65 · 03F03

Introduction

It has recently been asserted by Von Plato [354, p. 314] that "[Gentzen's] discovery of natural deduction and sequent calculus in 1932–33, with a full control over the structure of derivations, followed from the use of a tree form for derivations."[1] This claim is based upon an examination of Gentzen's inaugural dissertation and the work of Paul Hertz (1881–1940) of which Gentzen took especial account.[2] We can agree that the tree or tableau method as developed in the second third of the twentieth century stemmed directly

---

[1] Von Plato's boldface emphasis.

[2] Von Plato [352, pp. 240–244] also compared and discussed the newly-discovered printed version of Gentzen's [90] doctoral thesis with the published version of his [91] "Untersuchung über das logische Schliessen", which bears the same title, and sketches the detailed normalization proof which was omitted from the published version; he provides a transcription (Von Plato [352], pp. 245–257) as well of the third chapter of the dissertation in which it occurs.

See Abrusci [7], Schroeder-Heister [325], and Legris [161] for studies of Hertz's work in logic, and Schroeder-Heister [325] and Von Plato [352] especially for its importance for Gentzen's work.

© Springer International Publishing Switzerland 2016

F.F. Abeles, M.E. Fuller (eds.), *Modern Logic 1850-1950, East and West*, Studies in Universal Logic, DOI 10.1007/978-3-319-24756-4_3

from the work of Gentzen and those who built upon it, and from those such as Hertz, in particular his *Satzsystem* [108–110], upon whose work Gentzen built or took into account. Schroeder-Heister [357, p. 251], citing Hertz's [110], is very explicit in attributing "the first to consider *tree-like proof structures*" to Hertz, explaining that for Hertz [110, p. 463] a proof is a linear sequence of inferences and [110, p. 464] an inference system is a tree-like structure. It must be crucially noted, however, that this characterization of proof structures as "tree-like" is Schroeder-Heister's, not Hertz's, who uses terms such as "Reihe"—sequence—and "Kette"—chain.[3]

In his textbook *Formal Logic: Its Scope and Limits*, the first to present the tree method, Richard Carl Jeffrey (1926–2002) briefly set out the standard history of the method by writing [126, p. ix], that the "roots" of "Beth's method of semantic tableaux (or Jaakko Hintikka's method of model sets)...go back...to Jacques Herbrand...",[4] and [126, p. 227] that "the tree method derives from Beth's method of *semantic tableaux* [38, 40] and equally from Hintikka's method of *model sets* [118, 119]",[5] that is, from Evert Willem Beth's (1908–1964) [38] *The Foundations of Mathematics* and [40] *Formal Methods* and Jaakko Hintikka's (b. 1929) [118] "Form and Content in Quantification Theory" and [119] "Notes on Quantification Theory". Jeffrey [126, p. ix] tells us that he borrowed from Raymond M. Smullyan (b. 1919) the idea of one-sidedness that reduces tableaux to trees, and he hints that the roots of Beth's semantic tableaux and Hintikka's model sets (may) go back to Jacques Herbrand (1908–1931).

A somewhat more nuanced account of the development of the tree, or tableau method, was given by computer scientist John Alan Robinson (b. 1928), developer of the Resolution method which has been an essential tool of automated theorem proving that combines the features of Herbrand quantification and deductive tableaux,[6] in *Logic: Form and Function, the Mechanization of Deductive Reasoning*. There Robinson [318, p. 290] notes that neither Gentzen nor Herbrand who "unearthed essentially the same" cut-elimination, as had Gentzen, according to which we have the provability of any provable formula *A* by a proof that does not introduce any formulas that are not subformulas, in terms of their suitability, of the formula *A*, "went over the ground opened up by Gentzen with his sequent calculus and cut-elimination idea, and gave more direct expositions of the basic phenomenon." It remained, Robinson [318, p. 290] continued, for Beth and

---

[3]"Unter einem Beweis," says Hertz [110, p. 463], "aus einem Satzsystem T—die Sätze von T nenen wir *oberste Sätze des Bewises*—für einen Satz e verstehen wir eine Reihe von Schlüssen, deren letzterer e als Konklusion besitzt..." And, about the inference system, Hertz [110, p. 464] says: "In eienem Schlußsystem bezeichnen wir als *Kette* eine Reihe von Sätzen, die mit dem untersten beginnt und von da immer weiter zu einem jeweils darüberstehenden fortschreitet, bis sie mit einem obersten endet."

[4]These words were dropped from the third (1991) edition, along with any mention whatever of Herbrand.

[5]This account has been retained in the third (1991) edition of Jeffrey's textbook, but moved to the preface (p. xi) with slight modification in the expression: "derives from" was changed to "based on" and "and equally from" to "or, equivalently". In the fourth (2006), posthumously published, edition, John P. Burgess retained this historical comment as a block quote (p. xi) in his new preface.

It is an inaccurate account, however, since the core documents in which Beth developed his deductive tableaux and semantic tableaux were published earlier, *viz.* (Beth [35–37]), whereas the books (Beth [38, 40]) to which Jeffrey referred presented Beth's later accounts of the tableaux.

[6]For details, see esp. Abeles [2, 4].

Hintikka to "independently, hit upon"[7] semantic tableaux in which we present a cut-free proof of a given true sequent by attempting to build a counterexample which would block the further progression of the proof. And it was then that Smullyan took up his exploration of what Robinson [318, p. 290] called the "nooks and crannies" of the tableau method.

The free decomposition rules employed by Jeffrey in his *Formal Logic* are precisely the inference rules which Gerhard Karl Erich Gentzen (1909–1945) introduced in his classical sequent-calculus **LK**. In fact, Jeffrey's rules are a subset of the rules of **LK**. Gentzen [91] showed that true sequents can always be given cut-free proofs, by providing an elaborate apparatus for converting any arbitrary proof with many cuts into a cut-free proof of the same sequent (For an exposition of **LK** and a discussion of cut-elimination and its importance, see Takeuti [338].) Meanwhile, Herbrand, in his [106] *Recherches sur la théorie de la demonstration*, made the equivalent observation of the provability of any provable formula A by a proof that does not introduce any formulae that are not subformulae of the formula A (the subformula property), that it is easy to show that **LK⁻** (i.e. **LK** without cut) is precisely the tree method. A chain of Gentzen sequents would indeed look very much like an inverted Smullyan tree, or like a Hintikka tree lying on its side. In fact, as Beth showed in a talk given at the International Congress of Mathematicians in Edinburgh in 1958 and published in Beth [39], his semantic tableaux, when closed, are equivalent to proofs in Gentzen's systems **LK** of classical sequents and **NK** of *N*-sequents for natural deduction. Moreover, Beth [39] used this fact to prove the completeness of the semantic tableau method. The remainder of the standard history states that a tree is a one-sided Beth tableau; and since the notion of one-sidedness is due to Smullyan, the tree is called a *Smullyan tree*. Although the first textbook to use the Smullyan *tree* is Jeffrey's [126] *Formal Logic*, essentially an undergraduate text,

---

[7]Gentzen's motivation in developing natural deduction was to present a system for carrying out proofs which were closer than the axiomatic methods then available to the way in which mathematicians approached and constructed proofs. Specifically, he wrote, in his printed doctoral thesis (Von Plato [353], p. 672; [354], p. 321), that "Wir wollen einen Formalismus aufstellen, der möglichst genau das wirkliche logische Schliessen bei mathematischen Beweisen wiedergibt." In the published version (1934, 176), this has become: "Mein ersten Gesichtspunkt war folgender: Die Formalisierung des logischen Schließens, wie sie insbesondere durch Frege, Russell und Hilbert entwickelt worden ist, entfernt sich ziemlich weit von der Art des Schließens, wie Wirklichkeit bei mathematischen Beweisen geübt wird". In the English translation of the published version by Manfred Szabo [92, p. 288], this becomes: "The formal-ization of logical deduction, especially as it has been developed by Frege, Russell, and Hilbert, is rather far removed from the forms of deduction used in practice in mathematical proofs. Considerable formal advantages are achieved in return. I intended, first of all, to set up a formal system which came as close as possible to actual reasoning." As Anellis [12] has shown, Herbrand's goal was to elucidate what *being a proof* meant for Hilbert's axiomatic system, as found in the work in particular of Hilbert, his explicit references in this being Hilbert's [112–115], Hilbert and Ackermann's [116], John von Neumann's [351], along with both editions of Whitehead and Russell's *Principia* [358, 359]. The crucial point of contact between Gentzen's system and Herbrand's was characterized by van Heijenoort [344, p. 10] in terms of Herbrand's Théorème Fondamental (Herbrand [106], pp. 112–113; [107], p. 138) being, according to Gentzen [91, p. 409, n. 6], and as van Heijenoort [344, p. 10] noted, a special case (*Spezialfall*) of Gentzen's *verschäfter Hauptsatz* for his system **LK**.

An account of the history of the method of natural deduction, from Gentzen to Irving Marmer Copi (*né* Copilowish; 1917–2002), and of some of its chief drawbacks, in particular as concerns quantifier rules, is given by Anellis [13]. A somewhat different interpretation of this history is given by Pelletier [299].

the first graduate-level monograph is Smullyan's own *First-order Logic*, which appeared
in 1968. In the standard history, the latter book by Smullyan is (to again borrow the
phraseology of Robinson [318, p. 290]), an investigation of the "nooks and crannies" of
the semantic approach to cut-free proofs of sequents. This semantic approach was also
independently and almost simultaneously developed by Beth and Hintikka. In this story,
the real significance of Smullyan's book is that it gives the first unified and systematic
exposition of semantic tableaux which Smullyan called *analytic tableaux*.

The *falsifiability tree*, or *confutation tableau* as John Lane Bell (b. 1945) and Moshé
Machover (b. 1931) called it in their [31] *Course in Mathematical Logic*, presented in
these texts for testing the validity of proofs in propositional logic and decidable fragments
of first-order predicate logic, we shall show, although not historically connected directly
with Christine Ladd-Franklin's (1847–1930) [151] inconsistent triads or antilogisms, or
with (Charles L. Dodgson) Lewis Carroll's trees, are equivalent to those early methods.
All of these are instances of proof by contradiction, with Carroll's trees and the modern
falsifiability tree method or confutation tableaux, arraying proofs by contradiction in
diagrammatic form.

The actual story of the development of the tree method is somewhat more complicated
than portrayed in the sketches by Jeffrey and Robinson, but also much more interesting.

Hertz and Gentzen

Von Plato's [354] research is based in large measure upon early manuscripts and
publications by Gentzen that predate his more well-known two-part article, and focuses
upon the influence of Hertz's system in the evolution of Gentzen's systems of natural
deduction and sequent calculi. It shows that Gentzen had "tree forms for derivations" as
he developed the rules for his systems of natural deduction and sequent calculi. It is on
this basis that Von Plato concludes that Gentzen was the originator of the tree method.
Von Plato does not go deeper into the historical background of Gentzen's work on the one
hand than consideration of the general background for Gentzen's work than a discussion
of the axiomatic systems of Frege's *Begriffsschrift* [87], the first non-diagrammatic two-
dimensional linear technique, with material implication, e.g., defined as *A* implies *B*,
represented as:

(and which Peirce, before turning to graphical methods, wrote, algebraically, as $A -\!\!< B$),[8]
of Hilbert, in particular Hilbert and Ackermann's [116] *Grundzüge der theoretischen*

---

[8]Frege described his notation in §§1–12 of the *Begriffsschrift*, and displayed the [traditional] square
of opposition in §12, but defining both universal (A- and O-) propositions and existential (E- and I-)
propositions in terms of implication and negation, as opposed to contemporary usage in which existential
propositions are defined in terms of conjunction (with, e.g. "Some *A* is *B*" in the Peano-Russell notation,
as $\sim(\forall x)(Ax \supset \sim Bx)$ rather than, as in contemporary usage, $(\exists x)(Ax \cdots Bx)$). Moktefi and Shin [189, p.

Logik and Hilbert and Bernays' [117] *Grundlagen der Mathematik*), and of Russell and Whitehead (especially their *Principia* [358]), or on the other, of the role of Hertz. Neither does he trace the development of tableau methods that evolved in large measure from Gentzen's systems of natural deduction and sequent calculi. Anellis [10] traces the contemporary history of the tree method, from its tentative origins in the Gentzen sequent-calculus and the method of natural deduction, through its evolution and comprehensive development as the "Smullyan tree" or analytic tableau (1964–68) from Beth tableaux (1955–56) and Hintikka's theory of model sets (1953–55), and considers Jean van Heijenoort's (1912–1986) development of the falsifiability tree as a special case of the truth tree (1966–74) and work in the 1970s on proving the soundness and completeness of the tree method.[9] Anellis [9, 14, 22] examines in detail van Heijenoort's studies of the completeness and soundness of the tree method and its relation to other proof-theoretic tools, specifically, the various procedures for formal derivations in various calculi, propositional calculus, first-order calculus, second- and higher-order calculi, his comparison of the various systems, i.e. the axiomatic method, natural deduction, the Gentzen sequent calculus, Herbrand's method, and especially tableaux methods, most particularly the falsifiability tree method, i.e. Smullyan's analytic tableaux in which proofs are carried out by contradiction, and of the completeness and soundness of these methods for various classical and non-classical calculi. This work by van Heijenoort included general considerations of: the falsifiability tree method for sentential calculus and classical quantification theory and their properties, in particular the soundness and completeness of the method; intuitionistic propositional calculus and quantification theory and the study of related methods, including Beth tableaux; Grzegorczyk's method; Gentzen's system of natural deduction; Herbrand's method of natural deduction; the application of the falsifiability tree method to modal logics.[10]

The Historical Sources of Peirce's Diagrammatic Logic

Although there were numerous scattered efforts to deal mechanically, in a diagrammatical or graphical manner, with classification of logical relations, for assessing the validity of arguments, or computing the combinatorial count of possible relations between a set number of terms or concepts, we concentrate, for practical reasons, only upon those that

---

649] consider Frege's notation to be essentially diagrammatic and Moktefi and Shin [189, pp. 652–661] analyze it as such.

  Venn [346, p. 38; 349, p. 481] identified seven categories of notations for basic types of propositions, which, counted together, gave thirty-three varieties in all, devised in the period from Leibniz to Frege, and which included Peirce's illation. Not included were earlier notations, such as that of Vives, nor the diagrammatic methods of Leibniz, Lambert, Euler, or his own.

[9] See the bibliography in Anellis [10] for the references to the relevant works of Smullyan, Beth, Hintikka and van Heijenoort. For details of van Heijenoort's contributions, see Anellis [22].

[10] See the appendix "Proof-theoretic and Related Writings of van Heijenoort's in the *Nachlaß* [Box 3.8/86-33/1] (exclusive of research notes and unfinished work)" (Anellis [22], pp. 447–448) for a comprehensive listing. Van Heijenoort's [345] is a comparative study of the strengths and weaknesses of the members of the "family of formal systems"—the axiomatic method, natural deduction, Gentzen sequent calculi, and the Herbrand method.

specifically influenced Charles Peirce and played a direct or indirect role in Peirce's own efforts to develop visual tools for the mechanical classification and investigation of logical arguments and were within the direct line of the development of his historical background as he sought to carry out his work in devising visual methods of proof and classification of logical arguments. Peirce's earliest extant writings, beginning at least in 1859, utilized diagrams.

Nathan Houser [122, p. 3] therefore suggested that Peirce's initial interest in treating logical relations using a graphical syntax may have been piqued by his fascination with geometrical constructions, or from familiarity with the Euler diagrams that were familiar from, and prevalent in, the logic textbooks of the era. But it was only after the end of his short professorial career at the Johns Hopkins University that he turned definitively from algebraic logic to the development of graphical logic, inventing first his entitative graphs and then his more refined existential graphs. Houser [122, p. 3] noted that: "Probably the main influences for Peirce's interest in graphical logics were Clifford and Sylvester, with their use of chemical diagrams for representing algebraic invariants; Kempe, with his theory of logical form; and Clifford, with his graphical method." With this outline in mind, we turn, then, to a fuller account of the influences that stood behind Peirce's graphical logics.[11]

Peirce once accounted for his difficulties in being understood as explainable because "my damned brain has a kink in it that prevents me from thinking as other people think." In fact, Peirce, despite his facility with, and original contributions to algebraic logic, was primarily and essentially a visual thinker, and he explicitly associated his left-handedness and right-brain dominance with his preference for diagrammatic thinking [266]; and in [267] he explicitly stated that he never thinks in words, but in diagrams. De Waal [79, p. 24] asserts that Peirce held that mathematical reasoning is diagrammatic. In *New Elements of Mathematics*, Peirce indeed suggested that all of mathematics is essentially diagrammatic, writing (see Peirce [290], vol. II, p. 345) that a diagram is "any visual skeletal form in which the relations of parts are distinguished by lettering or otherwise, and which has some signification, or at least some significance." He further asserted (see Peirce [285], par. 2.778) that such a diagram "will either be geometrical, that is, such that familiar spatial relations stand for the relations asserted in the premises, or will be algebraical, where the relations are objects which are imagined to be subject to certain rules, either conventional or experimental." Under this interpretation he says a diagram, [290, vol. II, p. 345] can just as well constitute "a system of equations written under one another so that their relations may be seen at a glance."[12]

---

[11]Roberts [316, pp. 16–17] also refers to Peirce's preference for diagrammatic thinking, to the early influences of Clifford and Sylvester [316, pp. 17–20], and to his extended attention to Kempe [316, pp. 20–25].

See Wilson [361, p. 510] on chemical graphs and Wilson [361, pp. 512–515] on counting trees, including in particular the work of Cayley, Sylvester and Clifford. Wilson [361, pp. 520–521] is interested in the work of Kempe only with respect to his erroneous proof of the four color problem.

[12]Many more logicians, chief among them Frege, Hilbert and Russell, and even elsewhere Peirce himself (see, *e.g.* [23]), would dissent from this latter view, on the ground that axiomatic systems, such as Euclid's or Peano's, no matter how closely reasoned and how logically close one equation may be to those around it, are not formal deductive systems, if the equations were not derived from one another by deliberate employment of explicit inference rules.

Peirce's childhood study of chemistry, his graduate degree in chemistry from Harvard College's new Lawrence School of Science, and his lifelong interest in chemistry,[13] strongly influenced him in the use of diagrams to investigate the construction of molecules in organic chemistry, based upon the carbon ring which became the basis for chemical structure theory (see [136, 137, 139]).[14]

Benzene rings, Kekulé [139, p. 88]

$$\begin{array}{c} H_C = C^H \\ {}_4 \quad {}_3 \\ H^{C\ 5} \qquad {}_2\ C^H \\ {}_6 \quad {}_1 \\ {}_H C - C^H_H \end{array} \qquad \begin{array}{c} H_C - C^H \\ {}_4 \quad {}_3 \\ H^{C\ 5} \qquad {}_2\ C^H \\ {}_6 \quad {}_1 \\ {}_H C = C^H_H \end{array}$$

More critically, from the standpoint of investigatory techniques, Peirce's methodological approach to investigations, in almost any subject whatever, was formed in large measure by his laboratory work in chemistry at the Lawrence School under the direction of Justus von Leibig's (1803–1873) student, Rumford Professor Eben Norton Horsford (1818–1893), which consisted of identifying chemical compounds by undertaking an assay of the contents of unmarked bottles of chemicals.[15] See Fisch [85, pp. xvii–xviii]; see especially De Waal [79, pp. 4–5], which explicitly draws a connection between Horsford's teaching and Peirce's investigatory practices. Walsh [355, p. 83] elucidates the relation by explaining that, regardless of when Peirce took his profession to be a chemist or a geodecist, "it is clear that at heart he was a logician," and, referring to the *Elements of Logic* of Richard Whately (1787–1863) [357], that "[f]rom the moment he opened Whately's book he found it impossible to think of anything, even chemistry, except as an exercise in logic. Charles initially saw logic as a classificatory science like chemistry." To this end, it is not remarkable that one of the earliest of his publications in logic was the [210] "On a Natural Classification of Arguments".[16]

---

[13]Charles S. Peirce to Victoria Welby (1909) (see Hardwick and Cook [102], p. 114): "I was educated as a chemist, and as soon as I had taken my A.B. degree, after a year's work in the Coast Survey, I took six months under Agassiz in order to learn what I could of his methods, & then went to the laboratory. I had a laboratory of my own for many years & I had every memory of any consequences as it came out; so that at the end of two or three years, I was the first man in Harvard to take a degree[Sc.B.] in chemistry *summa cum laude*."

[14]Kekulé [136] explained that chemical structure is dependent upon the idea of atomic valence, with special reference to the tetravalence of carbon, and [137]—the year that Peirce received his undergraduate degree—that carbon atoms like to form chains, and in [138]—the year that Peirce wrote undelivered lectures on the logic of science—published his discovery of the benzene ring.

Peirce wrote about his early interest in chemistry in [265, 266]. The Peirce *Nachlaß* includes more than 20 undated manuscripts on chemistry, including computations for valencies of various compounds and atomic weights of elements. His earliest professional publication was on "The Chemical Interpretation of Interpenetration" in the *American Journal of Science and Arts* [200].

[15]The Peirce biography [48, p. 47] gives the name, incorrectly, as Horsfeld.

[16]Late in life, Peirce [265, 266] wrote about his interests and studies in chemistry. In Peirce [265], both his study of chemistry and his discovery of Whately's [357] *Elements of Logic* and his deep interest in logic are mentioned in tandem.

Peirce was also, as a logician and historian of logic, already familiar with the *arbor porphyriana* (tree of Porphyry) or *arbor prædicamentalis*, and its history, with Euler diagrams, which had their origin in the work of Leibniz, who variously employed rectangles, intersecting and concentric circles or ellipses, and lines.[17] Leibniz's line diagrams below (see, e.g. [165, pp. 292–321]; [164, VII, B, iv, pp. 1–10] under the title "De Formæ Logicæ comprobatione per linearum ductus"):

represent respectively, the propositions "All *B* is *C*", "Some *B* is *C*", "No *B* is *C*", and "Some *B* is not *C*", with syllogisms expressible by the appropriate combinations of these figures. These are given side by side with the circular diagrams that Euler later adapted from Leibniz's scheme of circles, and with the work of John Venn (1834–1923) in developing the Venn diagrams—which Venn [346, p. 1] called *Eulerian circles*.

Euler [83], *Lettres à une Princesse d'Allemagne* (carta CIII, 17 Feb., 1761) [*1788*, p. 101]

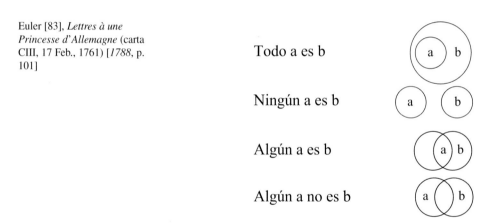

Venn made the diagrams feasible for a large number of terms by using intersecting ovals rather than Euler circles, and for greater complexity still, added shading to signal the existence of at least one item populating the shaded region.[18] Venn himself admitted that,

---

[17]See Baron [28, pp. 117–119] on Leibniz's diagrams. Baron [28] is remiss at providing complete adequate documentation.

[18]William Jesse Newlin (1878–1958) of Amherst College complained [191, pp. 539–540] that all available diagrams, Venn's included, which, he says [191, p. 540], creates "more confusion than ever", and were inconvenient or flawed, and in his [191] "A New Logical Diagram" offered what amounts very clearly to an adaptation of Dodgson's and Allan Marquand's (1853–1924) diagrams as presented by Marquand [184]. Newlin [191, p. 540] finds the latter as "com[ing] nearest" to being able to represent any number of classes properly. The originality of Newlin's work is, however, suspect, to the extent at least that he fails to cite Dodgson's work, and misidentifies Marquand, referring to him [191, p. 540] as "H. Marquand". Little is known about Newlin beyond the basic statistical data, also that he held an A.B. degree from Amherst, two Master's degrees, one in engineering, both from the Massachusetts Institute of

although theoretically capable of further expansion for dealing with ever greater numbers of terms, this diagrammatic technique becomes practically unmanageable beyond five terms. On the other hand, he imagines but few occasions which would call for analysis of arguments with so many terms.

Euler diagrams are based on Leibniz's diagrams. Peirce [287, par. 347–371] notably gave a critical exposition of Euler diagrams and of their subsequent historical impact, without, however, noting their origin in Leibniz's diagrams. However, the material presented in "Euler's Diagrams" as found in Peirce [287] was patched together by the editors from various scattered manuscripts as well as from Peirce's [237] entry on "Logical Diagram (or Graph)" for Baldwin's *Dictionary* and cannot, therefore, be regarded as a complete or unified and systematic discussion.[19] Without mentioning Leibniz, Peirce began in [237, p. 28; 287, par. 347] by stating that: "A diagram composed of dots, lines, etc., in which logical relations are signified by such spatial relations that the necessary consequences of these logical relations are at the same time signified, or can, at least, be made evident by transforming the diagram in certain ways which conventional 'rules' permit." For Peirce [237, p. 28], referring to chapter XI of Venn's *Symbolic Logic* [348, pp. 240–260], Venn diagrams are a "practical improvement" over Euler diagrams. In the manuscript "On Logical Graphs (Graphs)" Peirce [246] (Ms. #479; published in part in Peirce [287], par. 4.358–361), in addition to some random remarks, focused on a brief sketch of the history of logical diagrams, and dealt with Euler diagrams; he also

---

Technology. He taught philosophy and mathematics at Amherst, and, although he is listed in the Harvard cinquicentennial register as having earned a graduate degree from Harvard in 1906, the level of the degree, or its subject-area, are not specified, nor is it listed in his Amherst faculty vitae. However, soon thereafter, in 1907, he was promoted to Professor of Philosophy and Theology at Amherst.

See, e.g. Gardner [89] for a survey of the history of graphical methods in logic and their application to logic machines; Dürr [80] for very brief expository and historical treatments of Euler and Venn diagrams; Kuzichev [149] for a discussion of Venn diagrams; Edwards [81] for a popular account; and Davenport [71] for a brief treatment of the role of graphical methods in the history of logic. Kuzicheva [150] deals not only with Venn diagrams and Euler diagrams, but also with Lambert diagrams and De Morgan diagrams. The focus of Coumet [70] is primarily on the attitudes of logicians and philosophers towards diagrammatic methods in the period from Euler to Venn, rather than with the specifics of the technical advances in diagrammatic methods in that same period.

Moktefi and Shin [189, pp. 616–618] provide a critical analysis of Euler diagrams and the modifications to them presented by Joseph-Diez Gergonne (1771–1859) in his [93] "Essai de dialectique rationelle"; see also Faris [84] on Gergonne diagrams. Hammer and Shin [111] offer an analysis of Euler's diagrammatic system and modifications for improvements proposed by Venn and Peirce.

[19]The manner of construction of the contents of the Hartshorne-Weiss edition in the 1930s of Peirce's *Collected Papers*, in particular when working with the manuscript materials, was to selectively paste together, without regard to the chronology of the composition, parts of papers that seemed to belong together in terms of their subject contents, and, where more than one variant existed, to select the one that seemed to the editors to be the best version. In short, the procedure was often to "mix-and-match". The history of the Peirce *Nachlaß*, including a brief account of how the Hartshorne-Weiss edition was organized, was given by Houser [121]; see esp. p. 1262.

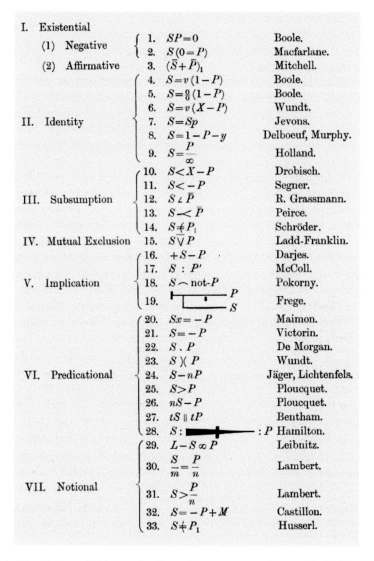

I. Existential			
(1) Negative	1.	$SP=0$	Boole.
	2.	$S(0=P)$	Macfarlane.
(2) Affirmative	3.	$(\bar{S}+\bar{P})_1$	Mitchell.
II. Identity	4.	$S=v(1-P)$	Boole.
	5.	$S=\frac{0}{0}(1-P)$	Boole.
	6.	$S=v(X-P)$	Wundt.
	7.	$S=Sp$	Jevons.
	8.	$S=1-P-y$	Delboeuf, Murphy.
	9.	$S=\dfrac{P}{\infty}$	Holland.
III. Subsumption	10.	$S<X-P$	Drobisch.
	11.	$S<-P$	Segner.
	12.	$S \angle \bar{P}$	R. Grassmann.
	13.	$S \prec \bar{P}$	Peirce.
	14.	$S \neq P_1$	Schröder.
IV. Mutual Exclusion	15.	$S \vee P$	Ladd-Franklin.
V. Implication	16.	$+S-P$	Darjes.
	17.	$S : P'$	McColl.
	18.	$S \frown \text{not-}P$	Pokorny.
	19.		Frege.
VI. Predicational	20.	$Sx=-P$	Maimon.
	21.	$S=-P$	Victorin.
	22.	$S . P$	De Morgan.
	23.	$S)(P$	Wundt.
	24.	$S-nP$	Jäger, Lichtenfels.
	25.	$S>P$	Ploucquet.
	26.	$nS-P$	Ploucquet.
	27.	$tS \parallel tP$	Bentham.
	28.	$S:$ ▬▬ $: P$	Hamilton.
VII. Notional	29.	$L-S \infty P$	Leibnitz.
	30.	$\dfrac{S}{m}=\dfrac{P}{n}$	Lambert.
	31.	$S>\dfrac{P}{n}$	Lambert.
	32.	$S=-P+M$	Castillon.
	33.	$S \neq P_1$	Husserl.

John Venn's Classification of Notations for the most common propositions of logic (Venn 1880a, 38; 1894, 481)

offered examples of two difficult problems [246, pp. 5–8, 21–22] using graphs, and, in another, undated manuscript also titled "On Logical Graphs" [282], proposed using a system of "curves convex inwards" as an improvement over Euler diagrams and logical algebra.

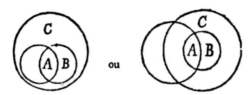

car alors, puisque la notion *A* a une partie contenue dans la notion *B*, la même partie se trouvera aussi certainement dans la notion *C* : d'où l'on obtient cette forme de syllogisme :

Quelque *A* est *B* :
Or Tout *B* est *C* :
Donc Quelque *C* est *A*.

An example of Euler diagrams for a syllogism

Leibniz himself referenced Ramón Lull's (Raimondus Lullius; 1234/35–1315) *Ars magna* (*ca.* 1370) and studied the categories from a logical perspective.[20] The tree of Porphyry, used for schematizing the relations between categories (for ordering of genus and species), was described in his commentary on Aristotle and discussed by Boëthius (Anicius Manlius Severinus Boëthius; *ca.* 480–*ca.* 520) in his *In Isagogen Porphyrii Commentorum* (see Boëthius [44]; see Hacking [100] for a history of the role of the tree of Porphyry through Euler and into the early nineteenth century). It can also be found in Peter of Spain's (Petrus Hispanius; Pedro Julio Rebello; Pope John XXI; *ca.* 1205/10–1277) *Summulae Logicales*, in which it was actually written out rather than merely described. Paulus Nicolettus Venetus's (Paolo de Veneto; 1372–1429) *Logica parva* (see [196], fol. 8r.) seems to be the first time the tree appeared in print.

It, together with Llull's *arbor scientiæ* in the *Ars magna* [169] were, as Nubiola [193, par. 5] noted, discussed by Peirce [286, par. 3.488, 285, par. 2.391; 288, par. 5.500]; in a "Lecture on Logic and Philosophy" (Ms. #342) of May 1879, Peirce (p. 8 of [293, pp. 7–9]) likens the tree of Porphyry to the biblical Jacob's ladder.[21]

Peirce's earliest known reference to Porphyry's logic, and to the tree of Porphyry, is in his "Upon Logical Comprehension and Extension", read at the American Academy of Arts and Sciences on 13 November 1867 (and published in Peirce [211]).[22] There, he says

---

[20]Schepers [324, p. 542] also refers in this regard to the *tabula prædicamentis* in the *Logica Hamburgensis* of Joachim Jungius (1587–1657) [132] and in the so-called *Elementa logica* [43] of Johann Heinrich Bisterfeld (1605–1655). Maróstica [183, p. 105] is among those who noted "first, that Llull, with his *Ars generalis* (1274–1308), influenced Leibniz in his conception of *ars combinatoria*, presented mainly in *Dissertatio de arte combinatoria* (1666), and in *On Universal Synthesis and Analysis* (1683). Second, that Leibniz influenced Peirce about the issue in several writings. Third, that Llull with his *ars combinatoria* (especially in his ternary period) influenced Peirce in his division and classification of signs."

[21]Identified (see comments in Peirce [293], 560*n*.7-9) as likely belonging to a lecture on "The Relations of Logic to Philosophy" delivered by Peirce to the Harvard Philosophical Club on 21 May 1879.

[22]For Peirce's source of Hispanus's *Summulae*, see Petrus Hispanus [301] and his [300]; for Porphyry, see Porphyry [305], a copy of the latter two belonging to Peirce's private library. Llull's *Ars magna* (earliest

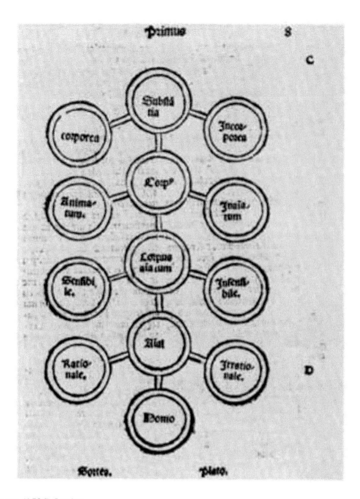

Paulus Venetus (1536) *Logica*

that "the tree of Porphyry involves the whole doctrine of extension and comprehension except the names. Nor were the scholastics without names for these quantities. The *partes subjectives* and *partes essentiales* are frequently opposed; and several other *synonymes* are mentioned by the Conimbricenses. It is admitted that Porphyry fully enunciates

---

printing [169]; see the [172] edition), and Paulus Venetus's [195] *Sophismata aurea perutilia*, a copy of which he owned, are mentioned by Peirce in his lesson plan of sixty lectures for a logic course he had planned out in the summer of 1883 ([219, 220]; see p. 487 of Peirce [293], 476–489). Peirce [213, 203*n*; 292, 263*n*.7], writing on the validity of the laws of logic, invokes the latter with respect to sophism 50 in a discussion of the Liar paradox.

Title page, Llull [171], *Arbor scientiæ*

the doctrine; it must also be admitted that the passage in question is fully dealt with and correctly explained by the mediæval commentators. The most that can be said, therefore, is that the doctrine of extension and comprehension was not a prominent one in the mediæval logic." His very brief discussion of the tree of Porphyry also occurs in his (1902b) article "Tree of Porphyry" for Baldwin's *Encylopaedia of Philosophy and Psychology*, where he notes [238, p. 14] that it appears in all "old logics" and "is supposed to illustrate the second chapter of the *Isagoge* of Porphyry, showing the genera on the trunk of the tree, and the specific differences on the branches, with substance at the top ...." Peirce himself had little to say, however, about Leibniz's diagrammatic work; most of what he says about Leibniz as a logician is embedded in an undated manuscript [282] on the history of logic which mentions the contributions of Scotus, Ockham, Cartesianism, Francis Bacon, Leibniz, and the Leibnizian logicians, Christian

Arbor naturalis et logicalis, Llull [170], *Logica nova*

Wolff and Johann Heinrich Lambert.[23] Usually a good historian, Peirce [287, par. 4.353] in this instance, takes up the argument, raised by the Leibnizian Johann Heinrich Lambert (1728–1777), in his [157, I, p. 28] *Architektonik*, of whether or not, as Johann Christian Lange (1699–1756) in his *Nucleus Logicæ Weisianæ* [160] claimed, Christian Weise (1642–1708) introduced Euler diagrams into the teaching of syllogistic; or if Friedrich Albert Lange (1828–1875) [159, p. 10] was correct, that they already occur in a work of Juan Luis Vives (1492–1550). In fact, however, the diagram in Vives's [350] *De Censura Veri* in his *Opera*, is closer to Leibniz's linear diagrams than to an Euler diagram, a large 'V' to represent the relation of inclusion among terms and premises in a syllogism

$$
\begin{array}{c}
C \\
\text{V} \\
A \\
\text{V} \\
B
\end{array}
$$

where C is the minor term or minor premise, A is the middle term or middle premise, and B is the major term or major premise. Peirce is correct in pointing out that an edition by Weise of the *Nucleus Logicæ* appeared [356] prior to Lange's [160] expanded edition. But Peirce also notes that what Weise produced is quite far from Euler's diagrams, and remarks that, given Weise's reputation in his own day, it is difficult to think that he

---

[23]Peirce's chief concern in regard to Leibniz was Leibniz's conception of logic taking the role of a *mathesis universalis*, as a *lingua characteristica* and *calculus ratiocinator*, and in the philosophical differences between Leibniz's philosophy of logic and that of Kant.

could have arrived at such a significant innovation without it being widely noticed. Peirce consequently concluded that Euler diagrams are indeed Euler's original invention.

Alfred Bray Kempe (1849–1922), Arthur Cayley's (1821–1895) one-time Cambridge University student, contributed to the logic of relations, applications of the logic of relations to geometry and foundations of geometry, and whose diagrams, together with analytical trees, had formed part of the inspiration for Charles Peirce's existential graphs for logic.[24] Among those having an influence upon Peirce's thought in developing diagrammatic tools, it was Kempe who had the most direct tangible direct influence that left a detectable residue in the *Nachlaß*. In the 1880s and 1890s, there was, as [98, p. 327] noted and detailed, substantial interaction between Peirce and Kempe, and we have seen, a correspondence ensued (see [141]), the earliest for which documentation is available is receipt by Peirce of Kempe's [140] "A Memoir on the Theory of Mathematical Form". The correspondence includes pages from Kempe's [144], with a note directing Peirce's attention to that article.

Cayley became interested in tree diagrams as a solution to and representation for combinatorial problems originating in differential calculus, and we thus find trees as representations for algebraic relations as early as 1857 in Cayley's work "On the Theory of Analytical Form Called Trees" [57]; this was followed up in his articles "On the Theory of Analytical Form Called Trees, with Applications to the Theory of Chemical Combinations" [61], and "On the Theory of Analytical Form Called Trees" [63] in the *American Journal of Mathematics*, where Charles Peirce, also having his [198] article "On the Logic of Number", was certain to have seen it. Hacking [110, p. 221] tells us that "[p]robably the first mathematical theorem about trees" is Cayley's 1857 paper.

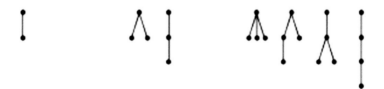

Trees, Cayley [57, p. 173]

Peirce, trained at Harvard as a chemist, was indeed partly influenced by this work, as well as by Kempe's diagrammatic method, using dots and lines and based upon chemical diagrams, as found in Kempe's [140] "A Memoir on the Theory of Mathematical Form" (see [316], pp. 20–25, [123], p. xlv, and [97], p. 140), and by William Kingdon Clifford's

---

[24]See Grattan-Guinness [98] on the interactions between Peirce and Kempe and [16] for details as background to the work of Tarski and his students.

(1845–1879) work on chemical diagrams for logic (see, e.g. Clifford [66–68]), in his own development of existential graphs for the logic of relations. Grattan-Guinness [97, p. 140] followed Peirce's own dating to assign 15 January 1889 as the point of departure for his conceiving, on the basis of his study of Kempe's work, the idea for developing his entitative and existential graphs. Also important in this development is "A Graphical Method of Logic" (see Peirce [233, 287], par. 418, pp. 468–470; see Murphey [190], pp. 196–197, Roberts [316], pp. 17, 22, 130, and Hawkins [104], p. 132), read to the National Academy of Sciences in November 17–18, 1896, referring to James Joseph Sylvester's (1814–1897) "On an Application of the New Atomic Theory to the Graphical Representations of the Invariants and Covariants of Binary Quantics", published in the *American Journal of Mathematics* and which Peirce was known to have seen (see [335]).[25] Despite an often friendly, but sometimes bitter and contentious rivalry between Peirce and Sylvester, particularly when in 1881–82 Sylvester (see [336], referring to Peirce's [214] "Description of a Notation for the Logic of Relatives . . . ", and Peirce's [218] reply) criticized Peirce's reduction of linear associative algebras to matrices in algebraic logic, and in particular criticized Peirce's interpretation of Sylvester's algebra of matrices as a special case of Peirce's algebra of relatives, Peirce credited Sylvester, *circa* 1877–78, with importing the method of graphs from chemistry into mathematics. Seibert [326] goes further, and asserts that Peirce's interest in logic in general and the hypothetico-deductive method in particular, predated even his reading of Whately's [357] *Elements Logic*, but arose as a result of his childhood work in chemistry.[26] In an uncompleted manuscript "A Problem of Trees" of uncertain date but tentatively traced to 1891, Peirce [228] himself pursued work on combinatorial trees, expanding on Sylvester's work at Johns Hopkins on combinatorics (for example, Sylvester's [337] "Note on the Graphical Method in Partitions". One might surmise that both the chemical diagrams adopted by Cayley, Kempe, and Peirce and the periodic table originating with Dmitrii

---

[25]Sylvester and Peirce were colleagues at the Johns Hopkins University, Peirce teaching there from 1879 to 1884, Sylvester from 1876 to 1883, and one of their mutual students, Christine Ladd-Franklin (while still a student, Christine Ladd), reported [155, p. 717] that: "Several of Professor Sylvester's students—understanding that the New Logic which Professor Peirce professed had connections with existing mathematics . . . joined his [Peirce's] class in logic . . . .". See also Parshall [194] on Sylvester at Johns Hopkins; see Houser [120, pp. liii–liv, lvi–lviii] for a summary of the issues of the Peirce-Sylvester controversy regarding Peirce's reduction of Sylvester's matrix algebra to an algebra of relations. Brent [48, pp. 140–141] looks at the personal aspects of the Peirce-Sylvester dispute. In a manuscript of 1903 for the Lowell lectures of 1903, Peirce [243] wrote bitterly of his conflict with Sylvester; remarking in particular to Sylvester's claim that some work shown him by Peirce, who, in turn, suspected that his work reduced to Cayley's theory of matrices, but was really nothing more than Sylvester's umbral notation. Later, Peirce adds that he discovered, with righteous satisfaction, that what Sylvester called "my umbral notation" had originally been published in 1693 by Leibniz.

Indicative of the effort by Peirce to treat contemporary mathematics from the standpoint of his algebraic logic, in an undated manuscript, Peirce [277] undertook to examine Cayley's [60] work on abstract geometry from the stand-point of the logic of relations.

For mention of Peirce's, Sylvester's and William Edward Story's (1850–1930) work on graph theory in connection with the four-color problem, see, e.g. Eisele [82, p. 55].

[26]See Andrews [8]. Picardi [302] explores the history of the interactions, influences, and similarities between logic and chemistry; Kedrov [134] details the concept of chemical element in its logical analysis, and Kedrov [135] is an historical-logical study of the development of the concept of element from Mendeleev, who began the charting of the table of chemical elements, to recent times.

Ivanovich Mendeleev (1834–1907) had their historical roots in, if not receiving a direct influence from, the "chemical sliderules" designed by William Hyde Wollaston (1766–1828) in 1814, which arranged chemicals on a scale with distances proportional to the logarithm of their equivalent or combining weights ([363]; see also [360]). Peirce mentions Wollaston in his [215] report "Note on the Theory of the Economy of Research" for the U. S. Coast Survey (see esp. p. 73, [293]), although not explicitly in connection with Wollaston's "chemical sliderules". The chemical slide rule is related, conceptually and historically, to the logical slide rules envisioned by Llull and built by Andrei [Jan] Khristoforovich Belobodskii (sometimes called Belobots'kyi; *fl.* mid-seventeenth cent.), producing a developed working model of a mechanical Euler slide rule for categorical syllogisms, and the logarithmic slide rules envisioned and built by John Napier (1550–1617) and his "bones" and William Oughtred (1574–1660) (see [73] for a brief history and account of slide rules and [52] for an extended history; see Gorfunkel' [95] for a treatment of Belobodskii).

Circular slide rule, *Quarta figura*. Llull, *Ars magna et ultima* (1501, 4)

These devices served as mechanical tools for computations, and their use was said by Augustus De Morgan (1806–1871) [73] to be considerably undervalued. In four undated manuscripts, presumably from various times in the 1890s, Peirce [278–281] undertook to evaluate Mendeleev's periodic table of the elements and proposed an alternative means of assigning valencies to the elements in Mendeleev's table. He thus may have arrived at his understanding of logical valency through his study of the periodic table of the elements and chemical valency.

Peirce would also have been very familiar with the *Cayley graph*, the directed graph in which each vertex represents a group element and edges are directed, as well as the diagrams of Kempe [140, 142–146], about which he wrote extensively.[27] Many of Arthur

---

[27]There are fourteen manuscripts listed in Robin [317]—Mss. 307 [244]; 308 [245], in which Peirce compares Kempe's graphs with his own existential graphs; 547 [276], an attempt to give an exposition of the main results of De Morgan and of Kempe; 622 [268], a sketch of the history of logic in the nineteenth century; 708 [235]; 709, 710, and 711 [270–272], in which Peirce discusses Kempe's [143] "On the Relation between the Logical Theory of Classes and the Geometrical Theory of Points"; 712 [273]; 713 [274]; in which Peirce praises Kempe's mathematical abilities and acumen in logic while bemoaning his formal training in logic; 714 [226] dealing with Kempe's [140] "A Memoir on the Theory of Mathematical Form"; and 715 [275], all of which deal with Kempe's work exclusively; 1170 [222], which includes, among other items, notes on Kempe's work in connection with Peirce's work from 1883 to 1909 for the *Century Dictionary*, notably on Kempe's terminology of logic and an index to Kempe's

Cayley's papers were brought to Peirce's attention by his former assistant, chemist Allan Douglas Risteen (1866–1932),[28] while they were working on the *Century Dictionary*. Kempe developed his diagrams in pursuing applications of the logic of relations to geometry and foundations of geometry. As explained by Wilson [361, p. 512]: "In 1874, Arthur Cayley observed that the diagrams corresponding to the alkanes all have a tree-like structure, and that removing the hydrogen atoms yields a tree in which each vertex has degree I, 2, 3 or 4 . . . ; thus, the problem of enumerating such isomers is the same as counting trees with this property," and remarks [361, p. 512] that Cayley's interest in tree counting problems dates at least to 1857, when working on a problem of Sylvester relating to differential calculus, enumerating rooted trees (see [57]). There, he notes [57, p. 173] that inspection of these figures that he provides "will at once show what is meant by . . . the terms *root*, *branches* (which may be either main branches, intermediate branches, or free branches), and *knots* (which may be either the root itself, or proper knots, or the extremities of the free branches)."[29]

The purpose of Kempe's work was to develop a graphing method to apply to geometry that would be based upon the relations between points, and in particular to use the theory of relations to axiomatize geometry. To this end, he developed a theory of relations graph-theoretically. Letting a point or "spot" represent a *unit*, one may obtain figures by considering pairs (i.e. dyads), triples (triads), *n*-tuples (*n*-ads), according to whether the points are connected. The edge between spots is either distinguished or not, depending upon the directedness of the edge. Thus, for example, in the dyad *ab*,

ab is distinguished, but *ba* is not. He then explains [140, pp. 3–4] that every collection of units has a definite form, as determined by the number of units and how the units, dyads, triads, etc. are distinguished or undistinguished. Thus, two collections having the same number of units but different distributions will have different forms, as, for example, will

---

theory of mathematical form,—covering the period from 1897 to 1909, which deal, in whole or in part, with Kempe's diagrams. In the manuscript "De Morgan's Propositional Scheme", Peirce [232, 253, Ms. #415] undertook to improve upon De Morgan's theory of propositions by expanding it, and to provide a graphical representation for it. In that same year, he also gave a graphical presentation of the relational proposition "Every mother loves some child of hers" [231].

[28]In a letter of 10 June 1891 [314]. Among the papers listed (Peirce [296], 407, *n*. 173:29–30) are (Cayley [57, 59, 62, 63]). Risteen, who had served as Peirce's assistant at the U.S. Coastal Survey and later achieved a doctorate from Yale University as a student of Josiah Willard Gibbs (1839–1903), is remembered chiefly for his [315] *Molecules and the Molecular Theory of Matter*. Two decades later, Peirce also wrote to Risteen regarding existential graphs. [269].

[29]Cayley [58, p. 265] made it clear that a *knot* is a "contour line which cuts itself: the point where this happens is in fact a knot, or geometrically the knot is a node or double point on the contour line", thus that the terms are synonymous. The latter term is now the one preferred.

the two tetrads

and

In his series of papers [140, 142–146] beginning with the (1886) "Memoir...",
Kempe, to a large extent in interaction with Peirce, worked out his theory of linear triads.
Peirce, in "Consequences of Common-Sensism" of *Pragmatism and Pragmaticism*, (see
[288], par. 5.505) called Kempe's "Memoir..." a "great memoir". Peirce's copy of the
"Memoir....." is heavily annotated (see [226], "Notes on Kempe's Paper on Mathematical Form"; see also [290, III, 440n.]).[30] In the "Memoir...", Kempe informally discussed
the various ways by which collections of various arities of mathematical atoms (units) can
be arrayed, and the graphical and geometrical, as well as tabular, representations of these
collections.

In general, Kempe can be said in the "Memoir..." to be searching for and describing
the combinatorial laws of algebra common to geometry, graph theory, group theory, and
linear and multilinear algebras. Algebras possessing one or more of these features are
called *primitive algebras* (see [140], p. 53). These laws are: commutation, association,
distribution, idempotence, and the law that $(ab) \neq a$ and $(ab) \neq b$. "Ordinary algebra",
by which Kempe [140, p. 53] meant Boolean algebra, is singled out as the primitive
algebra with idempotence. Kempe [140, p. 54] then introduces triads, as composed of
three units which give rise to primitive equations, that is, to equations expressing the
laws that characterize primitive algebras. The triad (abc) can represent a collection of
units, a, b, c, and can be interpreted as the pair (ab) of undistinguished atoms with
their product c. Where a, b, c are distinguished from l, m, n, the pairs which $(abc...)$
form with a, b, c is different from the pairs which $(abc...)$ forms with l, m, n (see
[140], p. 27). A quadrate algebra is one in which, for any two systems S_1 and S_2, all
combinations yield Cartesian products, and each product is itself an ordered pair taken

---

[30]See Anellis [16, pp. 281–288, 291] and Grattan-Guinness [98, pp. 332–335] for details.

as a unit rather than as a simple unit. The remainder of the "Memoir..." is devoted
to the mathematics of triads applied to geometry and algebraic logic. Kempe's next
paper "Note to A Memoir on the Theory of Mathematical Form" (1887) is very much
a reply to Peirce's comments and criticisms, and in particular to Peirce's letter to Kempe
of 17 January 1887. Likewise, Kempe's [146] "The Theory of Mathematical Form: A
Correction and Clarification" is a reply to the same letter and to similar comments made
by Peirce [234, 168*ff.*] in his paper "The Logic of Relatives". Thus, Kempe's "Note ...."
introduces, in an informal manner, some technical corrections to the "Memoir...",
largely concerning the differences between distinguished and undistinguished n-ads, in
the case of the former, where, e.g., $(n_1n_2 \ldots n_n) \neq (n_n \ldots n_2n_1)$. This was in response to
Peirce's letter and comments in Peirce's [235, 290, III, pp. 431–440] reply to Kempe.
Peirce's [235, p. 15] reply argued that Kempe's fundamental property of plane projective
geometry was too indefinite to be of use, but Peirce's [235] reply dealt primarily with the
nature of graphs developed by Kempe to represent n-ads. The detailed and formal work of
clarifying the characteristics of distinguished and undistinguished n-ads is carried out in
a rigorous and axiomatic manner, along with matrix-theoretic proofs, in Kempe's [143]
"On the Relation between the Logical Theory of Classes and the Geometrical Theory
of Points". It hinges on the fully developed concept of a *base system* of distinguished
and undistinguished linear triads whose form is fully determined by the logico-algebraic
laws that Kempe presents in that paper.[31] The remainder of this paper is given over
to demonstrating that the algebraic theory of forms has an interpretation in the class
calculus, and to presenting that interpretation. George Joseph Stokes (1859–1935) [334]
went on to assert that "between the mathematical theory of points and the logical theory
of statements, a startling correspondence exists. Between the laws defining the form of the
system of points, and those defining the form of a system of statements, perfect sameness
exists with one exception; namely the inclusion of the parallel postulate in the system of
points."[32] In a notebook with a miscellany of entries (the only entry dated being from

---

[31]These are (Kempe [143], p. 149):

K1: If $ap \cdot b$ and $cp \cdot d$ exist, then there is a q such that $ad \cdot q$ and $bc \cdot q$.
K2: If $ab \cdot q$ and $cp \cdot d$ exist, then there is a q such that $aq \cdot d$ and $bc \cdot q$.
K3: If $ab \cdot c$ and $a = b$, then $c = a = b$.
K4: If $a = b$, then $ac \cdot b$ and $bc \cdot a$ for any c.
K5 (Continuity): No entity is absent from the system which can consistently be present in the system.

[32]Descartes' work in *La Géométrie* (1637), published as an appendix to his *Discours de la méthode* (1637)
was intended to develop geometry as a *mathesis universalis* in which geometric figures are expressible
in algebraic terms. The Leibniz program of creating logic as a *lingua characteristica* and a *calculus
ratiocinator* by algebraicizing the syllogistic, sought to take the logic resulting therefrom as a *mathesis
universalis*. The most explicit early expression of the parallel between logic and algebra was given by
the Leibnizian mathematicians Jacob [Jacques] Bernoulli (1654–1705) and Johann [Jean] Bernoulli's
(1667–1748) [32] *Parallelismus ratiocini logici et algebraici*. Leibniz's work, other than a comparative
handful of his output, was otherwise unknown until the second half of the nineteenth century, with the
publication by Carl Immanuel Gerhardt (1816–1899) of the seven-volume *Mathematische Schriften* [163]
and the seven-volume *Philosophische Schriften* [164]. Thus, for example, Jevons tells us that Boole did
not learn of Leibniz's work until a year after completion of his [45] *Laws of Thought*, when he first
heard of it from Robert Harley (1828–1910). Jevons [130, I, p. xxi] wrote, referring to Harley's [103]
"Remarks on Boole's Mathematical Analysis of Logic" based upon his talk at the British Association for

24 December 1903), Peirce [247], after summarizing Kempe's [143] "On the Relation between the Logical Theory of Classes and the Geometrical Theory of Points", examined some possible permutations of linear triads.

Peirce (e.g. Peirce [270–272]) expressed particular concern for what he considered Kempe's careless use of the difference between *distinguished* and *undistinguished* terms. He attempted (in Peirce [270], pp. 3–4) to help Kempe by supposing that the claim that $ab\cdot c$ and $ba\cdot c$ are indistinguishable merely asserts that a and b have exactly the same relationship to the system, and not that $a = b$. If indeed Kempe had meant that $ab\cdot c$ and $ba\cdot c$ are indistinguishable means that $a = b$, then by his axiom or law of continuity (K5), c must fail to exist, even though K5 states that c does exist. Kempe's law of continuity, says Peirce [270, p. 4], "would seem to imply that there always is such an 'entity', although then it must be asked (Peirce [270], p. 3): "But what has all this to do with 'the distribution of linear triads through the system' which is all [Kempe's five laws] profess to define?"

Kempe in [144] "The Subject Matter of Exact Thought" expanded upon the axiom of continuity that he presented in "On the Relation between the Logical Theory of Classes and the Geometrical Theory of Points", according to which no entity is absent from the system which can consistently be present in the system, thus introducing what is apparently a bivalent truth-theoretic aspect for the equations of his system. Literally, "truisms" and "falsisms" are understood to be equations having the same algebraic value as other equations of the system. In fact, as Stokes [334, p. 7] pointed out, Kempe was primarily concerned here with the consistency of his system, and "truism" and "falsism" turn out to be the disguised laws of identity and contradiction respectively.

Peirce's general reaction to Kempe's work focused on technical issues concerning the problem of the symmetry and asymmetry of combinations of relations. Many of the technical details of Peirce's interactions with Kempe have their roots in the work on invariant theory and associated graph-theoretical methods carried out by Kempe and by Peirce's and Sylvester's Johns Hopkins student, Fabian Franklin (1853–1939), as well as of Charles's father, Benjamin Peirce (1809–1880).[33]

Clifford began his work on diagrammatic structures in an attempt to present a map of the solution by William Stanley Jevons (1835–1882) of the combinatorial problem for syllogisms with more than three terms. In particular, Jevons [127–129, vol. I, pp. 154–164; 130, vol. I, pp. 134–143] computed the number of types of compound statements in which three classes or terms were involved where the premises of a syllogism occur as a type of fourfold statement. Clifford then asked about the solution to the corresponding problem where four terms or classes are involved. Letting A, B, C, D be the terms or classes, and a, b, c, d be their respective negations, Clifford [67] carried out the enumeration

the Advancement of Science, held in Nottingham in August 1866, that: "It has recently been pointed out to me, however, that the Rev. Robert Harley did draw attention, at the Nottingham Meeting of the British Association, in 1866, to Leibnitz's anticipations of Boole's laws of logical notation, and I am informed that Boole, about a year after the publication of his *Laws of Thought*, was made acquainted with these anticipations by R. Leslie Ellis."

[33] Grattan-Guinness [98] notes that Peirce's interactions with Kempe, and especially Kempe's work on invariants and their graphs, played a crucial role in inspiring Peirce to develop his existential graphs, and employs the history of these interactions to trace the stages of Peirce's development of them. Walsh [355] focuses upon the details of the relations between Benjamin Peirce's and Charles Peirce's work.

and produced a diagram representing the solution, as follows (Clifford [67], p. 83; [69], p. 3):

```
        aBCD                    abCD                ABcd
         |                  aBcD \   / AbCD          |
 AbCD — ABCD — AbCD              X            aBcD — abcd — aBcd
         |                  aBcD /   \ aBCD          |
        ABcD                    ABcd                abCd
```

Together with Cayley's analytical trees, Kempe's and Clifford's diagrams formed a large part of the inspiration for Peirce's existential graphs for logic. Nathan Houser notes in particular the role of Cayley's tree diagrams, writing [124, p. xlviii] that, upon studying Cayley's tree diagrams and their application to chemistry "[i]t occurred to Peirce that Arthur Cayley's diagrammatic method of using branching trees to represent and analyze certain kinds of networks based on heritable or recurrent relations would be useful for his work on the algebra of the copula and his investigation of the permutations of propositional forms by the rearrangement of parentheses." Houser [124, p. xlviii] reports that, as a consequence of this work Peirce asked Risteen in early June of 1891 to add "trees" to the list of his areas of professional expertise. It was around this time, in the Spring of 1891, that Peirce wrote "On the Number of Dichotomous Divisions: A Problem of Permutations" ([227]; see [296], pp. 222–228),[34] in which he used binary trees, inspired by Cayley's work on analytic trees, to compute the number of propositional forms containing any number of copulas. In this manuscript, Peirce shows the result, a Catalan sequence, with the number of unique terms for combinations with respect to terms being $n+1$ for n-many "punctuations" or "separations". Given terms A, B, C, and Peirce's copula of illation (—<), e.g., we have $(A$—$<B)$—$<C$ and A—$<(B$— $<C)$, showing two copulas for three terms. Thus, we have $\frac{(2n+1)!}{n!(n+1)!}$ permutations and the number of separations being $\frac{(2n)!}{n!(n+1)!}$. In a work sheet (reproduced at [296], p. 225) to an early version of "On the Number of Dichotomous Divisions: A Problem of Permutations", alongside the computation of the Catalan sequence, binary trees are drawn to illustrate the sequence of numbers of propositional forms for the number of copulas involved.

Peirce's introduction to the work in matrix theory, graph theory and its combinatorics, and in general, to linear algebra and group theory, began under the auspices of his father's home teaching. In 1881, Charles Peirce published Benjamin Peirce's [214, 217] *Linear Associative Algebra*, adding to it his own advances in the form of notes and an appendix. He was apparently not immediately familiar with the Erlangen program of Felix Klein (1849–1925) [148], but was informed in his work on non-Euclidean geometry from at least 1881.[35] As Charles Peirce was an active member of the New York (now American)

---

[34]The second of two related manuscripts, Mss. ##73, 74, the other being "A Problem of Trees" [228], both listed as undated in the Robin catalog [317].

[35]Peirce refers to Klein's geometrical work, *e.g.*, in an undated manuscript that has been associated with his [217] "On the Logic of Number" as a possibly earlier version of part of [217]; see Peirce [293, 580*n*.300:9-22].

Mathematical Society and at least part of the time resided in New York City, he had occasion to personally hear Klein when on 30 September 1893 and again in October 1896 Klein lectured to the Society. Charles Peirce's contributions to his own edition of the *Linear Associative Algebra* was to demonstrate that the various linear and multilinear algebras could all be reduced to interpretations of the logic of relations. Starting from the concept of the "symbolical algebra" of George Peacock (1791–1858), the mathematicians who most contributed to the development of symbolical algebra and to the growth and maturation of algebra in the nineteenth century included: George Boole, Augustus De Morgan (1806–1871), Sylvester, Benjamin Peirce, his sons James Mills Peirce (1833–1906) and Charles, Hermann Günther Grassmann (1809–1877), and Cayley. Benjamin Peirce, Sylvester, and Cayley in particular, developed linear and multilinear algebras and matrix theory. It was on the basis of matrix algebra that Charles Peirce in unpublished manuscripts of 1893–1902 devised matrices for the truth-functional analysis of the binary connectives of propositional logic.

Peirce's manuscript of 1889, "Notes on Kempe's Paper on Mathematical Form" [226, 290, III, 440n.] was, as we have already remarked, heavily annotated, and highly critical of Kempe's efforts to "get rid of" triads, Peirce arguing that triads, or triple relatives (together with dyads, or dual relatives, and monads) are both necessary and sufficient for achieving the full expressive power of the logic of relations. In one two-page undated manuscript on Kempe Peirce [274] praised Kempe's mathematical powers and native instinct for doing logic, but was critical of "his sad want of training" in logic, and offered specific criticisms. Likewise, in the unfinished and fragmented article intended for *The Monist*, "The Bed-Rock beneath Pragmaticism", Peirce [261] cited Kempe's [140] "Memoir..." as an "invaluable, very profound, and marvellously strong contribution to the science of Logic" (see also [297], p. 208). In [261], Peirce presented his entitative and existential graphs, and explicitly compared the existential graphs to chemical graphs. Peirce's manuscript [235, pp. 5–6, 11, 15–19; 290, III, p. 436], a reply to Kempe's [146] reply to Peirce's [234] discussion in "The Logic of Relatives" of entitative graphs (see also [286], par. 468), is devoted to demonstrating that, despite differences in detail with Kempe's graphs and his own, and despite Kempe's claim to the contrary, tetrads can be rewritten as triads, but not as monads or dyads, and that, "accordingly, when we follow out the idea of Mr. Kempe's system, we arrive at a result that each graph consists of *monads, dyads,* or *triads.*" Starting from the tetrad,

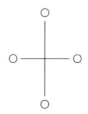

Peirce [235, pp. 6, 11; 290, III, p. 436] reduces the tetrad to two connected triads:

We may perhaps summarize the survey to the historical background to Peirce's work in logic and mathematics in general, and also insofar as applicable to his diagrammatic approach, by again quoting [120, p. xviii]: "In logic and mathematics, and even in philosophy, aside from his predecessors, the influence of Cayley, Sylvester, Schröder, Kempe, Klein, and especially Cantor stands out."

The Genesis of Charles L. Dodgson's Tree Method

An exposition of Lewis Carroll's, the pen name of Charles Lutwidge Dodgson (1832–1898), tree method, the earliest modern use of a truth tree to reason efficiently in the logic of classes, first appeared in William Warren Bartley's [30] edition of *Symbolic Logic, Part II, Advanced*, never previously published. The next publication on this topic was *Lewis Carroll's Method of Trees: Its Origins in 'Studies in Logic'* [1]. The Journal Editor, in a Note to this article, remarked, "The trees developed by Carroll in 1894, which anticipate concepts later articulated by Beth in his development of deductive and semantic tableaux, have their roots in the work of Charles Peirce, Peirce's students and colleagues, and in particular in Peirce's own existential graphs." [10, p. 22] And in a comprehensive article of his own, he suggested that "Perhaps this valuable contribution to proof theory [Dodgson's tree method] ought to be called the Hintikka-Smullyan tree method, or even the Dodgson-Hintikka-Smullyan tree . . . ." [11, p. 62]. A close examination of the methods Dodgson used in both parts of his symbolic logic books followed in *Lewis Carroll's Formal Logic* [3]. Recently, evidence that Dodgson's tree method was part of a visual proof system that he was developing appeared in *Toward A Visual Proof System: Lewis Carroll's Method of Trees* [6].

Dodgson created the first part of his visual proof system, a diagrammatic system, beginning in 1887 in a small book titled *The Game of Logic*. By 1896 he was able to use diagrams to represent syllogisms and map the set of valid syllogisms (he used only A, E, I propositions so there were just fifteen syllogisms in his system) to a set of twenty biliteral diagrams so that each of the valid syllogisms corresponds to one of the twenty diagrams and each of the twenty diagrams corresponds to one of the fifteen valid syllogisms. Using a set of rules to represent the premises of a syllogism in a triliteral diagram, followed by a second set of rules to transfer the information from the triliteral diagram to a biliteral diagram representing the syllogism's conclusion, he created a sound and complete diagrammatic proof system for syllogisms [56, pp. 27–31].

His system was capable of detecting fallacies, a subject that greatly interested him. He wrote, "[T]he Fallacy may be detected by the 'Method of Diagrams', by simply setting them [the propositions] out on a Triliteral Diagram, and observing that they yield no information which can be transferred to the Biliteral Diagram" [53, p. 81].

With a view to extending his proof method, Dodgson went on to expand his set of diagrams, eventually creating diagrams for eight sets (classes), and describing the construction of nine set and ten set diagrams. His set diagrams are self similar, i.e. each remains invariant when the scale is changed. For example, by placing a vertical line segment in each of the sixteen partitions of a four-set diagram, a five set diagram with 32 cells results.

Contrasting Dodgson's and Venn's diagrammatic systems, we know that Venn used his system as a proof method for problems like the following one whose object is to show the relation between a and b.

All a is either b and c, or not b.

If any ab is c, then it is d.

No da is bc. [348, pp. 116–117]

(Answer: no a is b.)

In both Dodgson's and Venn's systems, existential propositions can be represented. But Venn did not provide a method to do so until 1894 in the second extensively revised edition of his 1881 book, *Symbolic Logic* (actually he employed two representations: integers and horizontal line shading). The use of a small plus sign '+' in a region to indicate that it is not empty did not appear until 1894, and Dodgson reported it in his symbolic logic book [349, pp. 131–132; 53, p. 174]. However, Dodgson may have been the first to use it. A manuscript worksheet on logic problems, probably from 1885, contains a variant of a triliteral diagram that has a '+' representing a nonempty region. But in his published work, Dodgson preferred the symbol '1' for a nonempty region and the symbol '0' to designate an empty region [166].

Comparing Venn diagrams with Carroll diagrams we see that both are maximal, i.e. no additional logic information such as inclusive disjunctions can be represented by them. For a large number of sets Carroll diagrams are easier to draw because they are self-similar, discontinuous, and capable of being constructed algorithmically. Their regularity makes it simpler to locate and erase cells that must be destroyed by the premises of a syllogistic argument, a task that is difficult to accomplish in Venn diagrams for five or more classes. (For a more comprehensive view of Dodgson's work in logic, see Moktefi [188].)

Ultimately, Dodgson abandoned his diagrammatic proof method and created what he called his Method of Trees primarily to handle relations in his sorites and puzzle problems. In a similar but reversed way, as Carroll migrated from his logic diagrams to his tree method, Peirce migrated from his algebra of relations to existential graphs to represent relations ('relatives'). In his logic of relatives Peirce developed quantifiers as operators on propositional functions defined for given domains. Peirce published three articles on the logic of relations, the first one, "Logic of Relatives" in 1870. Here he extended Augustus De Morgan's theory of relations, which had appeared in [75, 76] in "On the Syllogism, No. IV, and on the Logic of Relations", and the same year in his book, *Syllabus of A Proposed System of Logic*. De Morgan had defined a relation as a mode of thinking two objects of thought together, i.e. a connection or lack of one. The name in the relation is the *subject*; the name to which it is in relation to is the *predicate*. He gave the example, 'in mind acting upon matter'. *Mind* is the *subject*, *matter* the *predicate*, *acting upon* is the *relation*. When names denote classes, the primary relation between them is that of containing and contained. He described this relation as mathematical in its character because a class is made up of classes, just as an area is made up of areas. When names denote attributes, the primary relation is that of containing and contained. He did not provide methods to handle relations [41–45], a task ultimately taken up by Peirce. In the manuscript "De Morgan's Propositional Scheme", Peirce [232] undertook to improve upon De Morgan's theory of propositions by expanding it, and to provide a graphical representation for it. In that same year, he also gave a graphical presentation of the relational proposition "Every mother loves some child of hers" [231].

In this article the connections between Dodgson's invention and the inspiration provided by one of Peirce's doctoral students, Christine Ladd-Franklin, together with a deeper insight into the development and significance of Dodgson's tree method constitute the subject of the narrative that follows.

In his book, *Symbolic Logic, Part I* which appeared in four editions in 1896, Dodgson, represented syllogisms as in this example:

No x are m′;

All m are y.

∴ No x are y′

in the form of conditional statements using a subscript form that is written symbolically as:

$xm'_0 \dagger m_1y'_0 \ \mathbb{P} \ xy'_0$ [54, p. 122] with the reverse paragraph sign signifying the connecting implication relation, which he defined as: the propositions on the left side "would, if true, prove" the proposition on the right side. Carroll [54, p. 119] Dodgson's algebraic notation is a modification of Boole's which he thought was unwieldy.

Influenced by Boole and his followers, Dodgson solved problems exemplifying the central problem of the symbolic logic of his time, known as the 'elimination problem', i.e. determine the maximum amount of information obtainable from a given set of premises. These problems often appeared in the form of a sorites, a linked set of syllogisms, or as a puzzle problem. To solve them he developed successively several techniques that he called the methods of subscripts, of underscoring, of diagrams, of barred premises, and finally, of trees.

Dodgson considered the tree method to be superior to the barred premises method. He wrote, "We shall find that the Method of Trees saves us a great deal of the trouble entailed by the earlier process. In that earlier process we were obliged to keep a careful watch on all the Barred Premises so as to be sure not to use any such Premis until all its 'Bars' had appeared in that Sorites. In this new Method, the Barred Premises all take care of themselves" [54, p. 287].

He took problems to solve from books and articles authored by important contemporary logicians, often citing multiple authors for a given problem like George Boole, Augustus De Morgan, W. B. Grove,[36] William Stanley Jevons, John Neville Keynes, John Venn, and Charles Sanders Peirce's students, Christine Ladd-Franklin and Oscar Howard Mitchell. Bartley presents these problems in chapter xxii of his reconstruction of part II

---

[36]Probably the biologist and taxonomist William Bywater Grove (1848–1938), specializing in microbiology and fungology.

of Dodgson's symbolic logic, a book that he never published. For Problems 7, 17, 18, 21, 22, and 23, *Studies in Logic* is listed as a source. All except one of the problems are from Ladd-Franklin's article [54, pp. 479–486].

One of them, number 2 on p. 52, in Ladd-Franklin's article, appears in Bartley's edition on p. 484 as problem 18. Although Dodgson changed the wording of the problem (which Ladd-Franklin had taken from p. 283 of Jevons's book, *Studies in Deductive Logic*), Dodgson cites only Peirce's *Studies in Logic* as his source for this problem. Some of Dodgson's solutions to these problems are included in a draft of his logic workbook now in the Princeton University Library. One of these pages, titled *Studies in Logic*, reproduced by Bartley in [54, pp. 486–487], contains Dodgson's solution to a problem proposed by W. B. Grove in *The Educational Times* from 1881 that Boole reproduced in his *Laws of Thought*, and given on p. 55 in Ladd-Franklin's article. (Either Dodgson or Bartley omitted the Ladd-Franklin citation because it does not appear in the list of sources for this problem.)

We know from Venn's review of *Studies in Logic* appearing in the October 1883 edition of *Mind*, soon after Peirce's book was published, that Peirce was well-known to the British symbolists, and that they were aware of Peirce's publications. Surely Dodgson would have read Peirce's Boolean algebraic approach to the logic of relatives in his "Note B" on pp. 212–228 in *Studies in Logic*.

The sale of Dodgson's library at his death included works on logic by Boole, Venn, Allan Marquand, Mitchell, Ladd-Franklin, Benjamin Ives Gilman, Peirce, John Neville Keynes, Rudolph Hermann Lotze (in English translation by Bernard Bosanquet), James William Gilbart (1794–1863), De Morgan, Bernard Bosanquet, Francis H. Bradley, John Stuart Mill, William Stirling Hamilton, William Whewell, and Jevons, among others. Bradley writes, "Some of these works presumably influenced his own writing; others he needed to consult in order to deal with Oxford adversaries, such as [John] Cook Wilson, who had studied with Lotze at Göttingen" [54, p. 31]. Dodgson corresponded with many British logicians, including James Welton (1854–1942), Thomas Fowler (1832–1904), William Ernest Johnson (1858–1931), Herbert William Blunt (1864–1940), Henry Sidgwick (1838–1900), John Alexander Stewart (1846–1933), and Bartholomew Price (1818–1898) [56, p. 6].

It seems reasonable to consider that Dodgson's reading of Ladd-Franklin's article in Peirce's *Studies in Logic* did influence the invention of his tree method. In her article, on p. 41, she gives a syllogistic rule to identify all valid syllogisms. To test an argument given as a syllogism, she describes a form that all valid syllogisms can be put into. If its triad (two premises and conclusion) has that proper form, what she calls an antilogism, i.e. an inconsistent set consisting of the two premises and the negation of the conclusion, the syllogism is valid, thus also known as an inconsistent triad.[37] Using a tree to test

---

[37] See also Ladd-Franklin [154, 156]. The antilogism was discussed by Eugene Shen (probably Youngding Shen; also known as Yu-Ting Shen; 1908–1992), who called it "the Ladd-Franklin formula" [329]. Ladd-Franklin [152, p. 583] and [153, p. 648, n. 1] notes that Josiah Royce (1855–1916) labeled her antilogism "inconsistent triad", and by I. Susan Russinoff [323], who placed it within the general context of the history of traditional logic and algebraic logic, while remarking [323, p. 463] that Ladd-Franklin failed to provide a rigorous proof of her "Rule of the Syllogism" in virtue of which the method of the antilogism is a proof of the validity or invalidity of arguments, and undertakes to establish it as a theorem. In vol. II, p. 75, of his three-volume *Logic* W. E. Johnson [131] takes credit for being the one, in his book to

the validity of a syllogism requires that an antilogism be assumed, i.e. the triad of the two premises and the negation of the conclusion is assumed to be an inconsistent set at the outset. Beginning at the root of the tree with the negation of the conclusion of the syllogism, and then proceeding via the rules of propositional logic to the bottom of the tree (leaves), either all the branches of the finished tree close, or at least one remains open. In the first instance, the syllogism is proved valid because there are no cases where the three propositions are true; in the latter instance the argument is proved invalid because an open path represents a set of counterexamples. In certain cases, Dodgson included the 'verification' of the tree as part of his method, defining it as, "[T]ranslating it [the tree] into Sorites-form that no Barred Premis will venture to make its appearance until all its Bars have been duly accounted for" [54, p. 287]. In effect, the verification process represents a linearization of the proof generated by the tree. Dodgson's tree verification anticipates the linear representation of a derivation tree proof given by Gerhard Gentzen.[38]

By interpreting a syllogistic argument in the form of a conditional, and setting up a test of the inconsistent triad of its propositions, both Ladd-Franklin and Dodgson changed the interpretation of the relationship of the conclusion to the premises of a syllogism (in classical logic, the premises entail the conclusion).[39] Dodgson's application of the antilogism in the form of his tree method produced a proof method for complex syllogistic arguments which, by modern standards, is sound, complete and decidable.[40]

We know Dodgson's created his tree method to solve complex soriteses, usually in the form of a puzzle problem, involving a very large number of premises, because he thought his other direct methods, particularly his method of subscripts, were inadequate. Even though Carroll considered his algebraic method to be superior to Boole's, he preferred his method of trees for working with a large number of premises. In all of these sorites/puzzle problems Carroll is dealing with propositions, not classes.

To use the tree method to test the validity of arguments, or equivalently the consistency of a set of suitable sentences, we list the basic inference rules for arbitrary propositions (sentences) S, T.

"introduce" the antilogism. Ladd-Franklin takes umbrage at having thus been "robbed" [156, p. 532], and goes on to explain why Johnson also failed to appreciate the full value of antilogism, not as a mere additional form of the syllogism, but as a replacement for the syllogism. She recommends to Johnson Shen's [329] article.

[38]In his early work, prior to 1935, Gentzen utilized derivation trees in natural deduction in the linear form of a sequent.

[39]In his 1927 paper Shen wrote, "The Antilogism was at first called by Dr. Ladd-Franklin the 'inconsistent triad'; apropos of it, the late Josiah Royce of Harvard was in the habit of saying to his classes: 'There is no reason why this should not be accepted as the definitive solution of the problem of the reduction of syllogisms. ...'"

[40]Complete details for constructing a tree to solve a sorites problem can be found in Abeles [1, pp. 27–29]; using a tree to test the validity of arguments, equivalently the consistency of a set of sentences, can be found in Abeles [3, pp. 152–155].

(1)	(2)	(3)	(4)	(5)	(6)	(7)	(8)	(9)
S	S → T	S ∧ T	S ∨ T	S ↔ T	~~S	~(S → T)	~(S ∧ T)	~(S ∨ T)
~S	~S T	S	S T	S ~S	S	S	~S ~T	~S
x		T		T ~T		~T		~T

(10)
~(S↔T)
~S   S
T   ~T

Rule 1. Contradiction	Rule 6. Double negation.
Rule 2. Conditional	Rule 7. Negation of conditional.
Rule 3. Conjunction	Rule 8. Negation of conjunction.
Rule 4. Disjunction	Rule 9. Negation of disjunction.
Rule 5 Biconditional	Rule 10. Negation of biconditional.

In rules 2, 4, 8, the symbols below the line are connected by ∨; in rules 5 and 10 there are two sets of these. In rules 3, 7, 9 the two symbols below the line are connected by ∧. Note that Dodgson considered the three statements: no x are y; some x exist and none of them are y; all x are not y, to be equivalent.

The tree method is a direct extension of truth tables and Dodgson had worked with an incomplete truth table in one of the solutions he gave to his Barbershop Problem in September 1894. Bartley writes, "The matrix is used . . . for the components; but the analysis and assignment of truth values to the compounds are conducted in prose commentary on the table" [54, 465n.]. (See Anellis [21] for a complete history of the truth table.)

Using truth tables to verify inconsistency is straight forward, but very inefficient, as anyone who has worked with truth tables involving eight or more cases knows. Instead, the truth tree method examines sets of cases simultaneously, thereby making it efficient to test the validity of arguments involving a very large number of sentences by hand or with a computer. To test the validity of an argument consisting of two premises and a conclusion, equivalently determining whether the set of the two premise sentences and the denial of the conclusion sentence is inconsistent, by the method of truth tables involving say, three terms, requires calculating the truth values in eight cases to determine whether or not there is any case where the values of all three terms are true. But a finished closed tree establishes that validity of the argument by showing there are no cases in which the three sentences are true. However, if any path in a finished tree cannot be closed, the argument is invalid because an open path represents a set of counterexamples.

It seems that Dodgson began to search the literature for problems involving complex relations that he could solve either by his subscript or tree methods. He found one of these in the weekly periodical, *Notes and Queries*, first published in 1849. In the issue of 14 January [77] (series 2, v IX, p. 25) titled "A Question in Logic", De Morgan wrote, "A great many persons think that without systematic study it is in their power to see at once all the relations of propositions to one another. I propose a case . . . whether you receive more than one answer . . . you will not receive many". Dodgson took up the challenge and sent his algebraic solution to his sister Louisa in a letter dated 15 March 1897 [Bartley 480-1].

He then went on to create a much more difficult problem modeled on it, calling it The Great-Grandson Problem. He solved it on 15 February 1897 and sent his solution to Cook Wilson the next day, remarking that it is a "new kind of problem" [54, p. 363] and that "My method of solution is quite new, & I greatly doubt if anyone will solve the problem"

[54, n. 362]. His new solution probably refers to the use of his Method of Cosmophases which he employed in his solution of De Morgan's problem. Dodgson defined the term 'Cosmophase', as "[t]he state of the Universe at some particular moment: and I regard any Proposition, which is true at that moment, as an *Attribute* of that Cosmophase" [54, p. 481].

Bartley provides many examples of sorites problems solved by the tree method in Book xii of part II of symbolic logic [54, pp. 285–319]. And several intricate puzzle problems solved by the tree method appear in Book xiii of part II of symbolic logic [54, pp. 326–338; 373–376].

We also know that Dodgson was proficient in proving theorems by the contradiction method in his many publications on geometry. Just as logic informed his geometric work, so geometry informed his logic writings. In his logic book, he used geometric notation and terms, e.g. the reverse paragraph symbol for the main connective of a syllogism, the implication relation, and the corresponding symbol ∴ for 'therefore'.

Dodgson's diary entries during the period 1894–1897 reveal the progress he made in discovering techniques to work with trees to solve these complex sorites problems. He first mentions his discovery on 16 July 1894 as an improvement over his diagrammatic method. (See Abeles [5] for an analysis of Dodgson's system of logic diagrams; also Moktefi and Shin [189] for an in-depth history of logic diagrams.)

"Today has proved to be an epoch in my Logical work. It occurred to me to try a complex Sorites by the method I have been using for ascertaining which cells, if any, survive for possible occupation when certain nullities are given. I took one of 40 premises, with 'pairs within pairs,' & many bars, & worked it like a genealogy, each term proving all its descendents. [sic.] It came out beautifully, & much shorter than the method I have used hitherto – I think of calling it the 'Genealogical Method'" [56, p. 34].

On 4 August, he connected the tree method with a scored sorites.

"I have just discovered how to turn a genealogy into a scored Sorites: the difficulty is to deal with forks. Say 'all a is b or c' = 'all A is b' and 'all α is c,' where the two sets A, α make up a. Then prove each column separately" [55, p. 158].

On 30 October, using a problem from a new edition of Keynes's book, *Studies and Exercises in Formal Logic*, he discovered how to navigate a tree representing a sorites with 21 premises having 10 attributes of which 8 are eliminated.

"Made a discovery in Logic, . . . the conversion of a 'genealogical' proof into a regular series of Sorites . . . .Today I hit on the plan of working each column *up* to the junction, then begin anew with the Premises just above and work into it the results of the columns, in whatever order works best . . . . This is the only way I know for arranging, as Sorites, a No. of Premises much in excess of the No. of Eliminands, where every Attribute appears two or three times in *each* column of the Table. My example was the last one in the new edition of Keynes" [56, p. 34].

The next day he wrote, "worked from 10 to 7 at my new discovery, & found a new 'dodge' in the genealogic method – suppose abc_0, I now do not write it

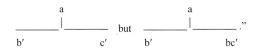

[55, p. 181] When an open branch is divided into two branches and a term, here b', appears in one of the branches and its negation is added to the other branch, we have an example of the use of the cut rule. Dodgson has anticipated a method that was not fully worked out until the 1930s [56, p. 35]. He wrote, "It is worthwhile to note that in *each* case, we tack on to *one* of the single Letters, the *Contradictory* of the *other* : this fact should be remembered as a *rule* . . . .We have now got a Rule of Procedure, to be observed whenever we are obliged to *divide* our Tree into *two* Branches" [54, p. 287].

He continued to discover new ways to improve his handling of trees, recording in his diary on November 12/13, 1896, "Discovered [a] method of combining 2 Trees, which prove $abc'_0 \dagger abd'_0$, into one proving $ab(cd)'_0$, by using the Axiom $cd(cd)'_0$" [55, p. 279].

Writing to his mathematically proficient sister, Louisa Dodgson, on 13 November 1896, where he answered questions she had raised about one of his problems that she was attempting to solve, we can see that Dodgson progressed from the use of his method of diagrams to his method of trees. He wrote,

"As to your 4 questions, . . . The best way to look at the thing is to suppose the Retinends to be Attributes of the Univ. Then imagine a Diagram, assigned to that Univ., and divided, by repeated Dichotomy, for all the Attributes, so as to have 2n Cells, for n Attributes. (A cheerful Diagram to draw, with, say, 50 Attributes! There would be about 1000,000,000,000,000 Cells.) If the Tree vanishes, it shows that every Cell is: *empty*" [56, p. 35].

Dodgson planned to do further work connecting his tree method with a generalization of his barred premises method, one that he named 'barred groups'. In an unpublished letter whose first page is missing, most likely from late 1896 or early 1897, probably to his sister, Louisa, he wrote,

"I have been thinking about that matter of 'Barred Groups' . . . . It belongs to a most fascinating branch of the subject, which I mean to call 'The Theory of Inference': . . . Here is one theorem. I believe that, if you construct a Sorites, which will eliminate all along, and will give the aggregate of the Retinends as a Nullity, and if you introduce in it the *same* letter, 2 or 3 times as an Eliminand, and its Contradictory the same number of times, and eliminate it *each* time it occurs, you will find, if you solve it as a *Tree*, that you didn't use all the Premises!" [56, p. 37]. Dodgson's death on 14 January 1898 ended this promising application of his tree method.

Peirce's Truth-Functional Analysis and the Origin of Truth-Table Matrices[41]

Peirce would no doubt have agreed with Charles Kidder Davenport (1900–1955) when he wrote [71, p. 145] of Euler diagrams that they have many "shortcomings". The chief of these difficulties, as Davenport [71, p. 145], was "the natural tendency that such graphs inculcate to interpret the propositions represented by them in terms of the relation of inclusion and exclusion among classes, . . . , thus placing unwarranted restriction on

---

[41] See Anellis [21], for full details. See also Anellis [17]. A general, if incomplete, chronological sketch of Peirce's work in logic is given in Roberts [316, pp. 129–135].

possible interpretations of the copula, adding that "[t]his of course rules out the subject-predicate relation which is implicit in the S is P proposition, and at the same time blurs the relation of a class to its members." Peirce's own copula of illation was intended to serve as class inclusion, material implication, set elementhood, the order relation, and the classical copula of existence, depending explicitly upon the context in which it occurred; and indeed, one of the criticisms leveled against Schröder's Subsumption (€) and Peirce's illation by Frege [88] and Russell [320] respectively was precisely that they conflated set membership with class inclusion. Venn diagrams, Davenport [71, p. 146] asserts, rendered obvious the "shortcomings" or "inconsistencies" of traditional logic; namely the logic of classes and associated Venn diagrams provide for the possibility of treating empty classes and limited universes of discourse . . . ," and most of the remainder of Davenport's work is given with examples, using Venn diagrams. It should be noted, however, that Boole was clearly aware that, given the null class, there were syllogistic inferences that were valid in Aristotle's syllogistic logic that were not in his own system, *and vice versa*, although he failed to undertake a comparison between his system and Aristotle's. Venn's chief goal in constructing his modification of Euler diagrams was to enable expressions for the null class, and he experimented with various means of indicating the existence of at least one element in a class to indicate existential import; his first effort in this direction was in his [347] paper "On the Diagrammatic and Mechanical Representations of Propositions and Reasonings". Frege meanwhile explored only the *Barbara*, *Felapton* and *Fesapo* syllogisms in the *Begriffsschrift*.[42] Although it did not require Venn diagrams to reach this recognition, Davenport is correct that the issue was first explicitly dealt with in a more than cursory or informal manner by Venn in his (1881) *Symbolic Logic*, who demonstrating, with the aid of his diagram [348, p. 112], that "*All X is Y*" is to be understood as there exists no class X such that "*X is not-Y*". Dodgson studied Venn's *Symbolic Logic* and as well explored the question of existential import [54, pp. 232–238]. He considered, for example, two possibilities which can justifiably be held: either I- and A-propositions assert existence, but E-propositions do not; or E- and A-propositions do, but I-propositions do not. He gives the example, citing Keynes [147, pp. 356–357] of *Darapti*: "All M is P. All M is S. Therefore some S is P", usually held by traditional logic to be valid, but then notes this is erroneous, since the syllogism is valid provided it is not the case that S exists but M and P do not. It was, however, Peirce who produced the first Boolean square of opposition, in 1865, having noted, in his Harvard Lecture VI on "Boole's Calculus of Logic" of 1865 [203, Ms. #100, March–April 1865; see 291, p. 228], part of an eleven-lecture series "On the Logic of Science", e.g., that the Law of Excluded Middle "cannot be expressed in Boole's system" if taken to mean "*A vel est B vel est non-B*" "because the existence of A is here implied. I mean that kind of existence which

---

[42]See Wu [364] for a brief survey of the history of the debates concerning existential import of propositions in Aristotelian vs. mathematical logic.

Frege, as [94, pp. 282–283] noted, singled out what amount to the syllogisms *Felapton* ("No *M* is *P*; All *M* is *S*: Therefore some *S* is not *P*"; Frege's [87] Formula 59) and *Fesapo* ("No *P* is *M*; All *M* is *S*: therefore some *S* is not *P*", Frege's [87] Formula 62) in the *Begriffsschrift* (Frege [87], §22) for translation; but only *Barbara* (Frege's [87] Formula 65) is universally valid when rewritten in his system. Nevertheless, Frege failed to notice that five of the seven rules of Aristotle's logic considered in the *Begriffsschrift* (Frege [87], §22) are invalid in his system. Frege gives his square of opposition in the *Begriffsschrift* (Frege [87], §12).

is implied in an affirmative copula, and in Logical Identity in general," and speaking there [203; see 291, p. 230] of the problem of existential import as "implication of Entity in Affirmatives", where an affirmative is defined "to be a proposition which implies the reality of the terms." In the earlier Lecture II [202, Ms. #95; see 291, p. 182] written in February–March, we find a square of opposition for the valid first three figures of the syllogism.[43] The motivation behind Peirce's earliest work in logic, as seen in his [209] "On an Improvement in Boole's Calculus of Logic", was to find a means to distinguish singular propositions from universal propositions. To this end, he came in his work on diagrams to use 'X' for a nonempty class (as occasionally had Venn) and '0' for the empty class ([246]; see Peirce [287], par. 4.359), with explicit rules following (Peirce [287], par. 4.36–4.366). In sum, Pierce in his study on "Euler's Diagrams" (see Peirce [287], par. 4.347–371), detected five faults which he undertook to remedy (see Peirce [287], par. 4.359–368).

Peirce began experimenting with diagrams and matrices, based upon the traditional square of opposition and Aristotelian syllogistic logic in 1865. In Lecture II of the Harvard Lectures "On the Logic of Science", as noted, he sought to order the first three figures of the syllogism in such manner as to display their relation to one another, while noting that *Darapti* and *Felapton* were excluded because "their premises [are] too broad" [202; see 291, p. 182]. In the same lecture [202; see 291, p. 182] he created the table

		E	A	
I	A	E	O	
A	I	O	E	
	O	I		

according to the scheme that universals are arrayed at the top, particulars at the bottom, affirmatives on the left, negatives on the right, and then giving the first figure (t, s, m), the second (t, m, s), and the third (m, s, t). The result is that the three figures are separate from one another in the table, the middle containing first figure, the top and bottom "wings" the second, and the side "wings" the third. Therefore, it is demonstrated that no inference is obtained in more than one way from the same "data". The same table,

---

[43]There is disagreement concerning the fourth figure of the syllogism, which has been variously attributed jointly to Aristotle's students Eudemus of Rhodes (*ca.* 350–*ca.* 290 B.C.) and Theophrastus of Ereseos (*ca.* 370/380–*ca.* 287 B.C.), or to Claudius Galen (*ca.* 129–199 A.D.) rather than to Aristotle. Łukasiewicz [179, pp. 38–42] is the chief exponent of the view that an unknown author, rather than Galen, simply compounded the syllogisms of the fourth figure from the first three figures. See Rescher [312], working from early medieval Arabic sources, for a now widely accepted defense of the view that the fourth figure is due to Galen. The validity of the fourth figure has not been a matter of dispute, but some logicians, Peirce among them, have argued that fourth-figure syllogisms are readily reducible by conversion, specifically, by transposition, to one of the original three figures. In Peirce [206, 291, p. 318], he called the syllogisms of the fourth figure "triangular", where the syllogisms of the other figures are "linear", and the triangular being convertible as moods of the linears. Likewise, in his Lowell Lecture I, he speaks [207, 291, p. 367] of the third figure as the "last" figure. He nonetheless takes these into consideration in his discussion of syllogisms in his [216] "On the Algebra of Logic". Peirce repeated in "One, Two, Three: Fundamental Categories of Thought and of Nature" [224, 294, pp. 242–243] that there are three figures of the ordinary syllogism", the fourth being indirect, readily reducible to moods of the others. Henle [105] surveys the arguments over whether the fourth figure is distinct from the others or readily reducible.

showing all moods of the syllogism, was produced the following year in the Lowell
Lecture I [207], Ms. #122 of September–October 1866; see Peirce [291, p. 368] and
again in the privately published "Memoranda Concerning the Aristotelian Syllogism"
[208; see 291, p. 507] distributed in November 1866 in conjunction with Peirce's Lowell
Institute lectures.[44] Peirce also created a closely related table in which he expressed the
relation among the valid syllogistic propositions, and another for enumerating all of the
moods of the syllogism, in the latter case, using boldface to denote the moods of the
first figure, italics for the second figure, and Roman for the third figure. He obtained,
in "An Unpsychological View of Logic . . ." [206, Ms. #109; see 291, p. 319], the
matrix:

	I	A	E	O
E	o	*E*	A	*i*
A	I	**A**	**E**	O
I	A	**I**	**O**	E
O	e	*O*	*I*	**a**

with a rule at the top, a case on the side, a result in middle, keeping in mind that there is
only one way to obtain a result from one syllogism.

In Lecture VIII, "Forms of Induction and Hypotheses" [204, Ms. #105; see 291,
p. 260] composed in April–May of 1865, he dealt diagrammatically with the three figures
separately. In Peirce's Lecture XI [205, Ms. #107; see 291, p. 302] composed in April–
May of 1865, the relation between the valid cases of the first three figures of the syllogism
were given as:

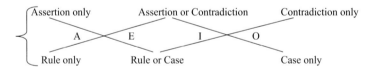

where A exists as assertion only, O as contradiction only, and E and I as either; and
E exists as rule only, I as case only, and A and O are neither particularly; and where,
as explained in the second version of a manuscript titled "An Unpsychological View of
Logic . . ." (Ms. #726, thought to be from 1865; see [291], p. 315), a *rule* is "a fact
expressed in a universal proposition, a *case* is a particular instance of the rule, and a *result*
being the conclusion obtained from these". We may readily compare Peirce's result with

---

[44]With the only difference in the latter instances being empty squares at the corners.

the modern or "Boolean" square of opposition and the traditional or "Aristotelian":

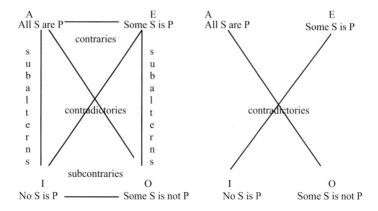

Peirce carried out his research on Aristotelian syllogisms and the relations among the moods and figure, then, from early 1865 through the end of 1866.

Hertz [108, p. 248] (see also [42], pp. 12–13), who made no reference to Peirce—nor, of course, could he have, since the manuscripts of Peirce which we have mentioned remained as yet unpublished—in discussing choices of axiomatic systems in obtaining proofs, presented two sets of diagrams; employing dots to represent the elements (or terms), a, b, c, ... of a proof and lines to represent implications of the form $a \to b$. He gave geometric illustrations of the possible unique sets of axioms where there is one axiomatic system available, and of possible sets of axioms where there are several unique sets of unique and several sets of non-unique sets of axioms, as well as the possible arrangements where there are unique "ideal" elements.

Davenport [71, pp. 153–154] argues for Dodgson's "transitional" graphical method as more suitable to the logic of relations than Venn's, while undertaking to "preserve the classical forms of propositions without such Boolean commitments as an empty class, and at the same time, to specify a universe of discourse within which can be generated all the relations between subclasses . . . ."

Peirce's interest in linear algebra, multilinear algebra, and matrix theory dates from his early education, fields to which both he and his father, Harvard mathematician Benjamin Peirce (1808–1880) contributed (see, e.g. [25, 96, 192, 355]). Charles Peirce was also familiar with the work in matrix theory, linear and multilinear algebras of Arthur Cayley and James Joseph Sylvester, and of Hermann Günther Grassmann (1809–1877). Charles Peirce's chief contributions were carried out in the notes and appendices he added to his reprint (1881) of his father's (1870) "Linear Associative Algebra", where he established that each of the algebras had a representation and could be classified in the logic of relatives, in much the same manner of the program undertaken by Felix Klein to demonstrate that geometries can be represented and classified according to groups. It is no doubt his familiarity also with non-commutative and nonassociative algebras, starting at least since 1873, with the non-Euclidean geometries of Lobachevskii,

Bolyai and Riemann, as well as the work of Klein,[45] that coalesced to encourage Peirce to undertake work in triadic logic. So, for example, in Peirce [229] he reviewed the English translation [174] by his former student George Bruce Halsted (1853–1922) of Lobachevskii's geometry [173].

We noted earlier that Dodgson had worked with an incomplete truth table [54, p. 31]. After remarking that Dodgson "applied "truth tables" to the solution to logical problems", Bartley went further, declaring that the "method was known to Boole, Frege, Peirce, and other nineteenth-century logicians too." This is misleading, presumably based upon a failure to recognize the crucial distinction, stressed by Shosky [331, pp. 12–13] between what he calls the truth table *technique* and the truth table *device*, the former being more properly understood as the truth-functional analysis of propositions, the latter as the matrix or tabular array of truth values for propositions. As shown in Anellis [17] and elaborated in Anellis [21], the first identifiable example of a complete truth table matrix array, which grew out of work on the truth-functional analysis of propositions, was that of Charles Peirce. Whether Peirce was influenced in this by Dodgson's partial truth table remains an open question.

What can be said with certainty is that Peirce's work in developing truth tables had its twin origins in the truth-functional analysis of equations for his algebra of relatives beginning in the early 1880s and in his penchant for diagrammatic thinking.

In his "On the Algebra of Logic: A Contribution to the Philosophy of Notation [223, pp. 188–189], under the name "icon of the first intention", of illation, '$a - < b$', we recognize material implication. The sign that Peirce chose, '$-<$', Peirce explained when he first introduced and employed it in his (1870) "Description of a Notation for the Logic of Relatives...", was designed to combine identity and the order relation, and could function, as the context required, either as an order relation, class inclusion, set elementhood, the copula, or the connective of material implication. And, aside from the greater typographical complexity of '$\leq$' because he found the latter alternative to be question-begging with regard to presupposing the relation it is intended to define, it was designed to distinguish it from the mathematical operators of equality and ordering.[46] It is the result of experimentation over a long period, which in later years included a pointing finger (☞). The development of notations for logical (and mathematical) operators with which Peirce experimented was part of his project of developing an *iconic* calculus or topological logic.[47] While working on this paper, Peirce undertook to develop a truth-

---

[45]See, *e.g.* Anellis [23, pp. 5–7].

[46]In the footnote where he introduced his illation sign, he explained [214, 318*n*.; 286, par. 47*n*.1; 292, 360 *n*. 1]: "My reasons for not liking the latter sign [is that] it seems to represent the relation it expresses as being compounded of two others which in reality are compilations of this. It is universally admitted that a higher conception is logically more simple than a lower one. Whence it follows from the relations of extension and comprehension, that in any state of information a broader concept is more simple than a narrower one included under it. Now all equality is inclusion in, but the converse is not true; hence inclusion is a wider concept than equality, and therefore logically a simpler one. On the same principle, inclusion is also simpler than being less than. The sign $\leq$ seems to involve a definition by enumeration, and such a definition offends against the laws of definition." In that same work, he also distinguishes algebraic from logical addition, by writing '+,' for the latter.

[47]Shin [330, p. 34] thus claims that "Peirce's invention of graphical systems was... a logical product of his theory of signs."

functional analysis of illation, and devised an abbreviated or indirect truth table. His first fully developed truth table in matrix form is found in his undated manuscript "An Outline Sketch of Synechistic Philosophy" [230] composed in 1893, we have an unmistakable example of a truth table matrix for the proposition expressing material implication, as

	t	f
t	t	f
f	f	t

In an untitled paper written in 1902 and placed in volume 4 of the Hartshorne and Weiss edition of Peirce's *Collected Papers* under the title 'The Simplest Mathematics" ([239]; see Peirce [287], 4.260–262), Peirce displayed the following table for three terms, x, y, z, writing **v** for *true* and **f** for *false*[48]:

x	y	z
v	**v**	**v**
v	**f**	**f**
f	**v**	**f**
f	**f**	**v**

By 1903, Peirce had also devised truth tables for triadic logic,[49] and these appeared in the edition of his works edited by Charles Hartshorne and Paul Weiss.

An *icon*, for Peirce (see, *e.g.* Peirce [285], par. 2.92; [287], par. 447, 4.531; [288], par. 5.73-74; [295], pp. 273, 277, 291, 304, 307, 460–461), is a sign which represents an object in virtue of resembling it in an essential manner, or sharing with it a distinguishing quality, as, *e.g.*, a pointing finger.

The term *topology*, in place of Leibniz's term *analysis situs*, originated in the work of Johann Benedikt Listing (1808–1882), first appearing in print in his [167] "Vorstudien zur Topologie", and widely known from the 1848 reprint (1848) where it is defined (1848, 6) as "the doctrine of the modal features of objects, or of the laws of connection, of relative position and of succession of points, lines, surfaces, bodies and their parts, or aggregates in space, always without regard to matters of measure or quantity"; but it did not begin to gain currency until used by Riemann [313, §2] and Poincaré [304]. Peirce was familiar with this work, and in his and James Mills Peirce's undated preface to their father's [197] *Elementary Treatise on Geometry*, they call it connective geometry [199; Ms. #94:09]. They explain (at [94:05]) that it "studies the relations of places" and deals with only a portion of the hypotheses accepted in other parts of geometry." As Burch [50, 51, p. 235] noted, borrowing Listing's [168] "Census Theorem", which in its newer version is the Euler-Poincaré formula, that describes the relationship of the number of vertices, the number of edges and the number of faces of a manifold. It has been generalized to include potholes and holes that penetrate the solid. As Burch [51, 233n.9] also noted, Peirce also borrowed Listing's [168] term *cyclosis*, or having a closed path; so that where any primitive term has cyclosis 0, if the term is constructible by these alone it also will have cylcosis 0. Other terms obtained from these primitives have cyclcosis 1.

[48]This is the work to which Lane [158] refers.

[49]Peirce's *Logic* (*Logic Notebook 1865–1909*) (R399, *1865–1909*); see Fisch and Turquette [86]. Peirce's original tables from Robin catalog Ms. #339, ([201], 00340, 00341, 00344), are reproduced as plates 1–3 at Fisch and Turquette [86, pp. 73–75]. Under the title "On Triadic Logic", the fragment Robin catalog, Ms. #339.00340-0344, dated 23 February 1909, was published in Haack and Lane [99, pp. 217–224].

Peirce's next step was to develop his X-box or X-frame notation to diagrammatically represent the possible combinations of truth-values for a dyadic relation. In his X-frame notation, the open and closed quadrants indicate truth or falsity respectively. For example, ⊠, the completely closed frame, represents row 1 of the table for the sixteen binary connectives, in which all assignments are false, and x, the completely open frame, represents row 16, in which all values are true.[50] The X-frame notation is based on the representation of truth-values for two terms as follows:

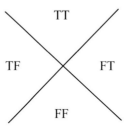

The full details of this scheme are elaborated by Peirce in his manuscript "A Proposed Logical Notation (Notation)" of *circa* 1903 [248, pp. 26–28].

Hubertus Gezinus Hubbeling (1925–1986) in his [125] "A Diagram-method in Propositional Logic", began by asserting that "[t]he diagram-method in logic has been propagated by various authors. The most famous attempts are those of Euler, Venn, and Peirce etc. Especially the so called Venn–Euler diagrams are well known." He adds that they "may be used in propositional logic with monadic functions." He goes on to illustrate this for classical propositional logic, but makes no effort to deal with the monadic predicate calculus, and states that a "completely formalized" calculus can be built with this method; but he rests content to give a few salient examples in the propositional calculus. He takes his diagrammatic method as an adaptation of, and improvement over, Venn–Euler diagrams. The chief "innovations" are (1) to use a system of pluses and minuses (rather than t and f or 1 and 0) for truth values, being apparently unaware of Leibniz's plus–minus calculus in his [162] *Non Inelegans Specimen Demonstrandi in Abstractis*; and (2) to replace the use of shading in the Venn–Euler diagrams with the system of pluses and minuses. Understandably, he also was unaware of Dodgson's use of a small '+' for a non-empty region on a worksheet of logic problems that Dodgson introduced late in 1884 or early 1885. Beyond that, Hubbeling's use of quadrants to represent the assignment of truth values is reminiscent of Peirce's X-frames, although there is no mention of these, or, for that matter, Peirce's [230] truth table, and no suggestion that Hubbeling was familiar with either.

---

[50]For details, see Clark [65] and Zellweger [367].

Peirce's Entitative and Existential Graphs[51]

Peirce spent much of his professional endeavors working on graphs and diagrams and in search of improvements over those, such as Euler's, which were already available. There are thirty-five extant manuscripts (Mss. ##479-514) in the Peirce *Nachlaß* that had been classified in the Robin Catalog [317] explicitly as being concerned with logical graphs. Of these, three (Mss. ##479, 481, 492) are devoted in large measure explicitly to Euler diagrams; most of the remainder are developments of Peirce's earlier entitative graphs and his later existential graphs. Many of these are undated; those that are dated range from the period 1896–1911. Peirce held his existential graphs to be his "chef d'oeuvre". And in the manuscript on "The Basics of Pragmaticism" [259], Ms. #280, dated to around 1905, after remarking that many errors and confusions made by philosophers arise as a result of their failure to accept or employ the logic of relations, he presents an elementary discussion of existential graphs, which he termed "quite the luckiest find that has been gained in exact logic since Boole". The claim that existential graphs explain logical fallacies also appeared in a two-notebook set [260] dating from the same time.

Existential graphs arose in part from Peirce's work on truth-functional logic and in the main out of his experimentation with graphical or diagrammatic methods for analyzing logical propositions and proofs.

The first extant published appearance of entitative graphs is found in his paper "The Logic of Relatives" [234, pp. 174–178, 185–189] in *The Monist*, where he employed it to elucidate the exposition of the algebra of relatives in as non-technical a manner as possible for philosophical readers, although he does not there call them such. Peirce [234, pp. 168–171; 286, par. 3.469-470] explicitly tells us that his system for graphically representing relational propositions was inspired by his study of chemistry, and refers [234, p. 168; 286, par. 3.470] to Kempe and Clifford; and in explicitly explaining logical relations in terms of chemical bonding, he in particular cites [234, p. 170; 286, par. 3.470] the chemical bonding theory of Julius Lothar Meyer (1830–1895).[52] Remarking [234, p. 174; 286, par.

---

[51]Zeman [365], Roberts [316], and Thibaud [339] are the standard sources for the exposition of existential graphs; Zeman [365] undertakes to present the system axiomatically; Thibaud [339] provides an exposition of the system by rendering it against the background of the algebraic logic as developed earlier by Peirce. Zeman [366] is a concise presentation; Shin [330] is an interpretive account set in the philosophical context of the issue of logic as calculus/logic as language distinction. Unfortunately, it is marred in its translation of β-graphs to first-order logic and in some unsound rules.

[52]Peirce does not give a reference for Meyer's work. In the first edition of *Die modernen Theorien der Chemie* [186], we find Meyer using atomic weights to arrange elements into families that share similar chemical and physical characteristics, while leaving a blank for an undiscovered element. His single major conceptual advance over his immediate predecessors was seeing valence, the number that represents the combining power of an atom of a particular element, as the link among members of each family of elements and as the pattern for the order in which the families were themselves organized. In "Die Natur der chemischen Elemente als Function ihrer Atomgewichte" [187] he gave a graphic display of the periodicity of atomic volume plotted against atomic weight. The editors of Peirce [286] insert a footnote (p. 297, *n*.*) to Peirce [286, par. 470] citing Seubert [327], without indicating that it is actually an anthology of papers by Meyer and Mendeleev that includes a reprint of Meyer [187].

3.475] that chemists employ Roman numerals to note the "adinity" of an atom, but that it is not really a necessity, he represents the relative copula with the diagram:

where the letter 'd' represents the monad "... is a mortal", and the letter 'h' represents the monad "... is a man", the line is the copula, and the antecedent is enclosed, the diagram expressing the proposition: "Anything whatever, if it be a man, is mortal". As he proceeds, his examples become more complex. In another source, the manuscript "Detached Ideas continued and the Dispute between Nominalists and Realists", Peirce [236, 253, Ms. #439] tells us that his system of existential graphs is a consequence of his study of the categories.[53]

In the fully developed system of entitative graphs, the surface of the graph was a sheet which represented a truth-theoretic plane, and the letters representing the terms of the calculus were connected by lines representing the relations between these terms. A "cut" in the sheet, depicted by a circle around a letter representing a term in the universe of discourse, indicated a hole in the sheet, and thus represented the negation, or falsity, of the encircled term. In his existential graphs, the next phase of his work, Peirce used a similar graphical technique to deal with quantified propositions. In the entitative graphs, P together with Q, i.e. their mere concatenation, means P or Q while in the existential graphs it would mean P and Q. Placing a cut around two concatenated terms P or Q which have each already been negated yields, by De Morgan's Laws, the proposition *not* (*not-P and not-Q*), i.e., *P or Q*:

---

[53]Peirce means his categories of Firstness, Secondness, and Thirdness, which is symptomatic of the Hegelian triadicism that was the philosophical basis of Peirce's metaphysics and epistemology from his student days. With respect to logic and language, for example, Peirce [229, 286, par. 3.422] wrote in 1892 in "Critic of Arguments. II. The Reader is Introduced to Relatives", a work he intended to publish as a logic textbook: "I will only mention here that the ideas which belong to the three forms of rhemata are firstness, secondness, thirdness; firstness, or spontaneity; secondness, or dependence; thirdness, or mediation," where by a *rhema* Peirce refers to Aristotle's term, typically understood as a predicate, but which in contemporary terms should more properly be understood as a propositional function, more particularly, an indeterminate function; see, *e.g.*, Steinthal [333, p. 234] and Luhtala [176] on the syntactic complexity of Aristotle's *rhema* taking the role of verb, predicate, and adjective. This doctrine of trichotomy is the basis for Peirce's Reduction Thesis, according to which all *n*-adic relations can be constructed solely from triadic relations. For an example, see Peirce [235, pp. 6, 11; 290, III, p. 436]. Peirce [244, 245] include arguments against Kempe's arguments, which Peirce considered "naïve", against admitting a third category into logic. For more on Peirce's triadicism or *teridentity* regarding the logic of relations, see, *e.g.*, Mertz [185], Herzberger [111], Burch [50, 51], Brunning [49] and Anellis [15].

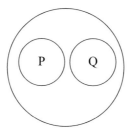

Peirce became dissatisfied with entitative graphs even before the issue of *The Monist* in which they appeared was printed, that is, while he was still checking the galley proofs, and he wrote to the journal's editor, Paul Calvin Carus (1852–1919) to describe his new system, the existential graphs, hoping to delay the publication of his paper so as to publish it with his existential graphs. Roberts [316, p. 27] quotes from Peirce's own account of this [259, pp. 21–21]:

> The writer described a system of logical graphs, since named "Entitative Graphs", [in [235]; but the ink was hardly dry on the sheets . . . when he discovered the far preferable system, on the whole, of Existential Graphs, which are merely entitative graphs turned inside out, and sent the gracious Editor a paper on the subject that could be squeezed into a single number by simply excluding everything else. But the Editor feared that so swift the advances of exact logic seemed to be, that, before the types were half set up, the second system might be superseded. However, eight years elapsed and not one jot or one tittle has in no wise passed from the system.

This is not to say, however, that Peirce did not continue to make refinements and improvements. In one undated, untitled manuscript [284, Ms. #513] a large portion of which contains a discussion of his algebraic logic and then turns to graphical methods (pp. 52–78), one will find a note in which Peirce expresses the judgment that "my cumbrous General Algebra with all its faults, seems preferable."

The only printed appearance of existential graphs in Peirce's lifetime was in a privately printed brochure [241] with a small distribution, prepared for his Lowell lectures, based on a manuscript [242]. This work provided transformation rules for these graphs. Beyond this one work, however, the logic community at large had to wait for Peirce's work on existential graphs until fragmentary manuscripts appeared in the fourth volume [287] of the edition in the early 1930s of Peirce's *Collected Papers* prepared by Charles Hartshorne (1897–2000) and Paul Weiss (1901–2002), writings which Willard Van Orman Quine (1908–2000) [311, p. 552] (as we shall subsequently see) largely savaged.

In the undated notebook "On Existential Graphs as an Instrument of Logical Research", Peirce [283, Ms. #498] wrote that he discovered these graphs late in 1896, but that he was practically there some 14 years before. This suggests that at least some of the undated manuscripts that deal with graphs could well date from the early 1880s while Peirce was still writing and publishing on his algebraic development of logic. The earliest record we have of existential graphs, what is called the *sheet of assertion*, occurs in connection with the α-graphs, which is to say existential graphs for propositional logic, upon which the other types of existential graphs, the β-graphs for quantified propositions, and the γ-graphs and tinctured γ-graphs for modal propositions, are based on as extensions, seems to be early in a thirty-one page manuscript notebook [249, pp. 2–3, Ms. #455]. This is the background against which the graph is drawn (it may be physically represented as a sheet of paper, or a chalk board, or, for more complex, layered graphs, a writing tablet, etc.), and serves to signify the universe of discourse with which we are concerned. Any signs or propositions inscribed upon it, e.g. "Socrates is mortal", are assertions that Socrates exists in this particular universe of discourse, and that Socrates is mortal. The universe of discourse, like De Morgan's "universe of a proposition, or of a name" [74, p. 380; 78, p. 2] that, unlike the fixed universe of all things employed by Aristotle and the medieval logicians which remained typical of traditional logic, "may be limited in any manner expressed or understood", may be changed as required.[54]

Alexander MacFarlane (1851–1913), with whom Peirce corresponded (see [181]), introduced in his [180, p. 42] *Principles of the Algebra of Logic*, without explicitly stipulating the fact, a representation of the universe of discourse by a square as an improvement over Venn diagrams. In his "The Logical Spectrum" [182, p. 286] he explicitly lets a square represent the universe of objects considered. This is, for the style, typical of modern Venn diagrams. It is not determinable when or whether Peirce's sheet of assertion might have been influenced by MacFarlane. In the interim, Allan Marquand, in "A Logical Diagram for n Terms" [184], introduced a square diagram for the universe of discourse, subdivided into $2n$ partitions, each representing the n terms of the universe of discourse, together with their possible combinations, positive and negative, thus, for terms P and Q, we have PQ, $\overline{P}Q$, $\overline{PQ}$, and with or without signifiers such as shading, to indicate existence or inexistence of terms for each combination of each quadrant. It is readily interpretable, however, as a tabular form, and likely precursor, of Peirce's X-frame. Other than the fact that Marquand's paper was accessible to Charles Dodgson in the Christ Church Library, there is no direct evidence that it was an inspiration for Carroll diagrams. Moktefi and Shin [189, pp. 635, 636] call both Marquand's and MarFarlane's "Venn-type" diagrams. Carroll diagrams unify the best features of those of Venn, Marquand, and MacFarlane. (See [5] for a complete discussion of the advantages of Carroll diagrams over Venn diagrams.) However, Venn-type diagrams became more popular, and when dealing

---

[54]Peirce (in Ms.493, an undated, unpaginated manuscript notebook) defined the *universe of discourse* as "aggregate of the individual objects which "exist", that is are independently side by side in the collection of experiences to which the deliverer and interpreter of a set of symbols have agreed to refer and to consider." The extensional conception of a universe of discourse, comprised of individuals and classes, was adopted by Peirce partially from De Morgan, but also partially from Oscar Howard Mitchell (1851–1889), who added the concept of *dimensionality* to De Morgan's universe [285, par. 2.536]. Underlying the semantic interpretation of a universe of discourse for Peirce was the ontological commitment to individuals and the classes to which they belong. The inhabitants of the universe of discourse may be physical, determined by experience, through the senses; or they may be imaginary, as populated by the contents of a work of art. As Peirce (with Christine Ladd-Franklin [298, p. 742; 285, par. 2.536]) wrote: "In every proposition the circumstances of its enunciation show that it refers to some collection of individuals or of possibilities, which cannot be adequately described, but can only be indicated as something familiar to both speaker and auditor. At one time it may be the physical universe of sense, at another it may be the imaginary "world" of some play or novel, at another a range of possibilities." As further expressed by Peirce [289, par. 6.351], therefore, "...I wish my description of what is true or false, to apply to what is not only true or false generally, but also to what is true or false under conditions already assumed. Whatever may be the limitations previously imposed, that to which the truth or falsity is limited may be called the *universe of discourse*. For example, at the mention of a certain name, every person initiated into the Eleusinian mysteries invariably experiences a feeling of awe. This is true. It is therefore true that every person initiated into the Eleusinian mysteries always experiences a sentiment of awe; not universally, but only under the limitations already understood before this is said."

with more than five terms, they employed intersecting ovals rather than circles. In any case, all of these diagram systems were intended for syllogisms, rather than for a logic of relations. Consider, *e.g.*, a Venn diagram of four intersecting ellipses, for dealing with four terms that appears in Venn's *Symbolic Logic* [348, p. 106]:

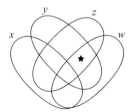

Working on existential graphs into the final decade of his life, Peirce, in preparing for his Lowell Institute lectures of 1903–04 [250], held that improvement in reasoning requires, first of all, a study of deduction, and that for this task, an unambiguous and simple system of expression is needed. The system in which reasoning is broken up into its smallest fragments by means of diagrams is the system of existential graphs which Peirce continued to develop in terms of fourteen conventions. In a very closely related set of notes for the Lowell lectures, he explained [251, Ms. #454] that this is a system for expressing any assertion with precision and is not intended to facilitate but to analyze necessary reasoning, i.e., deduction. The system is introduced by means of four basic conventions, called "principles", and four rules or "rights" of transformation. In the first draft for Lecture 3 of the Lowell lecture series of 1903–04, Peirce [254, Ms. #458] asserted that the system of existential graphs is the "simplest mathematics". Elsewhere [240], he defined the "simplest mathematics" as a two-valued system, a special case of which is Boolean algebra. With that in mind, he offered the α-graphs, and β-graphs, i.e. existential graphs for propositional logic, along with their transformation rules.

In other manuscript notebooks of 1903 for the Lowell lectures, Peirce [252, 253, Mss. ##455, 456] developed the α- and β-graphs, and in a manuscript for Lecture III [255, Ms. #460] we find the γ-graphs, which Peirce held to be required for the intelligibility of the categories of Firstness, Secondness, and Thirdness, and which enable the graphical concepts of *multitude*, *infinity*, and *continuity* to be expressible or representable. What is unique about γ-graphs, he further held, is that they permit abstractions, or mere possibilities and the laws of the subjects of discourse, inasmuch as they are designed to introduce modality into the logical system graphically. In a manuscript in connection with his third and fourth lectures, he wrote [256; Ms. #466] that existential graphs were initially introduced to illuminate the nature of pure mathematics, and then used in the discussion of the concept of multitude.

Work on the α-, β-, and γ-existential graphs continued as Peirce revised his notes for the Lowell lectures, developing and refining their "conventions" or formation rules and rules of passage, as well as on refining the concept of existential graphs and of their applications. Much of this work occurred in 1903, and was carried on until the end of 1913, when Peirce became too ill to continue writing. In a manuscript notebook begun on October 2, 1902 towards the third draft of the third Lowell lecture, Peirce [257, Ms. #457] presented a decision procedure for existential graphs in terms of an α-possibility that depends upon the expansion of quantified propositions as logical sums and products, and thus anticipates in graphical form Leopold Löwenheim's (1878–1957) [175] theorem for determining the validity of quantified formulas in a given domain.[55] Part of the manuscript in preparation for the fifth in the series of Lowell lectures [258; Ms. #468] dated December 4, 1903, is devoted to an effort to apply existential graphs to the principles of logic. Peirce had become involved with this question from the very outset of his career, and first dealt with it in his manuscript [212; Ms. #593]. Existential graphs, published in part as his [213] article "Grounds of Validity of the Laws of Logic", can be conceived precisely as a tool for determining the validity of arguments by visual inspection. In the manuscript "Analysis of some Demonstrations concerning definite Positive Integers (N)" of 1905–06, the bulk of which is devoted to the logic of relatives and its application, as the title suggests, to the definition and construction of the positive integers,[56] we also find in Peirce [262, Ms. #70:05] graphical representations of the formulas $(\bar{a} + \bar{b}) \cdot c$ and $c \, (\bar{a} + \bar{b}) \cdot c$ respectively. A few pages later [263, Ms. #70.07], a diagram is given for the quantified formula $\prod_{\gamma} \sum_{\alpha} q \cdot \alpha_n \cdot q_0 \beta u \cdot (\bar{q}_{\gamma} u + \bar{q}_{\alpha} u + \bar{q}_{\beta} u) \cdot q_{\gamma} u$. In the supplement to this manuscript on defining the properties of the integers [264, Ms. #70s:76], probably dating from August 9, 1906, Peirce diagrams the expression "I is S to X" where the line may be either straight or curved, and multiple lines, or *ligatures*, serve as branching for multiple relata., and denote existence. A line drawn heavily is the *line of identity*.

The Tableau Method: From Gentzen to van Heijenoort

The manuscripts for Part Two of *Symbolic Logic* in which Dodgson developed his tree method lay undiscovered until William Bartley happened upon it, publishing a brief account [29], and prepared and published the surviving text [54]. Quine's review of Carroll [54] was not entirely complimentary. Under such circumstances, Dodgson's work neither had, nor could have had, any role in the development of deductive tableaux and the variations which developed in the second third of the twentieth century at the hands of Gentzen, Beth, Smullyan, and van Heijenoort. Similarly, in the first third of the

---

[55]In his introduction to Löwenheim's [175], van Heijenoort [343, p. 228] says: "Löwenheim's work links up with that of Peirce and Schröder". It was left to Badesa [26, 27] and Brady [46] to fill in the details; see also Anellis [18].

[56]In *Laws of Form*, [332], George Spencer-Brown (b. 1923) claims to have created a new and simplified graphical system of logic—one which Bertrand Russell enthusiastically praised (see [322], p. 166), chiefly, we conjecture, for the author's claim that it is isomorphic to the systems of Nicod and of Sheffer, the latter of which Russell, in his introduction to the second edition of *Principia Mathematica*, with an abundance of hyperbole, essentially declared [321, p. xiii] to be the single greatest achievement in logic since the appearance of the first edition of the *Principia*. Spencer-Brown's system is based upon the primitive operation of *distinction* or *different* (logically identical with the Sheffer stroke, NOR), which Spencer-Brown undertook to employ to develop arithmetic and algebra. The operator $\leq$, in its design strongly resembles Peirce's sign of illation $\leqq$, in the manuscript "The Logic of Relatives: Qualitative and Quantitative" [225, 294, pp. 372–378]. A quotation from Peirce [294, p. 373], and Kauffman [133, pp. 88–93] asserts Spencer-Brown's calculus to be isomorphic to Peirce's existential graphs even though Spencer-Brown's: sign separates "outside" from "inside", and Peirce's sign is illation, or, as we would more typically call it, material implication.

twentieth century, the traditional logic, in which the syllogism played a seminal role, and the algebraic logic of De Morgan, Boole, Peirce, and Schröder, slowly fell into disuse among the mathematically-minded logicians, as attention shifted to the axiomatic methods of Frege, Hilbert, Russell, and Whitehead.[57] The bulk of Dodgson's work in logic that was known was treated largely as recreational. While such a notion might appeal to followers of the later Wittgenstein, for whom language games about the analysis of ordinary language were useful in an effort to root out nonsense in philosophy, it led Richard Bevan Braithwaite (1900–1990) to write, somewhat erroneously, [47, p. 174] that: "Carroll regarded formal and symbolic logic not as a corpus of systematic knowledge about valid thought nor yet as an art for teaching a person to think correctly, but as a game." It is in this sense that George Pitcher (b. 1925) [303, p. 592] was convinced of the deep influence which Dodgson had upon Wittgenstein, but it left little of interest to logicians working to develop axiomatic systems or related formal systems.[58]

Although parts of manuscripts presenting Peirce's existential graphs appeared in the early 1930s in the multi-volume *Collected Papers* (Peirce 1931–1935) of Peirce published by Hartshorne and Weiss, this work attracted little attention at that time, in part because of the judgment by reviewers such as Quine [309–311] who disparaged the value of Peirce's work, finding "considerable dross" along with "much gold" [309, p. 229], and in part because of the focus of attention at that time on the axiomatic method and related systems that arose as logicians sought more felicitous alternatives. Gentzen for example, developed both his Sequent Calculi and his method of natural deduction in search of a more natural way than the axiomatic method to undertake theorem proving,[59] Hilbert and Bernays' *Grundlagen* was just beginning to appear, Herbrand's work had just appeared, and the work of Frege and of Whitehead and Russell was still a major focus of attention, the second edition of *Principia Mathematica* having appeared only a decade previously. Moreover, the work of Frege, Russell, and Hilbert was being directly challenged by Gentzen [91, p. 176]. Quine was particularly dismissive of the value of Peirce's diagrammatic logic, writing about the material in "Book II. Existential Graphs" (Peirce [287], pp. 293–470) of the fourth volume [311, p. 552]:

> The other material on exact logic has to do with logical graphs. A series of extensions and modifications of EULER's scheme of diagrams leads PEIRCE to an elaborate scheme of his own, designed for the expression of propositions involving any manner of complexity in point of relational structure, quantity and even modality. The system is intended rather for the analysis of logical structure than for the facilitation of inference; because of its cumbersomeness it is

---

[57] See, *e.g.* Anellis [20] on the reactions of traditional logicians as they attempted to accommodate to the development of mathematical logic, in particular in the work of Bertrand Russell.

[58] As Pycior [308, p. 169], referring to Sewell [328, pp. 24–25], has noted, this accords well with the view of De Morgan and Carroll regarding the nature of the symbolical algebra:

> If neither prose nor poetry," she asks, "can provide the necessary structure for Nonsense, is there some other system by which language could be organized into an independent and consistent, if nonsensical, structure?" She finds such structure in game rules, or "an enclosed whole, with its own rigid laws which cannot be questioned within the game itself.

[59] Van Heijenoort [345, p. 7], for example, spoke of a "family of formal systems", whose members were: the axiomatic method; Herbrand quantification; Gentzen sequents; and natural deduction. The last three of these arose within a few years of one another, in the first half of the 1930s.

less suited to the latter purpose than is the algebraic form of logic. One questions the efficacy of PEIRCE's diagrams, however, in their analytical capacity as well. Their basic machinery is too complex to allow one much satisfaction in analyzing propositional structure into terms of that machinery. While it is not inconceivable that advances in the diagrammatic method might open possibilities of analysis superior to those afforded by the algebraic method, yet an examination of PEIRCE's product tends rather, apagogically as it were, to confirm one's faith in the algebraic approach.

As a consequence, Peirce's contributions were largely ignored and diagrammatic tools, insofar as they were taught in logic courses, tended to be relegated to introductory formal logic texts that restricted themselves to traditional logic.[60] The judgment in the middle of the twentieth century of the value of Peirce's work, as well as that of his immediate predecessors and colleagues, De Morgan, Boole, Marquand, Ladd-Franklin, and Schröder, was such that nothing of their original work was included in the source book on mathematical logic [343] of Jean van Heijenoort (1912–1968).[61]

Likewise, although Lane [158, p. 284] wrote that: "For many years, commentators have recognized that Peirce anticipated the truth-table method for deciding whether a wff is a tautology," and pointing specifically to Peirce's [223] "On the Algebra of Logic: A Contribution to the Philosophy of Notation", and to Peirce's [239] "Chapter III. The Simplest Mathematics (Logic III)" published in the Hartshorne-Weiss edition of Peirce [287, p. 213] as evidence that "it has long been known that he gave an example of a two-valued truth table",[62] Peirce's work on the truth-functional analysis of propositions and his development of truth tables suffered much the same fate. Even after Atwell Rufus Turquette (b. 1914) and Max Harold Fisch (1901–1995) wrote about Peirce's truth tables that Peirce developed for his trivalent logic in the period 1902–1909 [86, 341], in the process of which Peirce devised a truth table for the sixteen binary connectives of propositional logic which were subsequently rediscovered by William Glenn Clark and announced by Shea Zellweger (see [64, 65, 367]), it remained a commonplace in the literature to attribute truth tables to Jan Łukasiewicz (1897–1956) [177], Emil Leon Post (1897–1954) [306, 307], and Ludwig Wittgenstein (1889–1951) [362]. The situation changed when Shosky [331] announced his discovery of a hand-written truth table for p on the verso of a transcript by Bertrand Russell dating from 1912 and tables for $p \vee q$, $p \supset q$, and $\sim p \vee \sim q$ in lecture notes of April 1914 taken by Thomas Stearns Eliot (1888–1965) in Russell's Harvard University logic course (and reproduced in Shosky [331], p. 23). The truth table matrix given by Peirce [230] in his manuscript "An Outline Sketch of Synechistic Philosophy" for his illation connective $x \mathbin{-\!\!\!<} y$, and for which an indirect

---

[60]An example of such texts would be the *Logic: An Introduction* of Lionel Ruby (1899–1972) [319], such as one of the authors had used in a first logic course. A comparison with most introductory logic textbooks used a century earlier would show little difference in subject-matter and essential content, but only in style.

[61]For the view of the relegation of the algebraic logicians to comparative obscurity through much of the twentieth century, see, *e.g.* Anellis and Houser [24]. See Anellis [19] on the reciprocal antagonisms between the members of the Boole-Peirce-Schröder tradition and the Frege-Russell tradition in the early twentieth century.

[62]One may in this regard think of Łukasiewicz [178, §13] referring to Peirce's [223] as devising the matrix method for which "the truth of theses [in propositional logic] depends not on their content, but on their truth value" and the examples which Łukasiewicz immediately provides for negation, material implication, conjunction and disjunction. See also Beth [41, p. 73].

truth table was already given by Peirce [221] in the manuscript "On the Algebra of Logic" and the accompanying supplements, is exactly equivalent to that given by Russell in 1914 for $p \supset q$ (see [21] for a full account of Peirce's work in truth-functional analysis and [21], 94 specifically for his truth table matrix for illation). One of the few who did explicitly recognize Peirce as having presented a truth table in 1885 was Beth (see [41], 73, who, regrettably, does not, however, include [223] in his bibliography).

A generation after Quine, Thibaud [339, p. 2] offered as explanations for lack of interest in Peirce's graphical system the multiplicity and novelty in part of his terminology, in part the isolated and scattered discussions by Peirce of his existential graphs, and the elusive strangeness of the accounts given by Peirce—"elliptical and . . . opaque"— and seemingly far removed from the comparative familiarity of the algebraic notation. Thibaud's remedy is to begin his exposition with a forty-page discussion [339, pp. 9–49] of Peirce's algebraic treatment of propositional logic prior to embarking upon a treatment [339, pp. 49–68] of his α-graphs. This same pattern is undertaken in Thibaud's survey of the logic of relatives and the β-graphs, and of modal logic and the γ-graphs, prior to winding up with a brief discussion of the algebra of existential graphs. Thibaud's list of multiple reasons for the disinterest in these graphs may be seen as an elaboration of Quine's complaint of its "cumbersomeness" and complexity.

When we come to examine work of the twentieth-century creators of the tree method, we are led, nevertheless, to inquire whether their work had been informed, either directly or indirectly, by the work of Peirce, Dodgson, or Ladd-Franklin. Given that Dodgson's work on trees lay undiscovered until after Smullyan completed his work on analytic tableaux, and that his direct familiarity at the time was with the work of his contemporary colleagues, and of Beth and Hintikka most particularly. While the work of Peirce and the members of his logical "school" was largely neglected, we might surmise that the most likely reply must be in the negative. That leads in turn to the question of whether Gentzen or Beth, or Hintikka were familiar to any degree with the relevant work at least of Peirce, if not of Ladd-Franklin, while we may safely assume that none of these researchers could have known about the lost work of Dodgson prior to its rediscovery and publication by Bartley. As already remarked, Beth, one of the more historically sensitive of the trio of Gentzen, Beth, and Hintikka,[63] did in fact note [41, p. 73] that

---

[63] Approximately one-tenth of the pages of Beth [41], the entire Part I, are given over to the "historical background of research into the foundations of mathematics". Beth in his [33] "Hundred Years of Symbolic Logic" offers a quick survey of developments from the publication in 1847 of De Morgan's *Formal Logic* and Boole's *Mathematical Analysis of the Laws of Thought* to the centenary of their publication, and a contrasting comparison of that period with the preceding years, from Aristotle until 1847. In doing so, Peirce is mentioned [33, p. 334] as having "made quite important contributions", receiving the baton of symbolic logic from Boole and De Morgan and passing it on to Frege and Schröder; Peirce did this, Beth [33, p. 334] explained, by "thoroughly revis[ing] and extensively enlarg[ing]", along with Jevons and Schröder, Boole's work. He [33, p. 337] recognizes Peirce's contributions to the calculus of relations, without, however, providing any information about what they were, beyond his "very important" introduction in 1885 of "the general and existential quantificators," and in 1881 applying the new symbolic logic to attempt a logical analysis of the concept of number, making him, along with Richard Dedekind, "forerunners" of the logical study of foundations of mathematics [37, p. 338]. Beth in his [34] "The Origins and Growth of Symbolic Logic", however, said even less; he merely mentioned Peirce's name [34, p. 274] as a successor of Boole and De Morgan in logic, "also famous as a mathematician and philosopher", but had nothing whatever to say about his work.

Peirce provided a truth table in 1885. He also noted [41, p. 67] that Ladd-Franklin in 1883 gave "[a]n adequate treatment of the classical syllogism within the framework of the logical calculus"; but there is no hint here whatever of the antilogism or of what the "adequate treatment" consisted, and he failed even to say what would count as an "adequate" treatment. Beth's [33, p. 337] is no more helpful, telling readers only that: "In 1883, Mrs. Christine Ladd-Franklin gave the first adequate treatment of classical syllogism, by means of a symbolism, created *ad hoc* ... ;" ... "first adherents of symbolic logic" failed to successfully treat the classical syllogism because they dealt only with equations, whereas in the symbolism of Boole's logical algebra, existential equations could be expressed only by an inequality. Reading Beth on the history of this period of the development of logic, and especially of algebraic logic, one might conclude that, at least with respect even to the published writings other than Boole's, he had only minimal, possibly second-hand, information. Regarding graphical methods, Venn and Dodgson are entirely unmentioned in any of Beth's historical writings.

Beth, we know, communicated with Gentzen, and noted the link between Gentzen's work and his own tableaux.[64] A very elegant and succinct explanation of the "link" is given in [72, p. 13]: "The crucial idea" behind the semantic tableau is that systematically looking for a counterexample (countermodel) to a logical truth leads, in case of failure (to find the counterexample), to a proof of the logical truth, in upside down form. This proof will in the first instance be one in a Gentzen system, but it can easily be converted into other types of proof, e.g. in an axiomatic system." De Jongh [72, p. 15] further notes that the method is "very perspicuous" in demonstrating the decidability for the propositional calculus, when attempts to find a counter example fail, as such efforts will all either eventually succeed or fail, while for first-order logic, the method "gives a lot of insight", since the failure to find a counterexample leads to a proof, and demonstrates completeness. As Troelstra [340, p. 19] succinctly expressed it, considering a sequent $\Gamma \Rightarrow \Delta$, where both Γ and Δ are finite: "If the tableau closes at a certain finite depth, then the proof of the sequent, as a suitable Gentzen-style sequent, may be read off from the tableau. If the tableau does not close at a finite depth, then it implicitly contains a countermodel refuting the sequent $\Gamma \Rightarrow \Delta$." Once more establishing the "link" between Gentzen's work and Beth's tableaux, De Jongh [72, p. 15] explains that "Gentzen's subformula theorem is also an immediate consequence, since a tableau uses only subformulas" of the presumed logically true propositions involved. There is nothing in either Gentzen's work or Hertz's to justify a conclusion that either had any direct or indirect knowledge of the work of Peirce, his students, or of the technical work of Dodgson, and, in any event, Dodgson's work on trees would and could, of course, certainly have been unknown to them.

---

[64]Van Dalen [342, pp. 9–10] reproduces two letters from Gentzen to Beth, of 25 January 1937 and 6 February 1937, in which Gentzen notes some errors in Beth's discussion of the relative consistency proof for intuitionistic logic, resulting from the presence of free variables. Van Dalen [342, p. 11] then remarks that: "Later on, Beth's own work on tableaux would link up with Gentzen's proof calculi." Saying nothing explicitly of Beth's [37] "Semantic Construction of Intuitionistic Logic", in which the semantic tableau method is applied to intuitionistic logic, Van Dalen [342, p. 11] remarks that "Gentzen's natural deduction system, and the related sequence calculus, provided a faithful and elegant rendering of intuitionistic argument."

Conclusion

If Von Plato, in tracing the tree form of deduction, had articulated the view that it arose in the course of the evolution of the work of Hertz and Gentzen, by way of the work of Beth, Hintikka, and Smullyan, and entirely *independently* of the influence of Peirce, Marquand, Ladd-Franklin, or Dodgson, the authors should have no quarrel with that assertion.

References

1. F.F. Abeles, Lewis Carroll's method of trees: its origins in *Studies in Logic*. Mod. Log. **1**, 25–35 (1990)
2. F.F. Abeles, Herbrand's fundamental theorem and the beginning of logic programming. Mod. Log. **4**, 63–73 (1994)
3. F.F. Abeles, Lewis Carroll's formal logic. Hist. Philos. Log. **26**, 33–46 (2005)
4. F.F. Abeles, From the tree method in modern logic to the beginning of automated theorem proving, in *From Calculus to Computers: Using 200 Years of Mathematics History in the Teaching of Mathematics*, ed. by A. Shell-Gellasch, D. Jardine (Mathematical Association of America, Washington, 2006), pp. 149–160
5. F.F. Abeles, Lewis Carroll's visual logic. Hist. Philos. Log. **28**, 1–17 (2007)
6. F.F. Abeles, Toward a visual proof system: Lewis Carroll's method of trees. Logica Universalis **6**, 521–534 (2012)
7. V.M. Abrusci, Paul Hertz's logical works: contents and relevance, in *Atii del Convegno Internazionale di Storia della Logica, San Gimignano, 4–8 dicembre 1982*, ed. by V.M. Abrusci, E. Casari, M. Mugnai (CLUEB, Bologna, 1983), pp. 369–374
8. G.E. Andrews, 1988. J. J. Sylvester, Johns Hopkins and Partitions, in [Duren, Askey, & Merzbach *1988*], pp. 21–40
9. I.H. Anellis, La obra de Jean van Heijenoort en el campo de la lógica: sus aportaciones a la teoría de la demon-stración. Mathesis **5**, 353–370 (1989)
10. I.H. Anellis, Editor's note: a history of logic trees. Mod. Log. **1**, 22–24 (1990)
11. I.H. Anellis, From semantic tableaux to Smullyan trees: a history of the development of the falsifiability tree method. Mod. Log. **1**, 36–69; 263 (1990b)
12. I.H. Anellis, The Löwenheim-Skolem theorem, theories of quantification, and proof theory, in *Perspectives on the History of Mathematical Logic*, ed. by T. Drucker (Birkhäuser, Boston/Basel/Berlin, 1991a), pp. 71–83
13. I.H. Anellis, Forty years of "Unnatural" natural deduction and quantification: a history of first-order systems of natural deduction, from Gentzen to Copi. Mod. Log. **2**, 113–152 (1991)
14. I.H. Anellis, Jean van Heijenoort's contributions to proof theory and its history. Mod. Log. **2**, 312–335 (1992) (English translation of [Anellis *1989*])
15. I.H. Anellis, Review of [Burch *1991*]. Mod. Log. **3**, 401–406 (1993)
16. I.H. Anellis, 1997. Tarski's development of Peirce's logic of relations, in [Houser, Roberts, & Van Evra *1997*], pp. 271–303
17. I.H. Anellis, The genesis of the truth-table device. Russell: J. Bertrand Russell Stud. (n.s.) **24**(Summer), 55–70 (2004a); on-line abstract: http://digitalcommons.mcmaster.ca/russelljournal/vol24/iss1/5/
18. I.H. Anellis, Review [Brady *2000*]. Trans. Charles S. Peirce Soc. **40**, 349–359 (2004)
19. I.H. Anellis, Some views of Russell and Russell's logic by his contemporaries. Rev. Mod. Log. **10**(1/2), 67–97 (2004–2005); electronic version: "Some Views of Russell and Russell's Logic by His Contemporaries, with Particular Reference to Peirce", http://www.cspeirce.com/menu/library/aboutcsp/anellis/views.pdf
20. I.H. Anellis, Did the *Principia Mathematica* precipitate a "Fregean Revolution"? Russell **31**, 131–150 (2011); simultaneous published in: Nicholas Griffin, Bernard Linsky, & Kenneth Blackwell, (eds.), *Principia Mathematica at 100* (The Bertrand Russell Research Centre, McMaster University, Hamilton, ONT, 2011), pp. 131–150

21. I.H. Anellis, Peirce's truth-functional analysis and the origin of the truth table. Hist. Philos. Log. **33**, 87–97 (2012a); electronic preprint (cite as arXiv:1108.2429v1 [math.HO]): http://arxiv.org/abs/1108.2429

22. I.H. Anellis, Jean van Heijenoort's contributions to proof theory and its history. Logica Universalis **6**, 411–458 (2012b) (A much expanded, revised and corrected version of [Anellis *1992*])

23. I.H. Anellis, Charles Peirce and Bertrand Russell on Euclid. X Seminário Nacional de História da Matemática; 10 pp. typescript for invited talk, 24–27 March, 2013, Universidade Estadual de Campinas, São Paulo, Brazil

24. I.H. Anellis, N. Houser, The nineteenth century roots of universal algebra and algebraic logic: a critical-bibliographical guide for the contemporary logician, in *Colloquia Mathematica Societatis Janos Bolyai* **54**. *Algebraic Logic, Budapest (Hungary), 1988*, ed. by H. Andréka, J.D. Monk, I. Németi (Elsevier Science/North-Holland, Amsterdam/London/New York, 1991), pp. 1–36

25. R.C. Archibald, Benjamin Peirce's Linear Associative Algebra and C. S. Peirce. Am. Math. Mon. **34**, 525–527 (1927)

26. C.C. Badesa, *El teorema de Löwenheim en el marco de la teoría de relativos*, Ph.D. thesis, University of Barcelona, 1991; published: Barcelona: Publicacións, Universitat de Barcelona

27. C.C. Badesa (Michael Maudsley, trans.), *The Birth of Model Theory: Löwenheim's Theorem in the Frame of the Theory of Relatives* (Princeton University Press, Princeton, 2004)

28. M.E. Baron, A note on the historical development of logic diagrams: Leibniz, Euler, and Venn. Math. Gaz. **53**, 113–125 (1969)

29. W.W. Bartley, Lewis Carroll's lost book on logic. Sci. Am. **227**(1), 38–46 (1972)

30. W.W. Bartley, 1977. Editor's introduction, in [Carroll *1977*], pp. 3–42

31. J.L. Bell, M. Machover, *A Course in Mathematical Logic* (North-Holland, Amsterdam/New York/London, 1977)

32. J. Bernoulli, J. Bernoulli, *Parallelismus ratiocini logici et algebraici, quem, una cum thesibus miscellaneis, defendum suscepit par fratrum Jacobus & Joannes Bernoulli, ille præsidis, hic resondentis vices agens* (Basel, 1685); reprinted: Jacobi Bernoulli, Basileensis, *Opera*, vol. I (Geneva: Cramer, 1744), 211–224; English translation by Terry Boswell in: [Boswell *1990*], §2, 175–178

33. E.W. Beth, Hundred years of symbolic logic: a retrospect on the occasion of the Boole De Morgan Centenary. Dialectica **1**, 331–346 (1947)

34. E.W. Beth, The origin and growth of symbolic logic. Synthèse **6**, 268–274 (1948)

35. E.W. Beth, Remarks on natural deduction. Koninklijke Nederlandse Akademie van Wetenschappen, Proceedings, series A (= Indagations Mathematicae) **58**(17), 322–325 (1955a)

36. E.W. Beth, Semantic entailment and formal derivability. Koninklijke Nederlandse Akademie van Wetenschappen (n.s.) **18**(13), 309–342 (1955b)

37. E.W. Beth, Semantic construction of intuitionistic logic. Mededlingen van den Koninklijke Nederlandse Akademie van Wetenschappen (n.r.) **19**(11), 357–388 (1956)

38. E.W. Beth, *The Foundations of Mathematics* (North-Holland, Amsterdam, 1959); 2nd revised edn., 1964; reprinted: [Beth *1966*]

39. E.W. Beth, Completeness results for formal systems, in *Proceedings of the International Congress of Mathematicians, 14–21 August 1958 (Edinburgh, 1958)*, ed. by J.A. Todd (Cambridge University Press, Cambridge, 1960), pp. 281–288

40. E.W. Beth, *Formal Methods* (D. Reidel/Gordon and Breach, Dordrecht/New York, 1962)

41. E.W. Beth, *The Foundations of Mathematics* (Harper & Row, New York, 1966)

42. J.-Y. Béziau, *Universal Logic: An Anthology* (Birkhäuser/Springer Basel, Basel, 2012)

43. J.H. Bisterfeld, *Johannis Henrici Bisterfeldii . . . Elementorvm Logicorvm Libri tres: ad praxin exercendam apprimè utiles . . . Accedit, Ejusdem Authoris, Phosphorus Catholicus, Sev Artis meditandi Epitome: Cui subjunctum est, Consilium de Studiis feliciter instituendis* (Verbiest, Lugduni Batavorum, 1657)

44. Boëthius, (Samuel Brandt. ed.), Anicii Manlii Severini Boethii, *In Isagogen Porphyrii commentorum editionis primae et secundae. Corpu Scriptorum Ecclesiasticorum Latinorum*, no. 48 (F. Tempsky, Vienna, 1906)

45. G. Boole, *An Investigation of the Laws of Thought, on which are founded the Mathematical Theories of Logic and Probabilities* (Walton & Maberly, London, 1854); reprinted: New York: Dover Publishing Co., 1951

46. G. Brady, *From Peirce to Skolem: A Neglected Chapter in the History of Logic* (North-Holland/Elsevier Science, Amsterdam/New York, 2000)
47. R.B. Braithwaite, Lewis Carroll as Logician. Math. Gaz. **16**, 174–178 (1933)
48. J. Brent, *Charles Sanders Peirce: A Life* (Indiana University Press, revised & enlarged ed, Bloomington/Indianapolis, 1998)
49. J. Brunning, 1997. Genuine triads and teridentity, in [Houser, Roberts, & Van Evra *1997*], pp. 252–263
50. R.W. Burch, *A Peircean Reduction Thesis: The Foundations of Topological Logic* (Texas Tech University Press, Lubbock, 1991)
51. R.W. Burch, 1997. Peirce's reduction thesis, in [Houser, Roberts, & Van Evra *1997*], pp. 234–251
52. F. Cajori, *A History of the Logarithmic Slide Rule and Allied Instruments* (The Engineering News Publishing Co./Archibald Constable & Co., New York/London, 1909)
53. L. Carroll, 1887. *The Game of Logic*. Reprinted with *Symbolic Logic, Part I*, as *The Mathematical Recreations of Lewis Carroll*. New York: Dover, 1958
54. L. Carroll, in *Lewis Carroll's Symbolic Logic*, ed. by W.W. Bartley (Clarkson N. Potter, Inc., Publishers, New York, 1977); revised, expanded ed., 1986
55. L. Carroll, in *Lewis Carroll's Diaries: The Private Journals of Charles Lutwidge Dodgson (Lewis Carroll): The first complete version of the nine surviving volumes with notes and annotations/Vol. 9, Containing Journal 13, July 1892 to December 1897*, ed. by E. Wakeling (Lewis Carroll Society, Luton, 2005)
56. L. Carroll, in *The Logic Pamphlets of Charles Lutwidge Dodgson and Related Pieces*. The Pamphlets of Lewis Carroll, v. 4, ed. by F.F. Abeles (Lewis Carroll Society of North America, New York, 2010)
57. A. Cayley, On the theory of analytical form called trees. Philos. Mag. **13**, 172–176 (1857); reprinted: (Andrew Russell Forsyth, ed.), *The Collected Mathematical Papers of Arthur Cayley* (Cambridge: Cambridge University Press, 1890), vol. III, pp. 242–246
58. A. Cayley, On contour and slope lines. Philos. Mag. **18**, 264–268 (1859a); reprinted: (Andrew Russell Forsyth, ed.), *The Collected Mathematical Papers of Arthur Cayley* (Cambridge: Cambridge University Press, 1891), vol. IV, pp. 108–111
59. A. Cayley, On the theory of analytical form called trees, second part. Philos. Mag. **18**, 374–378 (1859b); reprinted: (Andrew Russell Forsyth, ed.), *The Collected Mathematical Papers of Arthur Cayley* (Cambridge: Cambridge University Press, 1891), vol. IV, pp. 112–115
60. A. Cayley, A memoir on abstract geometry. Philos. Trans. R. Soc. Lond. **160**, 51–63 (1870); reprinted: (Andrew Russell Forsyth, ed.), *The Collected Mathematical Papers of Arthur Cayley* (Cambridge: Cambridge University Press, 1893), vol. VI, pp. 456–469
61. A. Cayley, On the theory of analytical form called trees, with applications to the theory of chemical combinations. *British Association Report*, 257–305 (1875); reprinted: (Andrew Russell Forsyth, ed.), *The Collected Mathematical Papers of Arthur Cayley* (Cambridge: Cambridge University Press, 1896), vol. IX, pp. 427–460
62. A. Cayley, The theory of groups: graphical representation. Am. J. Math. **1**, 174–176 (1878); reprinted: (Andrew Russell Forsyth, ed.), *The Collected Mathematical Papers of Arthur Cayley* (Cambridge: Cambridge University Press, 1896), vol. X, pp. 403–405
63. A. Cayley, On the theory of analytical form called trees. Am. J. Math. **4**, 266–268 (1881); reprinted: (Andrew Russell Forsyth, ed.), *The Collected Mathematical Papers of Arthur Cayley* (Cambridge: Cambridge University Press, 1896), vol. XI, pp. 365–367
64. W.G. Clark, New light on Peirce's iconic notation for the sixteen binary connectives, in *Peirce's Contributions to Logic, Peirce Sesquicentennial Congress, 5–10 September 1989, Harvard University, Cambridge, MA. Organizers: N. Houser and D. D. Roberts; abstracts* (1989), p. 3
65. W.G. Clark, 1997. New light on Peirce's iconic notation for the sixteen binary connectives, in [Houser, Roberts, & Van Evra *1997*], pp. 304–333
66. W.K. Clifford, Remarks on the chemico-algebraic theory (Extract from a Letter to Mr. Sylvester). Am. J. Math. **1**, 126–128 (1878); reprinted: [Clifford *1882*], 255–257
67. W.K. Clifford, On the types of compound statements involving four classes. Memoirs of the Literary and Philosophical Society of Manchester (3 ser.) **6**, 81–96 (1879); reprinted: [Clifford *1882*], 1–13
68. W.K. Clifford, *Mathematical Fragments: Being Facsimiles of His Unfinished Papers Relating to the Theory of Graphs* (Macmillan, London, 1881)

69. W.K. Clifford, (Robert Tucker, ed., with an introduction by Henry John Stephen Smith), *Mathematical Papers by William Kingdon Clifford* (Macmillan, London, 1882)

70. E. Coumet, Sur l'histoire des diagrammes logiques, "figures géométriques". Mathématiques et sciences humaines **60**, 31–62 (1977); http://archive.numdam.org/article/MSH_1977__60__31_0. djvu and http://archive.numdam.org/ARCHIVE/MSH/MSH_1977__60_/MSH_1977__60__31_0/ MSH_1977__60__31_0.pdf

71. C.K. Davenport, The role of graphical methods in the history of logic. Methodos **4**, 145–164 (1952); reprinted: revised, expanded ed., Milano: Editrice la Fiaccola, 1952

72. D. De Jongh, 2008. Beth's main results in mathematical logic, in [Van Benthem, Van Ulsen, & Visser *2008*], pp. 12–16

73. A. De Morgan, Slide (or sliding) rule, in *The Penny Cyclopædia for the Diffusion of Useful Knowledge*, vol. XXII (Charles Knight & Co., London, 1842), pp. 129–134

74. A. De Morgan, On the syllogism, I: on the structure of the syllogism. Trans. Camb. Philos. Soc. **8**, 379–408 (1846); reprinted in [De Morgan *1966*], 1–21

75. A. De Morgan, On the syllogism no. IV: and on the logic of relations. Trans. Camb. Philos. Soc. **10**, 331–358 (1860a)

76. A. De Morgan, *Syllabus of a Proposed System of Logic* (Walton and Maberly, London, 1860b)

77. A. De Morgan, A question in logic. Notes Queries s. 2, **9**(211), 25 (1860c)

78. A. De Morgan, in *On the Syllogism and Other Logical Writings*, ed. by P. Heath (Yale University Press, New Haven, 1966)

79. C. De Waal, *Peirce: A Guide to the Perplexed* (Bloomsbury, London/New York, 2013)

80. K. Dürr, Des diagrammes logiques de Leonhard Euler et de John Venn, in *Proceedings of the Tenth International Congress of Philosophy (Amsterdam, August 11–18, 1948), Library of the Xth International Congress of Philosophy*, vol. I, Pt. 2 (North-Holland, Amsterdam, 1949), pp. 720–721

81. A.W.F. Edwards, *Cogwheels of the Mind: The Story of Venn Diagrams* (with a Foreword by Ian Stewart) (The Johns Hopkins University Press, Baltimore/London, 2004)

82. C. Eisele, 1988. Thomas S. Fiske and Charles S. Peirce, in [Duren, Askey, & Merzbach *1988*], pp. 41–55

83. L. Euler, 1768–72. *Lettres à une princesse d'Allemagne sur divers sujets de physique & de philosophie*. Saint Petersbourg: De |'Imprimerie de |'Académie impériale des sciences; Paris: chez Royez, 1787–88

84. J.A. Faris, The gergonne relation. J. Symb. Log. **20**, 207–231 (1955)

85. M.H. Fisch, 1982. Introduction, in [Peirce *1982*], pp. xv–xxxv

86. M.H. Fisch, A.R. Turquette, Peirce's triadic logic. Trans. Charles S. Peirce Soc. **2**, 71–85 (1966)

87. G. Frege, *Begriffsschrift, eine der arithmetischen nachgebildete Formelsprache des reinen Denkens* (Verlag von Louis Nebert, Halle, 1879); reprinted: (Ignacio Angelelli, Hsg.), *Freges Begriffsschrift und andere Aufsätze* (Hildes-heim: Olms, 2te. Aufl., 1964); English translation by Stefan Bauer-Mengelberg in [van Heijenoort *1967*], 5–82. English translation by Terrell Ward Bynum in Terrell Ward Bynum (ed. & trans.), *Conceptual Notation and Related Articles* (Oxford: Clarendon Press, 1972), 101–203

88. G. Frege, Kritische Beleuchtung einiger Punkte in E. Schröders *Vorlesungen über die Algebra der Logik*. Archiv für systematische Philosophie **1**, 433–456 (1895)

89. M. Gardner, *Logic Machines and Diagrams* (Dover, New York, 1958); Brighton: Harvester Press, 2nd edn., 1982

90. G. Gentzen, *Untersuchungen über das logische Schliessen*, Doctoral dissertation, printed, in the Bernays Archive, Eidgenössische Technische Hochschule Zürich, 1933; English translation by Jan Von Plato in [Von Plato *2008*], 245–257

91. G. Gentzen, Untersuchungen über das logische Schließen. Mathematische Zeitschrift **39**, 176–210, 405–431 (1935); reprinted: Darmstadt: Wissenschftliche Buchgesellschaft, 1974; English translation: [Gentzen *1964–65*]

92. G. Gentzen, 1964–65. (Manfred Egon Szabo, trans., with an introduction by Paul Bernays), Investigations into logical deductions. Am. Philos. Q. **1**, 288–306 (1964), **3**, 204–218 (1965); reprinted: [Gentzen *1969*], 68–131

93. J.-D. Gergonne, Essai de dialectique rationelle. Annales de Mathmatiques pures et appliquées **7**, 189–228 (1816–17)

94. D.A. Gillies, The Fregean revolution in logic, in *Revolutions in Mathematics*, ed. by D.A. Gillies (Clarendon Press, Oxford, 1992) (paperback edition, 1995), pp. 265–305
95. A.K. Gorfunkel', Andrei Belobotskii—poet i filosof kontsa XVII nachala XVIII v. Trudy Otdel drevnerusskoi literatury **XVIII**, 188–213 (1962)
96. I. Grattan-Guinness, Benjamin Peirce's *Linear Associative Algebra* (1870): new light on its preparation and 'Publication'. Ann. Sci. **54**, 597–606 (1997)
97. I. Grattan-Guinness, *The Search for Mathematical Roots, 1870–1940: Logics, Set Theories and the Foundations of Mathematics from Cantor through Russell to Gödel* (Princeton University Press, Princeton/London, 2000)
98. I. Grattan-Guinness, Re-interpreting 'λ': Kempe on multisets and Peirce on graphs, 1886–1905. Trans. Charles S. Peirce Soc. **38**, 327–350 (2002)
99. S. Haack, R. Lane (eds.), *Pragmatism, Old and New: Selected Writings* (Prometheus, Amherst, NY, 2006)
100. I. Hacking, Trees of logic, trees of porphyry, in *Advancements of Learning: Essays in Honour of Paolo Rossi*, ed. by J.L. Heilbron (Leo S. Olschki, Firenze, 2007), pp. 219–263
101. E. Hammer, S.-J. Shin, Euler's visual logic. Hist. Philos. Log. **19**, 1–29 (1998)
102. C.S. Hardwick, J. Cook (eds.), *Semiotic & Significs: The Correspondence between Charles S. Peirce and Victoria Lady Welby* (Indiana University Press, Bloomington, 1977); 2nd edn., Elsah, IL: The Press of Arisbe Associates, 2001
103. R. Harley, Remarks on Boole's mathematical analysis of logic, in *Report of the 36th meeting of the British Association for the Advancement of Science, held in Nottingham, August 1866, Transactions*, Sect. A (London, 1867), pp. 3–6
104. B.S. Hawkins Jr., Review of [Roberts *1973*]. Trans. Charles S. Peirce Soc. **11**, 128–139 (1975)
105. P. Henle, On the fourth figure of the syllogism. Philos. Sci. **16**, 94–104 (1949)
106. J. Herbrand, *Recherches sur la théorie de la démonstration*, Ph.D. thesis, University of Paris, 1930; http://archive.numdam.org/ARCHIVE/THESE/THESE_1930_110_/THESE_1930_110_1_0/THESE_1930_110_1_0.pdf; http://ar-chive.numdam.org/article/THESE_1930__110__1_0.djvu; also in *Prace Towarzystwa Naukowego War-szawskiego*, Wydział Ill, no. 33, 128 pp.; reprinted: [Herbrand *1968*], 35–353; English translation in: [Herbrand *1971*], 44–202; English translation of Chapter 5 in: [van Heijenoort *1967*], 525–581
107. J. Herbrand, in *Écrits logiques*, ed. by J. van Heijenoort (Presses Universitaire de France, Paris, 1968)
108. P. Hertz, Über Axiomensysteme für beliebige Satzsysteme. Teil I. Mathematische Annalen **87**, 246–269 (1922); English translation by Javier Legris: [Béziau *2012*], 11–29
109. P. Hertz, Über Axiomensysteme für beliebige Satzsysteme. Teil II. Mathematische Annalen **89**, 76–102 (1923)
110. P. Hertz, Über Axiomensysteme für beliebige Satzsysteme. Mathematische Annalen **101**, 457–514 (1929)
111. H.G. Herzberger, Peirce's remarkable theorem, in *Pragmatism and Purpose: Essays Presented to Thomas A. Goudge*, ed. by L.W. Summer, J.G. Slater, F. Wilson (University of Toronto Press, Toronto/Buffalo, 1981), pp. 41–58, 297–301
112. D. Hilbert, Neubegründung der Mathematik. Erster Mitteilung. Abhandlungen aus dem mathema-tischen Seminar dem Hamburgischen Universität **1**, 157–177 (1922); reprinted: [Hilbert *1932–35*], III, 157–177
113. D. Hilbert, Die logischen Grundlagen der Mathematik. Mathematische Annalen **88**, 151–165 (1923); reprinted: [Hilbert *1932–35*], III, 178–191
114. D. Hilbert, Über das Unendliche. Mathematische Annalen **95**, 161–190 (1925); English trans-lation by Stefan Bauer-Mengel-berg as "On the Infinite" in [van Heijenoort *1967*], 369–392; English translation by Erna Putnam & Gerald J. Massey, in Paul Benacerraf & Hilary Putnam (eds.), *Philosophy of Mathematics: Selected Readings* (Cambridge/London/New York/New Rochelle/Melbourne/Sydney: Cambridge University Press, 2nd ed., 1983), 183–201
115. D. Hilbert, Die Grundlagen der Mathematik. Abhandlungen aus dem mathematischen Seminar dem Hamburgischen Universität **6**, 65–85 (1928); English translation by Stefan Bauer-Mengelberg & Dagfinn Føllesdal: [van Heijenoort *1967*], 464–479
116. D. Hilbert, W. Ackermann, *Grundzüge der theoretischen Logik* (Springer-Verlag, Berlin, 1928)

117. D. Hilbert, P. Bernays, 1934–39. *Grundlagen der Mathematik*. Berlin: Springer Verlag, Bd. I, 1934; Bd. II, 1939; dual-language edition: (Claus-Peter Wirth, Jorg Siekmann, Michael Gabbay, & Dov M. Gabbay, eds.), *Grundlagen der Mathematik*, I/*Foundations of Mathematics*, I, Pt. A, Prefaces & §§1–2. London: College Publications, 2011
118. J. Hintikka, Form and content in quantification theory, in J. Hintikka, *Two Papers on Symbolic Logic*. Acta Philosophica Fennica **8**, 7–55 (1955a)
119. J. Hintikka, Notes on quantification theory. Societas Scientiarum Fennica, Commentationes Physico-Mathematicae **17**(12), 1–13 (1955)
120. N. Houser, Introduction, in [Peirce *1989*], pp. xix–lxx
121. N. Houser, The fortunes and misfortunes of the Peirce papers, in Michael Balta & Janice Deledelle-Rhodes (eds.), Gérard Deledalle (general ed.), *Signs of Humanity*, vol. 3 (Mouton de Gryuter, Berlin, 1992), pp. 1259–1268; http://www.cspeirce.com/menu/library/aboutcsp/houser/fortunes.htm
122. N. Houser, 1997. Introduction: Peirce as Logician, in [Houser, Roberts, & Van Evra *1997*], pp. 1–22
123. N. Houser, 2000. Introduction, in [Peirce *2000*], pp. xxv–lxxxiv
124. N. Houser, 2010. Introduction, in [Peirce *2010a*], pp. xi–xcvii
125. H.G. Hubbeling, A diagram-method in propositional logic. Logique et Analyse **8**, 277–288 (1965)
126. R.C. Jeffrey, *Formal Logic: Its Scope and Limits* (McGraw-Hill, New York, 1967); 3rd ed., 1991; 4th ed., (edited, with supplementary material, by John P. Burgess), Indianapolis/Cambridge: Hackett Publishing, 2006
127. W.S. Jevons, On the inverse, or inductive, logical problem. Proc. Lit. Philos. Soc. Manchester **XI**, 65–68 (1871a)
128. W.S. Jevons, On the inverse, or inductive, logical problem. Mem. Lit. Philos. Soc. Manchester (3rd ser.) **V**, 119–130 (1871b)
129. W.S. Jevons, *The Principles of Science, a Treatise on Logic and Scientific Method* (Macmillan & Co., London, 1874)
130. W.S. Jevons, *The Principles of Science, a Treatise on Logic and Scientific Method*, 2nd edn. (Macmillan & Co., London/New York, 1877)
131. W.E. Johnson, *Logic*, 3 vols. (Cambridge University Press, Cambridge, 1921–24); reprinted: 1940; reprinted: New York: Dover Publications, 1961
132. J. Jungius, *Logica Hamburgensis: hoc est, Institutiones logicae in usum schol. Hamburg. conscriptae, & 6 libris comprehensae* (Offermann, Hamburgi, 1638)
133. L.H. Kauffman, The mathematics of Charles Sanders Peirce. Cybern. Hum. Knowing **8**, 79–110 (2001)
134. B.M. Kedrov, Ponyatie khimicheskii element i ego logicheskii analiz. Filosofskie zapiski (Moscow), tm. **I**, 118–178 (1946)
135. B.M. Kedrov, *Razvitie ponyatiya elementa ot Mendeleeva do nashikh dnei. Oput istoriko-logicheskogo issledovaniya* (Nauka, Moscow, 1948)
136. A. Kekulé, Über die sogennante gepaarten Verbindungen und die Theorie der mehratomigen Radicale. Annalen der Chemie und Pharmacie **104**, 129–150 (1857)
137. A. Kekulé, Ueber die Constitution und die Metamorphosen der chemischen Verbindungen und über die chemische Natur des Kohlenstoffs. Annalen der Chemie und Pharmacie **106**, 129–159 (1858)
138. A. Kekulé, Sur la constitution des substances aromatiques. Bulletin de la Société Chimique de Paris **3**, 98–110 (1865)
139. A. Kekulé, Ueber einige Condensationsproducte des Aldehyds. Annalen der Chemie und Pharmacie **162**, 77–124 (1872)
140. A.B. Kempe, A memoir on the theory of mathematical form. Philos. Trans. R. Soc. Lond. **177**, 1–70 (1886)
141. A.B. Kempe, ca. 1886-. Correspondence with Charles S. Peirce; Robin catalog, Ms. #L232a; including pages from *Nature (December 18, 1890)*
142. A.B. Kempe, Note to a memoir on the theory of mathematical form. Proc. R. Soc. Lond. **48**, 193–196 (1887)
143. A.B. Kempe, On the relation between the logical theory of classes and the geometrical theory of points. Proc. Lond. Math. Soc. **21**, 147–182 (1889–90)
144. A.B. Kempe, The subject matter of exact thought. Nature **43**, 156–162 (1890)

145. A.B. Kempe, Mathematics. Proc. Lond. Math. Soc. **26**, 5–15 (1894)
146. A.B. Kempe, The theory of mathematical form: a correction and clarification. Monist **7**, 453–458 (1897)
147. J.N. Keynes, *Studies and Exercises in Formal Logic, including a Generalisation of Logic Processes in Their Application to Complex Inferences*, 3rd, rewritten and enlarged edn. (Macmillan, London/New York, 1894)
148. C.F. Klein, Vergleichende Betrachtungen über neuere geometrische Forschungen (Erlangen Programm), Inagural-Dissertation, Universität Erlangen, 1872; printed: Erlangen: Deichert, 1872; reprinted: Mathematische Annalen 43 (1893), 63–100
149. A.S. Kuzichev, *Diagrammy Venna: Istoriya i primeneniya* (Nauka, Moscow, 1968)
150. Z.A. Kuzicheva, Graficheskie metodu logiki Massov. Istoriia i Metodologi Estestvennyh Nauk **29**, 75–85 (1982)
151. C. Ladd-Franklin, 1883. On the algebra of logic, in [Peirce *1883a*], pp. 17–71
152. C. Ladd-Franklin, Explicit primitives again: a reply to Professor Fite. J. Philos. Psychol. Sci. Methods **9**, 580–583 (1912a)
153. C. Ladd-Franklin, Implication and existence in logic. Philos. Rev. **21**, 641–665 (1912)
154. C. Ladd-Franklin, The antilogism: an emendation. J. Philos. Psychol. Sci. Methods **10**, 49–50 (1913)
155. C. Ladd-Franklin, Charles S. Peirce at Johns Hopkins. J. Philos. Psychol. Sci. Methods **13**, 715–722 (1916)
156. C. Ladd-Franklin, The antilogism. Mind (n.s.) **37**, 532–534 (1928)
157. J.H. Lambert, *Anlage zur Architectonic, oder Theorie des Einfachen und des Ersten in der philosophischen und mathematischen Erkenntnis*, 2 Bde. (Johann Friedrich Hartnoch, Riga, 1771)
158. R. Lane, Peirce's triadic logic revisited. Trans. Charles S. Peirce Soc. **35**, 284–311 (1999)
159. F.A. Lange, *Logische Studien. Ein Beitrag zur Neubegründung der Formalen Logik und Erkenntniss-theorie* (J. Baedecker, Iserlohn, 1877)
160. J.C. Lange, *Nucleus logicæ Weisianæ sic auctus et illustratus, ut vera ac solida Logicæ peripatetico scholasticæ purioris fundamenta detegantur et ratione matematica per varias scholasticas præfigurationes huic usui inservientes ad ocularem evidentiam deducta proponantur* (Müller, Giessen, 1712).
161. J. Legris, 2012. Paul Hertz and the origins of structural reasoning, in [Béziau *2012*], pp. 3–10
162. G.W. Leibniz, ca. 1690. *Non Inelegans Specimen Demonstrandi in Abstractis*; in: Johann Erdmann (Hsg.), *God. Guil. Leibnitii opera philosophica quae extant Latina Gallica Germanica omnia* (Berlin: Eichler, 1839/40), II, 94–97; reprinted: (Leibniz-Forschungsstelle der Universität Münster, Hsg.), *Philosophische Schriften* Bd. 4: *1677–Juni 1690*, Th. A (Berlin: Akademie Verlag, 1999), 845–855
163. G.W. Leibniz, 1849–63. (Carl Immanuel Gerhardt, Hsg.), *Leibnizens mathematische Schriften*, 7 Bde. Berlin: Wiedmannsche Buch-handlung & Halle: Schmidt; Hildesheim: Georg Olms, 1971
164. G.W. Leibniz, 1875–90. (Carl Immanuel Gerhardt, Hsg.), *Leibnizens philosophische Schriften*, 7 Bde. Berlin: Wiedmannsche Buch-handlung & Halle: Schmidt
165. G.W. Leibniz, (Louis Couturat, ed.), *Opuscules et fragments inédits de Leibniz extraits de la Bibliothèque royale de Hanovre* (Félix Alcan, Paris, 1903); reprinted: Hildesheim: Georg Olms, 1961
166. J. Lindseth, Jon A. Lindseth Collection, L188
167. J.B. Listing, Vorstudien zur Topologie, Göttinger Studien, 811–875 (1847); reprinted: Göttingen: Vanden-hoeck und Ruprecht, 1848
168. J.B. Listing, Der Census räumlicher Complexe, der Verallgemeinerung des Euler'schen Satzes von den Polyedern. Ab-handlungen der Königlichen Gesellschaft der Wissenschaften zu Göttingen **10**, 97–182 (1861); reprinted: Göttingen: Dietrich, 1862
169. R. Llull, 1501 (Pere Posa, ed.), *Ars magna et ultima*. Barch[ino]ne: impressum per Petru[m] Posa . . . .
170. R. Llull, *Liber de logica nova* (Alonso de Proaza, València, 1512)
171. R. Llull, *Ars scientiæ* (F. Fradin, Lyons, 1515)
172. R. Llull, *Raymvndi Lvllii Opera ea qvae ad adinventam ab ipso artem vniversalem, scientiarvm artivmqve omnivm breui compendio, firmaq[ue] memoria apprehendendarum, locupletissimaq[ue] vel oratione ex tempore pertractan-darum, pertinent: vt et in eandem qvorvndam interpretvm*

scripti commentarii: qvae omnia sequens indicabit pagina: & hoc demum tempore coniunctim emendatiora locupletioraq[ue] non nihil edita sunt (Sumptibus haeredum Lazari Zetzneri, Argentorati [Strasbourg], 1617)

173. N.I. Lobachevskii [Lobatchewski],*Geometrische Untersuchungen zur Theorie der Parallellinien*, 2nd (photo-reproduction) edn. (Fincke, Berlin, 1887)
174. N.I. Lobachevskii [Lobatchewski], (George Bruce Halsted, trans.), *Geometrical Researches in the Theory of Parallels* (University of Texas, Austin, TX, 1891)
175. L. Löwenheim, Über Möglichkeiten im Relativkalkul. Mathematische Annalen **76**, 447–470 (1915). English translation by Stefan Bauer-Mengelberg in [van Heijenoort *1967*], 228–251
176. A. Luhtala, *On the Origin of Syntactical Description in Stoic Logic* (Nodus Publikationen, Münster, 2000)
177. J. Łukasiewicz, O logice trójwartościowej. Ruch filozofczny **5**, 169–171 (1920); English translation as "On the Principle of Excluded Middle", (Jan Woleński and Peter Simons, trans.), *History and Philosophy of Logic***8** (1987) 69
178. J. Łukasiewicz, 1938. 'Die Logik und das Grundlagenproblem', *Les Entretiens de Zürich sur les fondements et la méthode des sciences mathématiques 6–9 12 (1938), Zürich, 1941*, 82–100; English translation: (Ludwik Borkowski, ed.), *Selected Works* (Amsterdam/London: North-Holland, 1970), 278–294
179. J. Łukasiewicz, *Aristotle's Syllogistic from the Standpoint of Modern Logic*, 2nd enlarged edn. (Clarendon Press, Oxford, 1951)
180. A. MacFarlane, *Principles of the Algebra of Logic* (D. Douglas, Edinburgh, 1879)
181. A. MacFarlane, 1881–1883. L 263. Two letters to Charles S. Peirce, March 29, 1881, and May 5, 1883. Robin catalog, Ms. #L263
182. A. MacFarlane, The logical spectrum. The London, Edinburgh and Dublin Philos. Mag. J. Sci. **19**, 286–290 (1885)
183. A.H. Maróstica, *Ars Combinatoria* and time. Studia Logica **32**, 105–134 (1992)
184. A. Marquand, A logical diagram for *n* terms. The London, Edinburgh and Dublin Philos. Mag. J. Sci. **12**, 266–270 (1881)
185. D.W. Mertz, Peirce: logic, categories, and triads. Trans. Charles S. Peirce Soc. **15**, 158–175 (1979)
186. J.L. Meyer, *Die modernen Theorien der Chemie, und ihre Bedeutung für die chemische Statik* (Maruschke und Berendt, Breslau, 1864)
187. J.L. Meyer, Die Natur der chemischen Elemente als Function ihrer Atomgewichte. Justus Liebigs Annalen der Chemie, suppl. **7**, 354–364 (1870); reprinted: [Seubert *1895*], 9–17
188. A. Moktefi, Lewis Carroll's logic, in *Handbook of the History of Logic*, vol. 4: *British Logic in the Nineteenth Century*, ed. by D.M. Gabbay, J. Woods (Elsevier, North-Holland, Amsterdam/Boston/London, etc., 2008), pp. 457–505
189. A. Moktefi, S.-J. Shin, A history of logic diagrams, in *Handbook of the History of Logic*, vol. 11: *Logic: A History of its Central Concepts*, ed. by D.M. Gabbay, J. Pelletier, J. Woods (Elsevier North-Holland, Oxford/Amsterdam/Waltham, MA, 2012), pp. 611–682
190. M.G. Murphey, *The Development of Peirce's Philosophy* (Harvard University Press, Cambridge, MA, 1961); reprinted: Indianapolis: Hackett, 1993
191. W.J. Newlin, A new logical diagram. J. Philos. **3**, 539–545 (1906)
192. L. Nový, Benjamin Peirce's concept of linear algebra. Acta Historiæ Rerum Naturalium Necnon Technicarium, Special Issue **7**, 211–230 (1974)
193. J. Nubiola, 1995. The Branching of Science according to C. S. Peirce; talk presented at the 10th International Congress of Logic, Methodology and Philosophy of Science August 22, 1995, Florence, Italy; posted to Arisbe: The Peirce Gateway: May 28, 1998, http://www.cspeirce.com/menu/library/aboutcsp/nubiola/branch.htm
194. K.H. Parshall, America's First School of Mathematical Research: James Joseph Sylvester at The Johns Hopkins University 1876–1883. Arch. Hist. Exact Sci. **38**, 153–196 (1988)
195. Paulus Venetus, *Sophismata aurea et perutilia* (Nicolaus Girardengus, Pavia, 1483)
196. Paulus Venetus, *Logic parva* (Stefano dei Nicolini da Sabbio, Venitiis, 1536)
197. B. Peirce, 1837. *Elementary Treatise on Plane and Solid Geometry*. Boston: J. Munroe and Co.; 1841; Boston/Cambridge: J. Munroe and Co., 1855; New York: Collins, Brother, and company, 1851; Boston, W. H. Dennet, 1867; 1872

198. B. Peirce, "Linear Associative Algebra", with notes and appendix by C. S. Peirce. Am. J. Math. **4**, 97–229 (1881); reprinted: New York: D. Van Nostrand, 1882

199. J.M. Peirce, C.S. Peirce, n.d. Preface to Benjamin Peirce's *Elementary Treatise on Geometry*; Robin catalog, Ms. #94; *New Elements of Geometry* by Benjamin Peirce, rewritten by his sons, James Mills Peirce and Charles Sanders Peirce; Robin catalog, Ms. #94; n.p., n.d., pp. 1–6, 1–4 ("Preface"), 2 pp. ("Nota Bene"), pp. 1–398, (pp. 7, 31–33, 35, 69–70, 74–76, 78, 92–94, 166–168, 175, 182–183, 235 missing), with pp. xvi, xvii, xviii, xix, and pp. 37–150 from Benjamin Peirce's *Plane and Solid Geometry* mounted and ready for revision

200. C.S. Peirce, The chemical interpretation of interpenetration. Am. J. Sci. Arts (ser. 2) **35**, 72–82 (1863); reprinted: [Peirce *1982*], 95–100

201. C.S. Peirce, 1865–1909. *Logic (logic notebook 1865–1909)*, Robin catalog, MS#339. (Ms. #339.00340-344; 23 February 1909, published in [Haack and Lane *2006*, 217–224]

202. C.S. Peirce, 1865a. Lecture II, Harvard Lectures: "On the Logic of Science", Robin catalog, Ms. #95; February–March, 1865; published [Peirce *1982*], 175–189

203. C.S. Peirce, 1865b. Lecture VI: "Boole's Calculus of Logic", Harvard Lectures: "On the Logic of Science", Robin catalog, Ms. #100; February–March 1865; published: [Peirce *1982*], 223–239

204. C.S. Peirce, 1865c. Lecture VIII: "Forms of Induction and Hypotheses", Harvard Lectures: "On the Logic of Science", Robin catalog, Ms. #105; see April–May, 1865; published [Peirce *1982*], 256–271

205. C.S. Peirce, 1865d. Lecture XI, Harvard Lectures: "On the Logic of Science", Robin catalog Ms. #107, April–May 1865; published: [Peirce *1982*], 286–302

206. C.S. Peirce, 1865e. "An Unpsychological View of Logic to which are appended some applications to the theory to Psychology and other subjects"; Robin catalog, Ms. #109, May-Fall 1865; published: [Peirce *1982*, 305–321]

207. C.S. Peirce, 1866a. Lowell Lecture I; Robin catalog Ms. #122, September–October 1866; published; [Peirce *1982*], 358–375

208. C.S. Peirce, 1866b. "Memoranda Concerning the Aristotelian Syllogism"; privately printed, and distributed at the Lowell Institute, November 1866; reprinted: [Peirce *1982*], 505–514

209. C.S. Peirce, 1868a. "On an Improvement in Boole's Calculus of Logic" (Paper read on 12 March 1867). Proc. Am. Acad. Arts Sci. **7** (1868), 250–261; reprinted: [Peirce *1984*], 12–23

210. C.S. Peirce, 1868b. "On the Natural Classification of Arguments" (Paper read 9 April 1867). Proc. Am. Acad. Arts Sci. **7**, 261–287; reprinted: [Peirce *1932*], para. 2.461–516; [Peirce *1984*], 23–48

211. C.S. Peirce, 1868c. Upon logical comprehension and extension. Proc. Am. Acad. Arts Sci. **7**, 415–432; reprinted: [Peirce *1984*], 70–86; http://www.iupui.edu/~peirce/writings/v2/w2/w2_06/v206.htm

212. C.S. Peirce, 1868d. [A Search for a Method. Essay VI]; Robin catalog, Ms. #593; pp. 249–264

213. C.S. Peirce, 1869. Grounds of validity of the laws of logic: further consequences of four incapacities. J. Specul. Philos. **2**, 193–208; reprinted: [Peirce *1984*], 242–272

214. C.S. Peirce, 1873. Description of a notation for the logic of relatives, resulting from an amplification of the conceptions of Boole's calculus of logic. Mem. Am. Acad. **9**, 317–378; reprinted: [Peirce *1933a*], para. 45–135; [Peirce *1984*], 359–429

215. C.S. Peirce, 1876. Note on the theory of the economy of research. *Coast Survey Report*, 197–201; reprinted: [Peirce *1989*], 72–78

216. C.S. Peirce, 1880. On the algebra of logic. Am. J. Math. **3**, 15–57; reprinted: [Peirce *1933a*, para. 154–251]; [Peirce *1989*], 163–209

217. C.S. Peirce, 1881. On the logic of number. Am. J. Math. **4**, 85–95; reprinted: [Peirce *1933a*], para. 252–280; [Peirce *1989*], 299–309

218. C.S. Peirce, 1882. A communication from Mr. Peirce. *The Johns Hopkins University Circulars* no. 22 (April), 86–88; reprinted: [Peirce *1989*], 467–472

219. C.S. Peirce, 1883a. (ed.), *Studies in Logic by the Members of the Johns Hopkins University*, Boston: Little, Brown & Co.; reprinted, with an introduction by Max Harold Fisch and a preface by A. Eschbach: Amsterdam: John Benjamins Publishing Co., 1983

220. C.S. Peirce, 1883b. "Syllabus of Sixty Lectures on Logic"; Robin catalog Ms. #459, untitled manuscript; Summer 1883; published: [Peirce *1989*], 476–489

221. C.S. Peirce, 1883–1884. 'On the Algebra of Logic'; MS., n.p., n.d., 5 pp. of a manuscript draft; 12 pp. of a typed draft (corrected by CSP); and 2 pp. of fragments, *ca.* 1883–84; Robin catalog, Mss. #527 and #527s

222. C.S. Peirce, ca. 1883–1909. [Notes for Contributions to the Century Dictionary]; Robin catalog, Ms. #1170; n.p., n.d., 73 pp

223. C.S. Peirce, 1885a. On the algebra of logic: a contribution to the philosophy of notation. Am. J. Math. **7**, 180–202; reprinted: [Peirce *1933a*], 210–249, [Peirce *1993*], 162–190

224. C.S. Peirce, 1885b. "One, Two, Three: Fundamental Categories of Thought and Nature"; Robin catalog, Ms. #901; published: [Peirce *1993*], 242–247

225. C.S. Peirce, 1886. "The Logic of Relatives: Qualitative and Quantitative"; Robin catalog, Ms. #584; published: [Peirce *1993*], 372–378

226. C.S. Peirce, 1889. "Notes on Kempe's Paper on Mathematical Form"; Robin catalog Ms. #714; January 15, 1889, 12 pp. Harvard University; published: [Peirce *2010a*], 25–29

227. C.S. Peirce, 1891. "On the Number of Dichotomous Divisions: A Problem of Permutations"; Robin catalog Ms. #74; pp. 1–10 (p. 7 missing); plus 17 pp. of another draft; published: [Peirce *2010a*], 222–228, identified as dating to Spring 1891

228. C.S. Peirce, 1891(?). "A Problem of Trees"; Robin catalog Ms. #73, n.p., n.d., 4 pp. (incomplete or unfinished)

229. C.S. Peirce, 1892. The Non-Euclidean Geometry. The Nation **54** (11 February 1892), 116; reprinted: (Arthur Walter Burks, ed.), *Collected Papers of Charles Sanders Peirce,* vol. VIII (Cambridge, Mass.: Harvard University Press, 1966), 72; (Kenneth Laine Ketner & James Edward Cook, compilers and annotators), *Charles Sanders Peirce: Contributions to 'The Nation',* Part One: *1869–1893* (Lubbock: Texas Tech Press, 1975), 135–137; [Peirce *2010a*], 271–274

230. C.S. Peirce, 1893a. "An Outline Sketch of Synechistic Philosophy"; Robin catalog MS #946, n.p., n.d., 7 pp

231. C.S. Peirce, 1893b. Book II. Introductory. Chapter VII. "Analysis of Propositions"; Robin catalog, Ms. #410; n.p., 1893, pp. 1–18; 1–19 (of a secretary's inaccurate copy)

232. C.S. Peirce, 1893c. "De Morgan's Propositional Scheme"; Robin catalog, Ms. #415; n.p., 1893, pp. 297–313

233. C.S. Peirce, 1896. "A Graphical Method of Logic"; RC Ms. #915, 17–18 November 1896; published: [Peirce *1933b*], para. 418, 468–470

234. C.S. Peirce, 1897a. The logic of relatives. The Monist **7**, 161–217; reprinted in part: [Peirce *1933a*], para. 456–552

235. C.S. Peirce, 1897b. "Reply to Mr. Kempe (K)"; Robin catalog, Ms. #708, n.p., n.d., pp. 1–9, 5–7, and 5 pp. of another draft; published: [Peirce *1976*], 431–440

236. C.S. Peirce, 1898. "Detached Ideas continued and the Dispute between Nominalists and Realists (NR)"; Robin catalog, Ms. #439; n.p., 1898, pp. 1–35, with a variant p. 24

237. C.S. Peirce, 1902a. "Logical Diagram (or Graph)", in [Baldwin *1902*], vol. II, 28; reprinted: [Peirce *1933b*], para. 347

238. C.S. Peirce, 1902b. "Tree of Porphyry", in [Baldwin *1902*], vol. II, 713–714

239. C.S. Peirce, 1902c. "The Simplest Mathematics"; January 1902 ("Chapter III. The Simplest Mathematics (Logic III))"; Robin catalog, Ms. #431, January 1902; see [Peirce *1933b*], 4: 260–262

240. C.S. Peirce, ca. 1903. "On the Simplest Branch of Mathematics (SM)"; Robin catalog, Ms. #2., n.p., [c. 1903?], pp. 1–2; 1–5, incomplete, with an alternative p. 5

241. C.S. Peirce, *A Syllabus of Certain Topics in Logic* (Alfred Mudge & Son, Boston, 1903)

242. C.S. Peirce, 1903b. "Syllabus of a course of Lectures at the Lowell Institute beginning 1903, Nov. 23. On Some Topics of Logic (Syllabus)"; Robin catalog Ms. #478; pp. 168, with variants

243. C.S. Peirce, 1903c. "Lecture II"; Robin catalog, Ms #302; notebook, n.p.

244. C.S. Peirce, 1903d. "Lecture III: The Categories Continued"; Robin catalog, Ms. #307; notebook, 1903; published in part: [Peirce *1934*], para. 5.71n (p. 9); para. 5.82-87 (pp. 16–34)

245. C.S. Peirce, 1903e. "Lecture III; The Categories Defended"; Robin catalog, Ms. #308; notebook, 1903; published in part: [Peirce *1934*, para. 5.66-81, except 5.71n1 and 5.77n1 (pp. 1–12); 5.88-92 (pp. 48–53)

246. C.S. Peirce, 1903f. "On Logical Graphs (Graphs)", Robin catalog, MS #479; pp. 1–64; plus 30 pp. of several starts; published, with omissions: [Peirce *1933b*], para. 4.350–371, as "On Euler Diagrams"

247. C.S. Peirce, 1903g. Notebook, Miscellaneous notes; Robin catalog, Ms. #1584; the only dated entry: December 24, 1903

248. C.S. Peirce, 1903h. "A Proposed Logical Notation (Notation)"; Robin catalog, MS. #530, n.p., [ca. 1903], pp. 1–45; 44–62, 12–32, 12–26; plus 44 pp. of shorter sections as well as fragments

249. C.S. Peirce, 1903i. MS., notebook; Robin catalog, MS. #455; n.p., 1903, pp. 1–31

250. C.S. Peirce, 1903j. [Lecture I]; Robin catalog, Ms. #450; notebook, n.p., 1903, pp. 1–26

251. C.S. Peirce, 1903k. "Lectures on Logic, to be delivered at the Lowell Institute. Winter 1903–1904. Lecture I"; Robin catalog, Ms. #454, notebook, n.p., 1903, pp. 1–26

252. C.S. Peirce, 1903l. [Lecture II]; Robin catalog, Ms. #455; notebook, n.p., 1903, pp. 1–31

253. C.S. Peirce, 1903m. "Lowell Lectures, Lecture 2. Vol. 2"; Robin catalog, Ms. #456; notebook, n.p., 1903, pp. 40–66

254. C.S. Peirce, 1903n. Lowell Lectures. 1903. Lecture 3. 1st draught; Robin catalog, Ms. #458, notebook, n.p., 1903, pp. 1–33

255. C.S. Peirce, 1903o. [Lecture III]; Robin catalog, Ms. #460, notebook, 1903, pp. 1–22; published, in part: [Peirce *1931*], para. 1.15-26

256. C.S. Peirce, 1903p. "Useful for 3rd or 4th?"; Robin catalog, Ms. #466, notebook, n.p., 1903, pp. 1–28, unfinished, with two p. 19s, both of which leave text intact

257. C.S. Peirce, 1903q. Lowell Lectures of 1903. 1st Draught of 3rd Lecture; Robin catalog, Ms. #457; notebook, n.p., begun October 2, 1903, pp. 1–10

258. C.S. Peirce, 1903r. 468. CSP's Lowell Lectures of 1903. Introduction to Lecture 5; Robin catalog, Ms. #468; notebook, n.p., December 4, 1903, pp. 1–9

259. C.S. Peirce, 1905a. "The Basis of Pragmaticism (Basis)"; Robin catalog, Ms. #280, n.p., [ca. 1905], pp. 1–48, plus fragments

260. C.S. Peirce, 1905b. "The Basis of Pragmaticism" Ms. #284; two notebooks, 1905, pp. 1–48 (one notebook); 49–91 (second notebook)

261. C.S. Peirce, 1905c. "The Bed-Rock beneath Pragmaticism (Bed)"; Robin catalog, Ms. #300; pp. 1–65; 33–40; 38–41; 37–38; 40–43.7; plus 64 pp. of fragments running brokenly from p. 1 to p. 60; published in part: [Peirce *2010b*], 208–210

262. C.S. Peirce, 1905–06. Robin catalog, Ms. #70:05

263. C.S. Peirce, 1905–06. Robin catalog, Ms. #70:07

264. C.S. Peirce, 1905–06. Robin catalog, Ms. #70s:76

265. C.S. Peirce, 1909a. "Studies in Meaning (Meaning)"; Robin catalog Ms. #619, March 25–28, 1909, pp. 1–14, with 2 rejected pp

266. C.S. Peirce, 1909b. "Preface (Meaning Preface to the Volume)"; Robin catalog Ms. #631; n.p., August 24, 1909, pp. 1–4

267. C.S. Peirce, 1909c. "Preface (Meaning Preface to the Book)"; Robin catalog Ms. #632; August 24–29, 1909, pp. 1–27, plus fragments

268. C.S. Peirce, 1909d. "(Meaning Pragmatism)"; Robin catalog, Ms. #622; n.p., May 26-June 3, 1909, pp. 34–70 (p. 50 missing), 42–43, 51, and fragments

269. C.S. Peirce, 1911. "A Diagrammatic Syntax"; Robin catalog, Ms. #500; n.p., December 6–9, 1911, pp. 1–19. Letter to Risteen on existential graphs

270. C.S. Peirce, n.d.(a). "Note on Kempe's Paper in Vol. XXI of the Proceedings of the London Mathematical Society" [Kempe *1889–90*]; Robin catalog Ms. #709; n.p., n.d., pp. 1–6, plus 3 pp

271. C.S. Peirce, n.d.(b). "Notes on Kempe's Paper"; Robin catalog, Ms. #710; n.p., n.d., pp. 1–2, plus 7 pp

272. C.S. Peirce, n.d.(c). "Notes on Kempe's Paper"; Robin catalog, Ms. #711; n.p., n.d., 4 pp

273. C.S. Peirce, n.d.(d). "(Kempe)"; Robin catalog, Ms. #712; n.p., n.d., 1 p

274. C.S. Peirce, n.d.(e). "(Kempe)"; Robin catalog, Ms. #713; n.p., n.d., 2 pp

275. C.S. Peirce, n.d.(f). "Kempe Translated into English"; Robin catalog Ms. #715; n.p., n.d., 1 p

276. C.S. Peirce, n.d.(g). "Logic of Relatives"; Robin catalog, Ms. #547, n.p., n.d., 18 pp

277. C.S. Peirce, n.d.(h). "Comments on Cayley's "Memoir on Abstract Geometry" from the point of view of the Logic of Relatives"; Robin catalog, Ms. #546; n.p., n.d., 5 pp

278. C.S. Peirce, n.d.(i). "Argon, Helium, and Helium's Partner"; Robin catalog, Ms. #1036; n.p., [c. 1890?], pp. 1–5
279. C.S. Peirce, n.d.(j). "Chemistry"; Robin catalog, Ms. #1038; n.p., n.d., pp. 1–5
280. C.S. Peirce, n.d.(k). "Chemistry The Elements"; Robin catalog, Ms. #1039; n.p., n.d. (but a reference to Clarke 1897 on p. 1), 55 pp., including a sequence pp. 1–6
281. C.S. Peirce, n.d.(l). "A Proposal of a change in the Atomic Weights with a remark on the Periodicity of the Properties of the Elements"; Robin catalog, Ms. #1044; n.p., n.d., 8 pp
282. C.S. Peirce, n.d.(m). [Fragment on the History of Logic]; Robin catalog, Ms. #1000; n.p., n.d., 2 pp
283. C.S. Peirce, n.d.(o). "On Existential Graphs as an Instrument of Logical Research"; Robin catalog, Md. #498; notebook (Harvard Cooperative), n.p., n.d.
284. C.S. Peirce, n.d.(p). Undated, untitled mnauscript; Robin catalog, Ms. #513; pp. 27–98, incomplete and in some disorder, with missing sections and many alternatives and/or rejects
285. C.S. Peirce, in *Collected Papers of Charles Sanders Peirce*, vol. II, *Elements of Logic*, ed. by C.Hartshorne, P. Weiss (Harvard University Press, Cambridge, MA, 1932); 2nd edn., 1960
286. C.S. Peirce, in *Collected Papers of Charles Sanders Peirce*, vol. III, *Exact Logic (Published Papers)*, ed. by C. Hartshorne, P. Weiss (Harvard University Press, Cambridge, MA, 1933a); 2nd edn., 1960
287. C.S. Peirce, in *Collected Papers of Charles Sanders Peirce*, vol. IV, *The Simplest Mathematics*, ed. by C. Hartshorne, P. Weiss (Harvard University Press, Cambridge, MA, 1933b); 2nd edn., 1960
288. C.S. Peirce, in *Collected Papers of Charles Sanders Peirce*, vol. V, *Pragmatism and Pragmaticism*, ed. by C. Hartshorne, P. Weiss (Harvard University Press, Cambridge, MA, 1934)
289. C.S. Peirce, in *Collected Papers of Charles Sanders Peirce*, vol. VI, *Scientific Meta-physics*, ed. by C. Hartshorne, P. Weiss (Harvard University Press, Cambridge, MA, 1935)
290. C.S. Peirce, in *The New Elements of Mathematics*, 4 vols, ed. by C. Eisele (Mouton, The Hague/Paris, 1976)
291. C.S. Peirce, in *Writings of Charles S. Peirce: A Chronological Edition*, vol. 1: *1857–1866*, ed. by M.H. Fisch (Indiana University Press, Bloomington, 1982)
292. C.S. Peirce, in *Writings of Charles S. Peirce: A Chronological Edition*, vol. 2: *1867–1871*, ed. by E.C. Moore (Indiana University Press, Bloomington, 1984)
293. C.S. Peirce, in *Writings of Charles S. Peirce: A Chronological Edition*, vol. 4: *1879–1884*, ed. by C.J.W. Kloesel (Indiana University Press, Bloomington/Indianapolis, 1989)
294. C.S. Peirce, in *Writings of Charles S. Peirce: A Chronological Edition*, vol. 5: *1884–1886*, ed. by C.J.W. Kloesel (Indiana University Press, Bloomington/Indianapolis, 1993)
295. C.S. Peirce, in *The Essential Peirce: Selected Philosophical Writings*, vol. 2: *(1893–1913)*, ed. by Peirce Edition Project (Indiana University Press, Bloomington/Indianapolis, 1998)
296. C.S. Peirce, in *Writings of Charles S. Peirce: A Chronological Edition*, vol. 8: *1890–1892* (N. Houser, general ed.) (Indiana University Press, Bloomington/Indianapolis, 2010a)
297. C.S. Peirce, in *Philosophy of Mathematics: Selected Writings*, ed. by M.E. Moore (Indiana University Press, Bloomington/Indianapolis, 2010b)
298. C.S. Peirce, C. Ladd-Franklin, Universe, in *Dictionary of Philosophy and Psychology: Including Many of the Principal Conceptions of Ethics, Logic, Aesthetics, Philosophy of Religion, Mental Pathology, Anthropology, Biology, Neurology, Physiology, Economics, Political and Social Philosophy, Philology, Physical Science, and Education and Giving a Terminology in English, French, German and Italian*, vol. II, ed. by J.M. Baldwin (Macmillan, New York/London, 1902), p. 742
299. F.J. Pelletier, A brief history of natural deduction. Hist. Philos. Log. **20**, 1–31 (1998)
300. Petrus Hispanus, in *Compendiarius parvorum logicalium continens perutiles Petri Hispani tractatus priores sex & clarissimi philosophii Marsilij dialectices documenta*, ed. by K. Pschlaer (Vietor Singrenius, Vienna Austriæ, 1512)
301. Petrus Hispanus, *Summulæ logicales* (apud M. Valentinum, Venetiis, 1597)
302. E. Picardi, *La chimica di concetti: linguaggio, logica, psicologia* (Il Mulino, Bologna, 1994)
303. G. Pitcher, Wittgenstein, Nonsense, and Lewis Carroll. Mass. Rev. **6**, 591–611 (1965)
304. H. Poincaré, Sur l'analysis situs. Comptes redus de l'Académie des Sciences **115**, 633–636 (1892); reprinted: (René Garnay & Jean Leray, éds.) *Œuvres de Henri Poincaré/publiées sous les auspices de l'Académie des sciences*, tme. 6: *Géométrie* (Paris: Gauthier-Villars, 1953), 189–192
305. Porphyry, (Christian August Brandis, Hsg.), *Isagoge*, vol. IV of *Scholia in Aristotelem* (G. Reimeri, Berolini, 1836)

306. E.L. Post, *Introduction to a General Theory of Elementary Propositions*, Ph.D. thesis, Columbia University, 1920. Abstract presented in *Bulletin of the American Mathematical Society* **26**, 437; abstract of a paper presented at the 24 April meeting of the American Mathematical Society

307. E.L. Post, Introduction to a general theory of elementary propositions. Am. J. Math. **43**, 169–173 (1921); reprinted: [van Heijenoort *1967*], 264–283

308. H.M. Pycior, At the intersection of mathematics and humor: Lewis Carroll's *Alices* and Symbolical Algebra. Vic. Stud. **28**(1), 149–170 (1984)

309. W.V.O. Quine, Review of [Peirce *1932*]. Isis **19**, 220–229 (1933)

310. W.V.O. Quine, Review of [Peirce *1933a*]. Isis **22**, 285–297 (1934)

311. W.V.O. Quine, Review of [Peirce *1933b*]. Isis **22**, 551–553 (1935)

312. N. Rescher, *Galen and the Syllogism* (University of Pittsburgh Press, Pittsburgh, 1966)

313. B. Riemann, Theorie der Abelschen Functionen. Journal für reine und angewandte Mathematik **54**, 115–155 (1857); reprinted: (Heinrich Weber, Hsg.), *Gesammelte mathematische Werke und wissenschaftlicher Nachlass* (Leip-zig: B. G. Teubner, 1892), 88–142

314. A.D. Risteen, 1891. Letter to Charles S. Peirce, 10 June 1891; Robin Catalog, MS#L376

315. A.D. Risteen, *Molecules and the Molecular Theory of Matter* (Ginn & Co., Boston/London, 1895); reprinted: 1896

316. D.D. Roberts, in *The Existential Graphs of Charles S. Peirce*, ed. by T.A. Sebok (Mouton, The Hague, 1973)

317. R.S. Robin, *Annotated Catalogue of the Papers of Charles S. Peirce* (University of Massachusetts Press, Amherst, 1967); http://www.iupui.edu/~peirce/robin/rcatalog.htm

318. J.A. Robinson, *Logic: Form and Function: The Mechanization of Deductive Reasoning* (Elsevier, North-Holland, New York, 1979)

319. L. Ruby, *Logic: An Introduction* (J. B. Lippencott, Chicago/Philadelphia/New York, 1950); 2nd edn., 1960

320. B. Russell, Sur la logique des relations avec des applications à la théorie des séries. Revue de mathématiques/Rivista di Matematiche (Torino) **7**, 115–148 (1901)

321. B. Russell, Introduction to the second edition, in *Principia Mathematica*, vol. I, 2nd edn., ed. by A.N. Whitehead, B. Russell (Cambridge University Press, Cambridge, 1925), pp. xiii–xlvi

322. B. Russell, *The Autobiography of Bertrand Russell 1944–1967*, vol. III (George Allen and Unwin Ltd., London, 1967)

323. I.S. Russinoff, The syllogism's final solution. Bull. Symb. Log. **5**, 451–469 (1999)

324. H. Schepers, Leibniz' Arbeiten zur einer Reformation der Kategorien. Zeitschrift für philosophische Forschung **20**, 539–567 (1966)

325. P. Schroeder-Heister, Resolution and the origins of structural reasoning: early proof-theoretic ideas of Hertz and Gentzen. Bull. Symb. Log. **8**, 246–265 (2002)

326. C. Seibert, Peirce's childhood laboratory. Peirce Project Newsletter 3 (no. 2, Fall, 2000; ©2001), 9, 11

327. K. Seubert (Hsg.), *Das naturliche System der chemischen Elemente. Abhandlungen von Lothar Meyer, 1864–1869 und D. Mendekejeff, 1869–1871* (Verlag von Wilhelm Engelmann, Leipzig, 1895)

328. E. Sewell, *The Field of Nonsense* (Chatto and Windus, London, 1952)

329. E. Shen, The Ladd-Franklin formula in logic: the antilogism. Mind, n.s. **36**, 54–60 (1927)

330. S.-J. Shin, *The Iconic Logic of Peirce's Graphs* (MIT Press, Cambridge, MA/London, 2002)

331. J. Shosky, Russell's use of truth tables. Russell: J. Russell Archives (n.s.) **17**, 11–26 (1997)

332. G. Spencer-Brown, *Laws of Form* (Allen and Unwin, London, 1969)

333. H. Steinthal, *Geschichte der Sprachwissenschaft bei der Greichen un Römern. Mit besonderer Rücksicht auf die Logik* (Ferdinand Dümmler's Verlagsbuchhandlung, Berlin, 1863); reprinted: Cambridge/New York, etc.: Cambridge University Press, 2013

334. G.J. Stokes, The mathematical theory of inference. Am. J. Math. **7**, 1–8 (1900)

335. J.J. Sylvester, On an application of the new atomic theory to the graphical representations of the invariants and covariants of binary quantics,—with three appendices. Am. J. Math. **1**, 64–125 (1878)

336. J.J. Sylvester, 1882. Remarks on C. Peirce's Logic of Relatives, given before the Mathematical Seminary, Johns Hopkins University, April, 1882. *The Johns Hopkins University Circulars*, no. 15, 203

337. J.J. Sylvester, 1883. Erratum. *The Johns Hopkins University Circulars*, no. 21, 46
338. G. Takeuti, *Proof Theory*, 2nd revised edn. (North-Holland, Amsterdam/New York/Oxford, 1967)
339. P. Thibaud, *La logique de Charles Sanders Peirce: De l'Algèbre aux Graphes* (Éditions de l'Université de Provence, Aix-en-Provence, 1975)
340. A.S. Troelstra, 2008. Beth's contribution to intuitionism, in [Van Benthem, Van Ulsen, & Visser *2008*], pp. 19–21
341. A.R. Turquette, Peirce's icons for deductive logic, in *Studies in the Philosophy of Charles Sanders Peirce* (2nd Series), ed. by E.C. Moore, R.S. Robins (University of Massachusetts Press, Amherst, 1964), pp. 95–108
342. D. Van Dalen, 2008. Beth and the emergence of intuitionistic logic, in [Van Benthem, Van Ulsen, & Visser *2008*], pp. 9–11
343. J. Van Heijenoort,*From Frege to Gödel: A Source Book in Mathematical Logic, 1879–1931* (Harvard University Press, Cambridge, MA, 1967)
344. J. Van Heijenoort, 1968. Préface to [Hebrand *1968*], pp. 1–12
345. J. Van Heijenoort, *El desarrollo de la teoría de la cuantificación* (Universidad Nacional Autónoma de México, Mexico City, 1976)
346. J. Venn, On the various notations adopted for expressing the common propositions of logic. Proc. Camb. Philos. Soc. **4**, 36–47 (1880a); reprinted: [Venn *1894*], 477–504
347. J. Venn, On the diagrammatic and mechanical representations of propositions and reasonings. The London, Edinburgh and Dublin Philos. Mag. J. Sci. (ser. 5) **10**, 1–18 (1880b)
348. J. Venn, *Symbolic Logic* (Macmillan & Co., London, 1881)
349. J. Venn, *Symbolic Logic*, 2nd edn. (Macmillan & Co., London, 1894); reprinted: Bronx, N.Y.: Chelsea Publishing Co., 1971
350. L. Vives, *Opera Io. Lodocivi Vivis Valentini* (Apvd Nivolavm Episcopivm, Basileao, 1555)
351. J. Von Neumann, Zur Hilbertschen Beweistheorie. Mathematische Zeitschrift **26**, 1–46 (1927)
352. J. Von Plato, Gentzen's proof of normalization for natural deduction. Bull. Symb. Log. **14**, 240–257 (2008)
353. J. Von Plato, Gentzen's logic, in *Handbook of the History of Logic*, vol. 5: *Logic from Russell to Gödel*, ed. by D.M. Gabbay, J. Woods (North-Holland, Amsterdam, etc., 2009)
354. J. Von Plato, Gentzen's proof systems: byproducts in a work of genius. Bull. Symb. Log. **18**, 313–367 (2012)
355. A. Walsh, *Relations between Logic and Mathematics in the Work of Benjamin and Charles S. Peirce*, Ph.D. thesis, Middlesex University, 1999; published: Boston: Docent Press, 2012
356. C. Weise, *Christiani Weisii nucleus logicae: succinctus regulis sufficientibus tamen exemplis in compendio exhibens quicquid a primis disciplinae auditoribus disci vel requiri potest, in eorum gratiam, qui majori logicae, iam dudum publicatae* (Zittaviae: Apud Joh. Casparum Meyerum, Leipzig, 1691)
357. R. Whately, *Elements of Logic* (James Munroe, Boston, 1845)
358. A.N. Whitehead, B. Russell, *Principia Mathematica*, 3 vols. (Cambridge University Press, Cambridge, 1910–13)
359. A.N. Whitehead, B. Russell, *Principia Mathematica*, 3 vols., 2nd edn. (Cambridge University Press, Cambridge, 1925–27)
360. W.D. Williams, Some early chemical slide rules. Bull. Hist. Chem. **12**, 24–29 (1992)
361. R.J. Wilson, Graph theory, in *History of Topology*, ed. by I.M. James (North-Holland, Amsterdam, 1999), pp. 503–530
362. L. Wittgenstein, (Charles Kay Ogden, transl., with an introduction by Bertrand Russell), *Tractatus logico-philosophicus/Logisch-philosophische Abhandlung* (Routledge & Kegan Paul, London, 1922)
363. W.H. Wollaston, A synoptic scale of chemical equivalents. Philos. Trans. **104**, 1–22 (1814); http://www.scs.illinois.edu/~mainzv/HIST/bulletin_openaccess/num12/num12%20p24-29.pdf
364. J.S. Wu, The problem of existential import (from George Boole to P. F. Strawson). Notre Dame J. Formal Log. **10**, 415–424 (1969)
365. J.J. Zeman, *The Graphical Logic of C. S. Peirce*, Ph.D. thesis, University of Chicago, 1964; available online at: http://www.clas.ufl.edu/users/jzeman/ ©2002

366. J.J. Zeman, Peirce's logical graphs. Semiotica **12**, 239–256 (1974)
367. S. Zellweger, 1997. Untapped potential in Peirce's iconic notation for the sixteen binary connectives, in [Houser, Roberts, & Van Evra *1997*], pp. 334–386

I.H. Anellis
Peirce Edition Project, Institute for American Thought, Indiana University-Purdue University at Indianapolis, Indianapolis, IN 46202, USA

F.F. Abeles (✉)
Departments of Mathematics and Computer Science, Kean University, Union, NJ 07083, USA
e-mail: fabeles@kean.edu

Logic and Argumentation in Belgium: The Role of Leo Apostel

Jean Paul Van Bendegem

Abstract To understand present-day research in logic and argumentation theory in Belgium, it is necessary to highlight the unique contribution of Leo Apostel. The first part of the paper deals with his main intellectual influences, namely Chaïm Perelman, Rudolf Carnap, and Jean Piaget. In the second part the Signific Movement and the Erlangen School are discussed, leading up to the present situation and thus to the promised understanding.

Keywords Logic · Argumentation theory · Leo Apostel · Chaïm Perelman · Rudolf Carnap · Jean Piaget · Gerrit Mannoury · Paul Lorenzen · Diderik Batens

Mathematics Subject Classification (2000) 03-03 · 01-02 · 01A60

Introduction

As so often happens, there can be a huge distance between the original conception of a paper and the final form it takes, and this contribution is no exception to that rule. My aim was to write about the Significs Movement, an important but unfortunately not that well known philosophical project in the early part of the twentieth century in the Netherlands. It was however impossible to write this history without bringing into the picture how that movement came to be known in my home country, namely Belgium. Which led me to the figure of Leo Apostel. His role has been crucial in the development of logic and argumentation theory in Belgium, ignoring for the moment his work in philosophy of science and in philosophy in general. This in its turn made it necessary to discuss the chief intellectual influences on his thinking, namely Chaïm Perelman, Rudolf Carnap and Jean Piaget. But the story would be incomplete if I did not pay attention to at least one other project that Apostel studied and was influenced by, namely the Erlangen School.

This paper is partially based on a presentation at the conference "Lebenswelt and Logic: The Erlangen School as Heir to Logical Empiricism", organised by Gerhard Heinzmann and Pierre Wagner, and held at the Université de Nancy in November 2008. I was asked specifically to talk about the influence of the Erlangen school in Belgium and therefore the focus of my presentation was on Paul Lorenzen's work and its reception in Belgium through Leo Apostel. Here I have largely left out this specific element (although some references will of course be made) and focused on the Signific Movement but obviously the two should be integrated into a more encompassing framework.

© Springer International Publishing Switzerland 2016
F.F. Abeles, M.E. Fuller (eds.), *Modern Logic 1850-1950, East and West*, Studies in Universal Logic, DOI 10.1007/978-3-319-24756-4_4

As I ended up with a more general picture of how logic and argumentation theory were developed in Belgium in the second half of the twentieth century, it seemed appropriate to have a look at the present-day situation.

So this paper has finally become a contribution to the development of logic and argumentation theory in Belgium through the work of Leo Apostel, with special attention to the Significs Movement and the Erlangen School and to the present-day situation. The structure is such that I will focus first on Apostel himself in section "Some Notes on Sources" with a few remarks about the sources used, as this was and still is not an entirely trivial task. Then I present a short biography in section "Short Biography" and some elements of his philosophical methodology in section "Leo Apostel's philosophy: A Few Notes on his Methodology". This is continued in section "Leo Apostel's Philosophy: The Main Influences" by a sketch of the main influences on Apostel's philosophical thinking. Then in section "The Signific Movement" I can discuss the introduction of the Significs Movement and, to a lesser extent, the Erlangen School in Belgium. Finally in section "Present-Day Situation or the Legacy of Leo Apostel" a few elements of the present-day situation will be outlined.

It is, I'm afraid, not an exaggeration to claim that most of this history has remained so far largely unwritten but needs to be written in order to understand how deep Apostel's influence has been on what is the present-day logico-philosophical community in Belgium. And, along the way, it will turn out that Irving Anellis himself to whom this volume is dedicated, also appears in the picture.

Some Notes on Sources

At present there is not yet a decent, sufficiently detailed biography of Leo Apostel. In 1989 under the editorship of Fernand Vandamme and Rik Pinxten, a three volume work was published, entitled *The Philosophy of Leo Apostel*, conceived along the lines of the well known Schilpp volumes. The first volume, *Descriptive and Critical Essays*, brings together a set of papers related to his work, the second, *The Philosopher Replies*, contains what the title says, and the third volume bears the title, *A life history*. The latter volume is the result of a series of conversations with Ingrid Van Dooren and, in that sense, relies to a large extent on Apostel's memory at that time. In other words, some care must be taken in assessing the correctness of historical details. This will force me in some cases to write in a rather hypothetical way concerning certain events.

As far as publications are concerned, the situation is much better. There is a complete list of all the writings of Leo Apostel, both academic and non-academic, that can be found at the following web address: http://logica.ugent.be/. Thanks to the efforts of Jenny Walry (1935–2008), a *compagnon de route* for many years and a philosopher-writer herself, we also have a complete set of the publications themselves. In addition, selections of papers and some posthumous works were published after Apostel's demise, in a form that made them more accessible to both the academic audience and the general public. Some of the posthumous work that has not been published can be found at the website mentioned above. Finally, the website also contains a short biography.

There are some general characteristics of the writings of Leo Apostel one has to take into account, and these are:

- A large part of the work is written in Dutch. This is not a coincidence, but a deliberate choice. Apostel was not the only philosopher at that time—at the University of Ghent one should mention (at least) Etienne Vermeersch (born 1932), Jaap Kruithof (1929–2009) and Rudolf Boehm (born 1927)—who held the idea that a philosopher has obligations not only towards the academic world but also toward society at large. This is comparable to the French idea of the 'intellectuel' who participates in societal debates, though there are clear differences, e.g., as far as philosophy itself is concerned (see further about the philosophical approach of Leo Apostel).[1] Of course, this inevitably means that part of his work was not internationally accessible.
- He did not particularly seek to publish in major journals. One must not conclude from this statement that he was philosophically isolated in Flanders or Belgium. Far from it. He had a quite active correspondence with academics all over the world or, using modern parlance, he was part of an extensive, world-wide network. It is fair to say that he preferred to discuss philosophy face to face rather than via an intermediary channel such as a journal, where its importance is often correlated with the time it takes to have a paper published, at least in his time and life. In addition, he did not want to waste time to write papers in the different formats required by different journals.
- Another element that seriously complicated matters is that Apostel did not like to read proof copies. To reread a text meant in his case almost certainly to rewrite it. And thus the process never came to an end. It is a true pity, short of a tragedy, that this also happened to his magnum opus, the two volume book, entitled *Matière et Forme* [1] on, but not solely, a philosophical foundation for a realist interpretation of causality.

These particularities explain the perhaps somewhat odd situation that, although the person is known or even well known, this need not imply that his works are equally well known, though this need not imply that his influence would have been minimal, quite the contrary, as the short biography in the next section shows (and, of course, also in the final section that deal with his influence on the generations after him, up to the present.)

Short Biography

Leo Apostel was born in Antwerp, Belgium on 4 September 1925. He studied philosophy at the Université Libre de Bruxelles (ULB) and graduated in 1948. His thesis was entitled *Questions sur l'Introspection* (*Questions concerning introspection*). After graduation

---

[1]This is a recurring theme in Belgian academic and philosophical life (and it can be extended to society at large in some cases): both "Northern" and "Southern" influences are present. We have both the image of the philosopher as an academician, rather isolated from society, often of his/her own choice, and doing extremely profound things in his/her ivory tower, and of the philosopher as the intellectual who engages him- or herself in everyday life, a typical example being of course Jean-Paul Sartre. More often than not, we end up being both.

he became an assistant of Chaïm Perelman, then professor of logic at the ULB. In the academic year 1950–1951 he travelled to the USA and was a CRB[2] fellow at the University of Chicago and of Yale, where he studied with on the one hand Rudolf Carnap and on the other hand Carl Gustav Hempel (aka Peter Hempel). From 1952 to 1956 he became a researcher with a grant of the N.F.W.O. (Nationaal Fonds voor Wetenschappelijk Onderzoek, National Fund for Scientific Research) and in March 1953 he became "Doctor in de Wijsbegeerte" at the ULB with a doctoral thesis on *La Loi et les Causes* (*Laws and Causes*). The academic year 1955–1956 was spent at the Université de Génève, as a member of the Centre International d'Epistémologie Génétique, the famous center founded by the well known genetic psychologist Jean Piaget. He remained a foreign member of that Center until the death of Piaget in 1980 and in 1983 he became 'Doctor honoris causa' at the Université de Génève. Starting from 1956 he became a professor at the ULB, teaching general courses in philosophy, logic and epistemology and in 1957 he became professor at the Rijksuniversiteit Gent (now the University of Ghent). During the academic year 1958–1959 he was guest professor at the Pennsylvania State University. He retired from the VUB—in 1969–1970 the Dutch-speaking part of the ULB founded its own institution, the Vrije Universiteit Brussel and Apostel joined this new institution—in 1974 and he retired from the Rijksuniversiteit Gent in 1979. He died on 10 August 1995, 3 weeks before his 70th birthday.

This short biography already introduces the three most important influences on Apostel's philosophical thinking: Chaïm Perelman, Rudolf Carnap and Jean Piaget, who will be discussed in the sequel. The mere fact that these three names occur here together already indicates that Leo Apostel is hardly to be considered a logical empiricist *pur sang*, but as someone who was looking for a synthesis of logical empiricism with other views and approaches. But that list of three would be incomplete if I did not mention explicitly the Signific Movement, the original topic of this paper, as explained in the introduction. It is important to mention that Apostel stayed in Amsterdam during 1950–1951, where he met Evert Willem Beth, Gerrit Mannoury and L. E. J. Brouwer. No doubt the last name mentioned is the most famous one, and most likely the first name mentioned will be the second most famous one but I am, alas, pretty sure that the name mentioned in the middle always needs some, if not a serious amount of background and explanation. That Leo Apostel valued the work of these three philosophers-logicians-mathematicians quite highly, is demonstrated by the fact that, in 1964, thanks to his efforts, Beth became 'Doctor honoris causa' of the University of Ghent (but unfortunately he was already too ill to collect it). And among his first (international) publications, see his [2], is an abstract about Brouwerian algebras.

---

[2]Commission for Relief in Belgium. This later turned into the Belgian American Educational Foundation (BAEF) and continued to award grants to Belgian researchers, such as the Fulbright-Hays grant, for short and longer stays in the USA.

Leo Apostel's Philosophy: A Few Notes on his Methodology

It is a nearly impossible task to summarize someone's methodological views in a few statements but the four items that follow are constitutive in the sense that any fuller treatment must incorporate at least these four aspects:

- Throughout his intellectual life, Leo Apostel had a clear view on the use of logic in order to achieve clarity, insight, and understanding. (As will become clear in the next section this is mainly due to the influence of Rudolf Carnap.) Although this need not imply that logic has to be seen solely as a tool, it rarely if never was a goal on its own for him. This explains his pluralistic, very open-minded attitude towards formal logical systems. The question was, while studying a philosophical problem, which logic(s) would be most suited to achieve the required clarity. Often, in fact, he developed his own systems.[3]
- Analysis requires synthesis to be complete. Although it would be wrong to attribute the idea of a unity of science to Apostel, either in the sense of the Diderot-d'Alembert *Encyclopédie* or of the Wiener Kreis *Encyclopedia of Unified Science*, he did seek to synthesize the broken fragments that constitute our personal and shared world view. In later life this took a concrete form in the *World Views Project*, where the aim was to bring together as many disciplines as possible, both from the natural sciences and the humanities.[4]
- No philosophy without science. Apart from the fact that this ties in nicely with the logical empiricist view, this commitment to 'check off' philosophical theories, ideas and suggestions with the relevant sciences implies that philosophy must be seen as an open-ended activity, as long as the sciences themselves do not reach an endpoint, a highly unlikely event. And it does seem odd to reflect, e.g., on time and space and not consult the physical sciences to see what contribution they can make. Although, of course, the philosophical conclusions might be that the physicists are on the wrong track.[5]
- Philosophy can be as scientific as science itself. This does not mean that philosophy is to be 'reduced' (in some meaning of the verb) to science but that philosophical problems can be treated as rigorous as it is done in the sciences. It does not mean either that the philosophical method has to be an exact copy of the scientific method—the counterarguments are easy to find—but that the search for a comparable method is meaningful. Apostel [3] presents the clearest defense of this view.

---

[3]Very often he was ahead of his time because he did not want to simplify matters so that a logical framework could be designed. If the philosophical problem dictated the need to incorporate certain features in the formal framework then that needed to be done. Needless to say that often the result was a rather complex logical system, and therefore rather difficult to work with.

[4]See Apostel and Van der Veken [34], Apostel et al. [35]. Both of these books are in Dutch but there is also a book in English, Apostel et al. [36]. This book is electronically available at: http://www.vub.ac.be/CLEA/pub/books/worldviews.pdf.

[5]I emphasize this point because I do not want to restrict the interaction between philosophy and science to a form of naturalistic exploration whereby the scientific practice itself is left untouched and the philosopher's task is first of all to understand that practice, not to criticize it, let alone improve it in some particular direction.

From these four elements a couple of consequences follow that I want to mention briefly:

- His writings appear very often to be rather eclectic. His already mentioned magnum opus *Matière et Forme* is a perfect example: a huge number of logical approaches are combined with one another to arrive at an overarching theory for a realist epistemology of causality and very diverse philosophical systems are discussed in an attempt to arrive at a form of synthesis. Nevertheless, eclecticism would be a very misplaced label for the final aim—whether successful or not is another matter—which was to arrive at a global theory.
- His need and desire for synthesis all too often led to 'grand' projects that in practice transformed themselves into unfinished projects. Of course, this in itself need not be a drawback as it invites others to continue the effort but it is not always easy to understand well what his aims were with a particular project. A perfect example was his grand project for a *Design of a Natural Philosophy*, which was partly published [4], partly made available, as mentioned, on the website posthumously.[6]
- The size of such synthesizing projects often made it necessary to be completely focused on it during a particular period. So at different times he had different interests. Again this may have created to the outside world the opposite impression of not being focused at all but such was not the case. After all, to explore a house, one has to do it one room at a time.[7]

In the short biography I mentioned three names that have been highly influential for Apostel's thinking. Let me sketch a few additional elements for each one of them.

Leo Apostel's Philosophy: The Main Influences

Chaïm Perelman

The first major intellectual figure in the life of Leo Apostel was the philosopher Chaïm Perelman (1912–1984). No doubt Perelman is best known for his work in argumentation theory, most notably through the [5] book on argumentation, jointly written with Lucie Olbrechts-Tyteca (1899–1987), where the concept of 'la nouvelle rhétorique' ('the new rhetoric') was launched. The main object of this approach can be formulated as follows. We all know the frequently claimed opposition between argumentation and logic. The latter aims for truth (and nothing but the truth), the former deals with attempts to convince an audience, whereby truth need not necessarily play an important role but efficiency does. Such an oppositional view must lead to the idea that in argumentation theory the main focus must be to identify all the mistaken forms of reasoning, hence the extensive study of fallacies, even all too often simply identified with argumentation theory. This

---

[6]See http://logica.ugent.be/leo_apostel/natuurfilosofie/index.html.

[7]Other interests (that I will not discuss any further in this paper) were Freemasonry, spirituality for atheists, anarcho-socialism, and postmodernism (together with Jenny Walry, mentioned already), to name but a few.

rough picture is sufficient to illustrate the curious fact that it is standardly formulated, one is tempted to say, through a *via negativa*: one lists primarily what one should *not* do. Perelman's and Olbrechts-Tyteca's aim was to provide this theory with a proper 'positive' foundation and hence the qualification 'nouvelle' was and still is justified.

An important consequence—and this will prove to be of importance when discussing the Signific Movement—is that the relation between mathematical reasoning and, e.g., juridical argumentation is not necessarily one of opposition but rather, to use Stephen Jay Gould's term—a matter of different *magisteria*.[8] Where certainty is present, mathematical-logical reasoning will do perfectly, where it is not, argumentation-cum-rhetoric steps in. The following quote makes this quite clear:

> For centuries many good minds have found in the artificial language of mathematicians an ideal of clarity and univocity that natural languages, with their lesser development, should strive to imitate. From this viewpoint, any ambiguity, obscurity, or confusion is considered an imperfection that can be eliminated not only in theory but also in fact. Because of the univocity and the exactness of its terms, scientific language is held to be the best instrument for demonstration and verification, and one would like to impose these characteristics on all language. (130, from the English translation[9])

The ultimate mistake would clearly be to impose mathematical-logical forms of reasoning where they would only create a false sense or an illusion of certainty.

Of equal importance is the fact that the framework for the new rhetoric is of a social nature: any argument is always presented before an audience that will or will not be convinced by a set of arguments. It is thus closer to a dialogue than a monologue. Although Perelman himself did not get involved with developments in so-called dialogue logic (and its variations: dialogical logic and dialectical logic—see further when the Erlangen School is discussed), this was a connection that Apostel would explore. After all, once we have a (positive) theory about argumentation, what would prevent us to try to formalize that theory (see [6] for one of the best examples of this enterprise)? Apostel's interest in both logic and argumentation and especially the relations between them—the use of formal methods to understand how arguments function and the use of argumentation theory to make the hidden assumptions visible in logical and mathematical reasoning—has been a continuous philosophical theme throughout his intellectual life.

---

[8] Stephen Jay Gould introduced this notion in his [37], where the main theme was the relation between science and religion. One might wonder how far the parallel can be drawn between the couple science-religion on the one hand and the couple logic-argumentation on the other hand. But that requires a separate treatment.

[9] Perelman and Olbrechts-Tyteca (1969) [5]. The original text: "Le langage artificiel des mathématiciens fournit, depuis des siècles, à beaucoup de bons esprits, un idéal de clarté et d'univocité que les langues naturelles, moins élaborées, devraient s'efforcer d'imiter. Toute ambiguïté, toute obscurité, toute confusion sont, dans cette perspective, considérées comme des imperfections, éliminables non seulement en principe, mais encore en fait. L'univocité et la précision de ses termes feraient du langage scientifique l'instrument le meilleur pour les fonctions de démonstration et de vérification, et ce sont ces caractères que l'on voudrait imposer à tout langage."

Rudolf Carnap

Was Perelman the major influence in Apostel's intellectual growth, then his maturity was assured by that philosopher he admired most, namely Rudolf Carnap (1891–1970). Before his stay with him in the USA, Apostel was already deeply impressed with Carnap's first major philosophical effort, namely his famous 1928 book [9] *Der Logische Aufbau der Welt*. I will not try to develop here in what ways Carnap's views influenced Apostel as this is a complicated and delicate matter (and I even doubt whether I am the best suited person to do so) but just indicate that the impact was not merely on the philosophical-intellectual level but also on the personal level. In his already mentioned autobiography ([7], third volume), he states the following:

> When I met Carnap, I felt he was everything I had imagined he should be: he was deep, clear, considerate, modest, giving all his attention to his students and his work, explaining with care the difficult things he was doing. (24)

and

> Carnap invited me to do my Ph.D. thesis with him in Chicago. I was honoured by his suggestion, but declined it with regret. The brilliant Carnap, whom I highly respected as a thinker and as a human being, did not lead a happy life at his university. In staff-meetings I often experienced how less creative people but with greater administrative power, attacked him. I knew my weakness and was not sure whether I would be able to endure this stressful atmosphere. Moreover Carnap overwhelmed me. I had to admire him from afar. My relation to him in a certain sense could have become self-destructive. (24–25)

The easiest qualification would be that Carnap turned Apostel into a logical empiricist but that would indeed be too easy. It is true of course that he saw himself as influenced by that particular school of thought but more as a starting point. Apostel's work on causality can serve as a perfect example: although he is relying on logical empiricists' ideas, at the same time he considerably deviates from it, if only by insisting on the importance of the ontological status of causal processes. A better qualification, but definitely not a final one, is to say that he kept faithful to Carnap's methodological views, expressed by Apostel himself thus:

> One of the admirable characteristics of R. Carnap, I think (others may consider it a weakness) was that he allowed for alternatives. Confronted with an alternative, he would try to find occasions where one horn of the dilemma applied, and others where the other was valid. (25)

Many of Carnap's writings exemplify this particular feature. If in any discussion around a specific topic, say in the philosophy of mathematics, where opposing positions were formulated, then he would typically look for a larger framework wherein both positions could be expressed, usually in some appropriate formal language, in such a way that arguments from all players concerned, could at least be translated into one another, thereby increasing the probability that a (partial) agreement could be reached, thus helping the discussion forward without a necessity to reach a final resolution. This last statement makes clear how Perelman's and Carnap's views became integrated in his thinking. But the picture would not be complete without the third major influence, namely Jean Piaget.

Jean Piaget

It is safe to assume that Jean Piaget (1896–1980) is primarily known for his groundbreaking studies on the development of thinking in children. His famous four-stage model of cognitive development—starting with the sensorimotor stage, from the pre-operational stage to the concrete operational stage, to culminate in the formal operational stage—has been quite influential (although, of course, now transformed and integrated into other frameworks). Space, time, number, causality, and moral judgment were topics of investigation through an elaborate series of ingenious experiments to demonstrate the different stages a child went through.[10] Famous are, e.g., the demonstrations that young children do not immediately grasp the idea of conservation, when the same amount of water is transferred from one vessel to another, especially in the case when one is thin and high and the other is almost flat and large.

It would however be unfair to characterize Piaget 'merely' as a psychologist. He was a biologist as well and, above all, a philosopher. The extension of his psychological work into the so-called genetic epistemology ('épistemologie génétique') is a clear transition from science into philosophy. The famous already mentioned center that Piaget established in 1955, the Centre d'épistémologie génétique, was described by Apostel in his autobiography as follows:

> Here was a group of philosophically minded interdisciplinary researchers who tackled major epistemological problems: in a sense Piaget both combined and went beyond the programs of logical empiricists (creation of a scientific philosophy) and intuitionists (providing constructivism with empirical foundations). (46)

So Apostel joined the center during the academic year 1955–1956 and that stay had a lasting impact on his work, primarily in relation to how logic(s) can be justified or, more precisely, how logical and mathematical rules can be justified.[11] He was always interested in the possibility of an empirical or constructive approach to the matter. This also explains why Apostel invited Paul Lorenzen to a conference held at the center. This should not be a surprising statement since the latter had worked on so-called operational logic ('Operative Logik', see Lorenzen [11]) which seemed, at least at first sight, to be connected with Piaget's work. I will return to the influence of Lorenzen and the Erlangen School when I will discuss the importance of the Signific Movement in the next section but let me just note here that the idea of dialogue logic was not yet present in his work and would only be developed later in collaboration with Kuno Lorenz. All that being said, it is surely no exaggeration to claim that Apostel was solely responsible for the introduction of this important variant of constructive logic in Belgium.

In later years he would continue to visit the center for short(er) stays and it is worth mentioning that he produced some very important contributions during those years, especially in the second half of the sixties, see, e.g., Apostel [12–15], related to the work of Piaget.

---

[10]All references can be found at: http://www.fondationjeanpiaget.ch/fjp/site/bibliographie/index.php.

[11]It is worth mentioning that in those early years Apostel collaborated, among others, with a young mathematician, namely, Benoit Mandelbrot before he became famous with his research on fractal phenomena.

The reader must have noticed that these three major influences all occurred rather early in his career. One should obviously not derive from this observation that in later years he was no longer influenced by other philosophers. But Perelman, Carnap and Piaget became the background for this future work and thus one can state that these three remarkable scholars had a life-long impact on his thinking. As I wrote in the introduction, we have now finally reached the stage where I can situate and explain the importance of the Signific Movement. So let us have a closer look at this curious movement that is all too often ignored in the history of philosophy, logic and mathematics.[12]

The Signific Movement

Its Introduction

It is a curious phenomenon that, on the one hand, the number of publications about this movement is not small (though not very large either), yet on the other hand it remains largely unknown in the broader philosophical field. Nevertheless its impact has been important, both in the Netherlands and in Belgium, and it is not easy to understand what happened to produce the situation we know today: important yet unknown. Before providing the necessary information about this movement, let me mention how Apostel came into contact with them by quoting once again the biography:

> In 1954 J. Piaget sent his *Traité de Logique* to Perelman, asking for his support against the severe attacks on it by E. W. Beth. Perelman proposed that I write a comment on it, which I did . . . and at about the same time, Piaget read "Logique et Preuve".[13] (46)

It is incidentally this comment by Apostel that led to an invitation by Jean Piaget to come to Geneva to Piaget's center. In order to situate Beth in this context, it is necessary to go back one generation and mention the original founders of the Signific Movement ('Signifische Beweging' in Dutch). Or more precisely: that person who would be responsible for its genesis, though not herself a member: Victoria Lady Welby. Little information was available about this intriguing philosopher, outside of *academia*, who played such a vital role in philosophy in the second half of the nineteenth century. That however has now changed, due to the considerable efforts of Susan Petrilli who in her [16] has brought together not yet published manuscripts, letters and notes of Lady Welby, together with additional texts and documents demonstrating her intellectual influence. This has now been complemented with Petrilli et al. [17] where a group of authors, including myself, see Van Bendegem [18], has been invited to further reflect about this remarkable woman. I will not use the available space here to present an extensive biography but simply use a diagram, see below, from Petrilli's [16] book that speaks for itself.

---

[12]And I should include semiotics as well as their views are more often associated with that domain than with philosophy. There are some exceptions but few and far between such as Nöth [38] and Cobley [10].

[13]See Apostel [39].

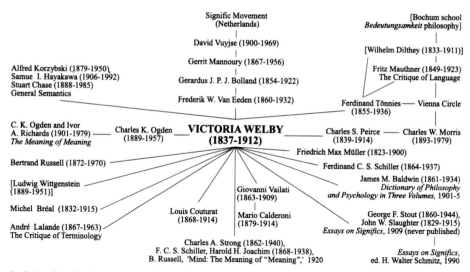

Irradiation of significs: Some important influences and connections, direct and indirect. The square brackets refer to relations hypothesized in the literature on Welby.

Although as stated she never held a position at a university, she was clearly in the center of an intellectual network, bringing together practically everybody of any importance in that period. A first interesting connection can be made here with (part of) the work of Irving Anellis, to whom this volume is dedicated, as it is sufficient to note that in the diagram many of the philosophers and mathematicians appear who have been his favorite objects of study. I just mention C. S. Peirce and Bertrand Russell as the most prominent examples.

Let me now focus on the upper middle branch, that includes the name of Gerrit Mannoury (1867–1956). But first a few words about Frederik van Eeden (1860–1932), who appears in between Mannoury and Welby. Although primarily known as a literary author in the Dutch-speaking world, there is also van Eeden—the philosopher, author of an essay, published in 1897, that bears some curious resemblances with Wittgenstein's *Tractatus* [50], *De redekunstige grondslag van verstandhouding* (*The logical foundation of communication* or *Logical basis of mutual understanding* or *Logical foundations of understanding*[14]) and van Eeden—the initiator of psychoanalysis in the Netherlands, and van Eeden—the political activist, responsible for the Thoreau-inspired 'Walden experiment' (see [19][15]).

---

[14]There exists, as far as I know, no English translation of van Eeden's treatise (and if absence of evidence is evidence of absence, it is badly needed). We do have at present a German translation, see van Eeden [40], entitled *Logische Grundlage der Verständigung*. As to the English titles, the first version is the one used by Gerrit Mannoury in his [41], the second version is Brouwer's, see Brouwer [42] and the third one is to be found in Bergmans [43], apparently to find a good translation is a non-trivial task.

[15]This book is unfortunately written in Dutch. I have not been able to find English papers and/or books that document this project in a substantial way.

He introduced Welby's, and thereby Peirce's ideas in the Netherlands. All these ideas were a source of inspiration for Mannoury. Apostel in the often mentioned autobiography states that:

> G. Mannoury showed me that one can look at mathematical foundations from an anthropological point of view. Mathematics is created by human beings, thus one must look at its psychological presuppositions. (20)

In more than one way, the ideas of Gerrit Mannoury were rather extraordinary for that time (and I would add perhaps today as well). Summarized in a few statements:

- He stressed, as the above quote underlines, that a constructivist approach is necessary to understand mathematics,
- He defended the idea of language as a kind of organism that should not be reduced to some ideal case but recognized as it is,
- For all forms of knowledge, including mathematics, he emphasized its social nature,
- Mathematics, science, and politics cannot be thought separately as there are mutual influences between all of them.

What these ideas led to can be perfectly well illustrated by the idea of the so-called 'language-steps'[16]:

> All language means, used by members of a society, can be divided up in language steps, that differ gradually from one another. This occurs on the basis of regularities related to the use and, above all, to the interconnections and resulting stability of the meaning of signs and speech acts. With every step these regularities become more important; in parallel run a stepwise social differentiation and stability of meaning. The resulting language steps serve specific communicative aims and the formulation of specific contents; furthermore they shape the social relations between individuals, who, while communicating, use a particular language step. (352, my translation).

The important and revolutionary implication is that the better people understand one another, e.g., one mathematician talking with another mathematician, the smaller the group will be because of all the requirements that need to be satisfied to make such a clear and unambiguous communication possible. To understand these requirements it is necessary to see the mathematical process as a genuinely social process.

To complete Mannoury's profile, one must add the fact that he was also politically active, being a communist—which incidentally strengthens the similarities between him and Apostel, who also promoted forms of political activism as an outcome of philosophical reflection. It becomes obvious that, firstly, Mannoury was in a sense a counterpart to Victoria Lady Welby but this time within *academia* and on the continent and, secondly, why Apostel was so impressed by this fascinating personality. Strangely enough, there are not that many direct traces to be found in the works of Apostel that refer

[16] I quote from the excellent historical survey, Schmitz [44] as it still stands on its own as a presentation of Signific thinking in full detail. Unfortunately the book exists only in the original Dutch version. But then again, fortunately, this is not the sole contribution of Walter Schmitz to the study of Welby and Significs, see, e.g., Schmitz [45].

directly to Mannoury.[17] As the next subsection shows his interests went in a somewhat different direction.[18]

Its Almost Disappearance, Introducing the Erlangen School and Strict Finitism

However, in the years following his stay in Amsterdam, he showed more interest in the work of Evert Willem Beth (1908–1964) and L. E. J. Brouwer. Although apparently the collaboration with Beth was not always easy, witnessing this quote from the autobiography:

> ... I worked on the idea of proving the probability of consistency and completeness. I tried to explain this to Beth, who was not enthusiastic given his doubts about the foundations of probability. Probability theory came for him after deductive logic in the logical order, and was not to be used in the metatheory of the latter. (20)

To a certain extent, the consequence of this shift in interest was that the social element in Mannoury's thinking disappeared into the background. As far as Brouwer is concerned, I have tried to sketch this evolution to see the socially embedded, real-life mathematician turn into the abstract 'creative subject' of Brouwer, who develops the whole of mathematics in its (individual) head (see the already mentioned Van Bendegem [18]). Add to this the observation that his original ideas were not straightforwardly expressible in formal-logical terms, and it becomes understandable why Apostel, so deeply influenced by Carnap, continued his search for logical models, looking at the work of Beth, Brouwer, Lorenzen and the Erlangen School.

A rather neat way to see this shift in research and to prepare the ground for the last section of this paper, I will have a look at a few elements that have to do with the teaching of Leo Apostel. We, as his students, were extremely lucky that the course notes were always up to date. In that sense, they were and are a good reflection of his research themes that occupied him in parallel with this teaching. It also means that we, at the undergraduate level, were very well informed of what the new developments in the field were. In the first part of the handbook, written by Apostel and Vandamme, see [21], the last chapter, entitled 'Logika en natuurlijke taal' ('Logic and Natural Language'), was out of the ordinary. In this short chapter attention is paid to the problem of the implication connective. The final section of this chapter carries the title 'Taal en logika als een dualistisch systeem' ('Language and logic as a dualistic system'). The authors emphasize that language is always a matter of a listener and a hearer or speaker and audience, and that language and

---

[17]Interestingly enough in Perelman and Olbrechts-Tyteca [5], p. 202, there is a single reference to Mannoury.

[18]An additional element that needs to be mentioned but that will not be further discussed in this paper is the influence of Mannoury's Marxist views. Although he was not deeply impressed by Hegel's work, nevertheless he was sensitive to the idea of dialectical processes with the intrinsic inconsistencies that are part and parcel of those processes. It would be a bold statement to claim that here lies Apostel's early interest in the formalization of such processes, using what we today know as paraconsistent logics in all their variations.

logic should reflect this duality. Within logic they see dialogue logic as an answer to this problem of duality. It is worthwhile to present one paragraph *in extenso*:

> Thinking in practice runs mainly, if not entirely, along the lines of interaction with others and oneself. Hence the question should be asked whether we would not obtain a more adequate and a more efficient logic (. . .) by looking at logic from this interactional perspective. We see several important initiatives in this direction, such as Hamblin (1970) [22], Lorenzen (1959) [23],[19] Lorenz (1968) [24]. [my translation[20]]

Although there is a clear connection to Signific thinking, especially the duality of speaker and hearer, no names such as Mannoury are mentioned.

In 1980 a second part was published by both authors, *Formele Logika. Deel II: Niet-Klassieke Systemen en Toegepaste Logika* (*Formal Logic. Part II: Non-Classical Systems and Applied Logic* [49]). Striking in this first edition, is the absence of dialogic(al) logic in whatever form. But in the first part of the book, entitled 'Pedagogiek, logika en aktietheorie' ('Pedagogy, logic and action theory'), attention is paid to the famous INRC-group[21] that came out of the Piaget studies, which was connected to the first designs of an action logic by Henrik von Wright.[22] Apart from an intriguing chapter on logic, language and music, in the last chapter 'Naar een formele dialektiek voor de sociale wetenschappen' ('Towards a Formal Dialectic for the Social Sciences') the authors deal with contradictions and introduce the idea of a formal dialectics. Names involved with formal logic that are mentioned here, are: Leo Apostel himself, Stanisław Jaskowski, and Richard Routley. Again no reference to Significs.

Finally, in 1986, the third edition appeared, which was seriously revised and extended. One of the major changes is that now a separate section of the book, almost 50 pages, is completely devoted to dialogue logic. Among the references one finds the following books and papers: Else Barth [20], "A New Field: Empirical Logic", L. Carlson [8], *Dialogue Games*, an extensive set of books and papers by J. Hintikka ranging from 1957 to 1981, and Erik Krabbe [25], "Formal Systems of Dialogue Rules". Strangely enough, now names such as Lorenzen and Lorenz are missing, but the basic description of what dialogue logic is, is attributed to Krabbe, but the main reason is that his presentation is transparently written, ideally suited for a handbook for students. Further editions of the book were done by Fernand Vandamme on his own, since at that time Leo Apostel had already retired and spent his time on other philosophical projects.

---

[19]This should be Lorenzen [23], but the mistake is easily explained as the book contains the proceedings of a conference that was held in Warsaw in 1959.

[20]The original text: "Het denken in de praktijk verloopt hoofdzakelijk, zo niet volledig, langs de lijnen van interaktie met anderen en zichzelf. Vandaar dat de vraag zich stelt of we niet een veel adequatere en efficientere logika (. . .) zouden verkrijgen, door de logika vanuit dit interaktioneel aspekt te benaderen. We zien verschillende belangrijke initiatieven in die richting: bijv. Hamblin (1970) [22], Lorenzen (1959) [23], Lorenz (1968) [24]."

[21]The symbols stand for: identity (I), negation (N), reciprocation (R), and correlation (C). These operations can be combined and, given specific relations between the four elements, such as $N = RC$, a group structure results, more precisely a Klein 4-group.

[22]The best known work of course is Von Wright [46]. See also Apostel [47], where he further developed this action logic (of course in combination with several other approaches).

This development away from Significs and into different forms of dialogue/dialogical logic, present in the handbooks, is also confirmed by the research itself. The already mentioned Apostel [6] is a perfect example as, one, this article is a presentation of a general theory of argumentation, and, two, it appeared in a volume, edited by Barth and Martens. To complete the picture, let me add that (1) Else Barth is a pupil of Beth, (2) Erik Krabbe a pupil of Barth, (3) Beth, together with Brouwer and A. Heyting had started the series *Studies in Logic and the Foundations of Mathematics*, published by Elsevier, that would become a standard for many years, and (4) the journal *Synthese* would change hands from Mannoury (and others) to Jaakko Hintikka around 1963–1964.[23] Put all these elements together and it becomes understandable why (unfortunately) the Signific Movement slowly disappeared into the background. There is however one element that stands apart and that is the discussion about strict finitism.

One of the members was the mathematician-statistician David Van Dantzig (1900–1959).[24] One of his particular interests and inspired by signific ideas was the idea of strict finitism, i.e., the view that not only the actual infinite has no role to play in mathematics—Brouwer's intuitionism had already undertaken that task and shown the possibility—but also the potential infinite should be eliminated. That being said, he wrote just one paper in 1956 [26] on the topic with the absolutely clear title: is $10^{10^{10}}$ a natural number? Of course, the answer is supposed to be negative. Although strict finitism was and still is not a particularly well received view in the foundational studies in mathematics, van Dantzig apparently had a dramatic effect on Apostel and I quote once more from the biography (note the use of the word 'infected'):

> ... and met the famous mathematician and radical finitist D. Van Dantzig. I was infected by finitism. (20)

I will not go into the details about strict finitism itself—I refer the reader to Van Bendegem [27] and [28]—but use the available space to sketch a curious connection between this topic and Irving Anellis. My Ph.D. started out with Leo Apostel as supervisor and, as I was already interested in the specific status of infinity in mathematics, leaning strongly towards strict finitism, Apostel encouraged me to take it as my subject. It is only at that time that I got to know the Signific Movement and, more particularly, van Dantzig. Literature about strict finitism was quite rare so my search (in those days) brought me almost everywhere. It may sound like an oddity but in Belgium exactly one philosopher, namely Herman Roelants, a mathematician-philosopher of the Catholic University of Louvain (Katholieke Universiteit Leuven) had ensured that the library had a subscription to the curious journal *Philosophia Mathematica*. And so Joon Fang appeared in my life. I ended up on the editorial board of the journal and we communicated about strict finitism and similar themes. When I read the obituary in *Modern Logic* of Fang, written by Irving Anellis, see Anellis [29], the circle was made round. Incidentally, there

---

[23]Let me add here the importance of the single issue that appeared in 1956, volume 10, number 1 that would deal with all the issues for 1956 up to 1958. 61 (!) papers, all of them the outcome of three conferences about significs held in the Netherlands in 1950 (in's Graveland), 1951 and 1952 (both in Amersfoort). Nearly all people mentioned in this paper were present at one or more of these meetings: Apostel, Mannoury, Piaget, Perelman, and Lorenzen.

[24]For his biography, see Albers [48], unfortunately only available in Dutch.

is also a connection with the Erlangen School, although not a positive one:

> Fang … came to dislike strongly his old friend Friedrich Kambartel, whom he had known when
> he [Fang] gave his first colloquium lecture at Münster in 1960 and Kambartel was a young student
> at Münster of Frege expert Friedrich Kaulbach and of Heinrich Behnke …. (147)

Friedrich Kambartel, later on, became a member of the Erlangen School!

As I have now made the transition from yesterday to today, at least for my own personal situation, let me broaden the picture and have a brief look at the legacy of Leo Apostel today.

Present-Day Situation or the Legacy of Leo Apostel

In this final section I briefly list a number of groups that have come about due to the efforts, usually in collaboration with others of course, of Leo Apostel. First and foremost, the *Centre National de Recherches de Logique—Nationaal Centrum voor Navorsingen in de Logica* (CNRL-NCNL, see http://www.logic-center.be/) must be mentioned. This center for research in logic was founded in 1955 by Philippe Devaux, Joseph Dopp, Robert Feys, Maximilien Freson, Chaïm Perelman, Franz Crahay, and Leo Apostel and still exists today. In principle, it brings together all logicians (in a fairly extensive meaning of the term) in Belgium. This center is also responsible for the publication of the journal *Logique & Analyse*. The journal was founded in 1958 and one of the editors-in-chief was Paul Gochet from the Université de Liège, until I took over about 20 years ago. The journal still adheres to its original aim, namely: "*Logique et Analyse* itself subscribes to no particular logical or philosophical doctrine, and so is open to articles from all points of view, provided only that they concern the designated subject matter of the journal." By 'designated subject matter' is meant the logical-analytical analysis of philosophical problems, understood as broadly as possible. Over the past 50 years, a wide range of papers on dialogue logic and argumentation theory (especially in the early period) have been published. I just mention the special issue of 1963 where the four issues of that volume were published as a single book on argumentation, containing, e.g., "Methodisches denken" by Paul Lorenzen.

Secondly comes the *Communication & Cognition* group, originally founded in 1967 by Marc De Mey and Fernand Vandamme. Its aim was clearly interdisciplinary: to bring together researchers from as many disciplines as possible to collaborate on all kinds of philosophical questions, relevant to their discipline, such as measurement, observation, theory formation, …. Although this group was mainly based at the University of Ghent, its influence was really world-wide as over the years they organized a series of conferences: *The Cognitive Viewpoint* (1977), *Theory of Knowledge and Science Policy* (1979), *Logic of Discovery and Logic of Discourse* (1982), *Evolutionary Epistemology* (1984), *George Sarton Centennial* (1984, organized in association with 4S—Society for the Social Studies of Science—and EASST—European Association for the Study of Science and Technology), *CC20* (1987), and *Argumentation, Logic and Cognition* (1989). They also published a journal *Communication & Cognition* and a book series, including the conference proceedings. At present, neither the group nor the journal exist anymore.

The department of philosophy at the University of Ghent also published its own journal, originally *Studia Philosophica Gandensia*, then *Philosophica Gandensia*, then *Philosophica* for short, still existing today, see http://logica.ugent.be/philosophica/, and that has published on topics related to the work of Apostel over the years. I just mention the three special issues on pragmatics and philosophy, edited by Apostel and Asa Kasher in 1981 and 1982 [30], including authors such as Daniel Vanderveken, Teun Van Dijk, and Ruth Manor.

Presently in Ghent the *Center for Logic and Philosophy of Science*, see http://logica.ugent.be/, is basically involved with research in logic and philosophy of science. Its director for many years was Diderik Batens who is now retired. Batens, a direct pupil of Leo Apostel, has continued his work. Although he was never particularly involved with dialogue logic and similar approaches, he remained close to Apostel's themes as he started out with doctoral work on the logic of induction, "Carnapian style", see Batens [31], moving to relevant logic, as developed by Alan Anderson and Nuel Belnap, which led him to the study of paraconsistent logic(s), working together with, e.g., Newton da Costa, Robert Meyer, and Graham Priest. Out of this resulted the grand research project dealing with so-called *adaptive logics*, see http://logica.ugent.be/adlog, as a research programme, where the basic idea is that, as long as there are no "problems", one can reason along the lines of classical logic, but, as soon as problems appear, e.g., in the form of inconsistencies, one switches to a paraconsistent logic. Worth mentioning is that in this work the name of Jaskowski reappears, thus making a connection with early work on dialogue logic. At present, the core figures of the Ghent research group are Joke Meheus and Erik Weber.

At the Vrije Universiteit Brussel, two groups are active. On the one hand, the *Center Leo Apostel*, CLEA, see http://www.vub.ac.be/CLEA/, led by its director Diederik Aerts, a physicist-philosopher. At present one of their main themes is the applicability of quantum mechanical models, outside of physics, e.g., in psychology or sociology. They still reflect the necessity of interdisciplinary research.

The second group is the *Center for Logic and Philosophy of Science*, see http://www.vub.ac.be/CLWF/, led by myself. The scope of this group ranges from logic, philosophy of mathematics, philosophy of science and the relations between science and art. My own research, as already indicated, started with strict finitism but also with an interest in dialogue logic. In 1985, I was the editor of a special issue on that topic, containing contributions from Paul Lorenzen, Friedrich Gethmann, Kuno Lorenz, Else Barth, Richard Sylvan, Jaakko Hintikka and myself, see Van Bendegem [32]. Later on, a second major topic became the philosophy of mathematical practices. As I have argued in Van Bendegem [18], I see this research as a continuation of the signific project, of course reinterpreted and adapted to present-day circumstances but in spirit still very close to Mannoury and Apostel. As far as the next generation is concerned and related to this particular project, let me just mention Bart Van Kerkhove, Patrick Allo, Karen François, and Kathleen Coessens.

In Conclusion

Perhaps it is a statement too trivial to make but it must be clear that what I have presented here is an initial and incomplete presentation of the importance of the work of Leo Apostel (with in addition a focus on the Signific Movement and the Erlangen School). A fuller story must of course include what has happened at the already mentioned Katholieke Universiteit Leuven and, to a lesser extent, at the University of Antwerp. And that would only cover the Dutch-speaking part of Belgium. I have only mentioned some of the researchers in the French-speaking part of the country and Brussels, namely obviously Chaïm Perelman and Paul Gochet but many need to be added, such as Jean Ladrière, Michel Meyer (a direct pupil of Perelman), Marcel Crabbé and Michel Ghins. As it happens, Gochet in Gochet [33] presented a first survey of philosophy in Belgium for the second half of the twentieth century. In short, most of the work remains to be done. However, I do believe that the overview presented here makes clear why certain topics are now important topics in parts of the philosophical research in Belgium and how Leo Apostel can be seen as directly responsible for this development. It is hence befitting that this legacy should be acknowledged as I hope to have done in this first attempt.

References

1. L. Apostel, *Matière et Forme. Introduction à une épistémologie réaliste*, 2 vols (Centre de Recherches de Logique, Brussels, 1974)
2. L. Apostel, Measure in Brouwerian algebras. J. Symb. Log. **20**(1), 90–91 (1955)
3. L. Apostel, Can metaphysics be a science? Studia Philosophica Gandensia **1**, 7–95 (1963) (reprinted in L. Apostel, *Zoektocht naar eenheid in verscheidenheid (Search for Unity in Diversity)* (VUB-PRESS, Brussels, 2002)
4. L. Apostel, *Natuurfilosofie. Voorbereidend werk voor een op de fysica gebaseerde ontologie* (Natural philosophy. Preliminary work for an ontology based on physics) (VUBPRESS, Brussels, 2000)
5. C. Perelman, L. Olbrechts-Tyteca, *Traité de l'argumentation. La nouvelle rhétorique* (Editions de l'ULB, Brussels, 1958) (English translation by J. Wilkinson, P. Weaver, *The New Rhetoric. A Treatise on Argumentation* (University of Notre Dame Press, Notre Dame, 1969))
6. L. Apostel, Towards a general theory of argumentation, in *Argumentation: Approaches to Theory Formation*, ed. by E.M. Barth, J.L. Martens (Benjamins, Amsterdam, 1982), pp. 93–122
7. F. Vandamme, R. Pinxten (eds.), *The Philosophy of Leo Apostel*, 3 vols (Communication & Cognition, Ghent, 1989)
8. L. Carlson, *Dialogue Games* (Reidel, Dordrecht, 1983)
9. R. Carnap, *Der Logische Aufbau der Welt* (Felix Meiner Verlag, Leipzig, 1928) (English translation by R.A. George, *The Logical Structure of the World. Pseudoproblems in Philosophy* (RKP, London, 1967))
10. P. Cobley (ed.), *The Routledge Companion to Semiotics and Linguistics* (Routledge, London, 2001)
11. P. Lorenzen, *Einführung in die operative Logik und Mathematik* (Springer, Heidelberg, 1955)
12. L. Apostel, Psychogenèse et logiques non classiques. In *Psychologie et Epistémologie Génétique. Thème Piagétiens, hommage à Jean Piaget* (Dunod, Paris, 1966), pp. 95–106
13. L. Apostel, Epistémologie de la linguistique. In: *Logique et Connaissance Scientifique. Encyclopédie de la Pléiade* (Gallimard, Paris, 1967), pp. 1055–1096
14. L. Apostel, Logique et dialectique. In: *Logique et Connaissance Scientifique Encyclopédie de la Pléiade* (Gallimard, Paris, 1967), pp. 357–374
15. L. Apostel, in *Syntaxe, Sémantique et Pragmatique*. Logique et Connaissances Scientifique. Encyclopédie de la Pléiade (Gallimard, Paris, 1967), pp. 290–311

16. S. Petrilli, *Signifying and Understanding: Reading the works of Victoria Welby and the Signific Movement* (Mouton de Gruyter, Berlin, 2009)
17. S. Petrilli, F. Nuessel, V. Colapietro (guest editors), On and beyond significs: centennial issue for Victoria Lady Welby (1837–1912). Spec. Issue Semiotica **196**(1/4) (2013)
18. J.P. Van Bendegem, Significs and mathematics: creative and other subjects. Semiotica **196**(1/4), 307–323 (2013)
19. M. Mooijweer, *De Amerikaanse droom van Frederik van Eeden (The American Dream of Frederik van Eeden)* (De Bataafsche Leeuw, Amsterdam, 1996)
20. E. Barth, A new field: empirical logic. Synthese **63**(3), 375–388 (1985)
21. L. Apostel, F. Vandamme, *Formele Logika. Deel I: Klassieke Systemen* (De Sikkel, Kappelen, 1975)
22. C.L. Hamblin, *Fallacies* (Methuen, London, 1970)
23. P. Lorenzen, Ein dialogisches Konstruktivitätskriterium, in *Infinitistic Methods*. Proceedings of the Symposium on Foundations of Mathematics (Warszawa 1959) (Pergamon, Oxford, 1961) , pp. 193–200 (reprinted in P. Lorenzen, K. Lorenz (eds.), *Dialogische Logik* (Wissenschaftliche Buchgesellschaft, Darmstadt, 1978), pp. 9–16)
24. K. Lorenz, Dialogspiele als semantische Grundlage von Logikkalkülen. Archiv für mathematische Logik und Grundlagenforschung **11**, 32–55 and 73–100 (1968) (reprinted in P. Lorenzen, K. Lorenz (eds.), *Dialogische Logik* (Wissenschaftliche Buchgesellschaft, Darmstadt, 1978), pp. 96–162)
25. E. Krabbe, Formal Systems of Dialogue Rules. Synthese **63**(3), 295–328 (1985)
26. D. Van Dantzig, Is $10^{10^{10}}$ a natural number? Dialectica **9**(3/4), 273–278 (1956)
27. J.P. Van Bendegem, Strict finitism as a viable alternative in the foundations of mathematics. Logique et Analyse **37**(145), 23–40 (1994) (date of publication 1996)
28. J.P. Van Bendegem, A defense of strict finitism. Constructivist Foundat. **7**(2), 141–149 (2012)
29. I. Anellis, Joon Fang of Jaean – a retrospective. Mod. Log. **3**(2), 145–155 (1993)
30. L. Apostel, A Kasher (eds.), Pragmatics and Philosophy. Three Spec. Issues of Philosophica **27–28** (1981), **29** (1982)
31. D. Batens, *Studies in the Logic of Induction and in the Logic of Explanation: Containing a New Theory of Meaning Relations* (De Tempel, Brugge, 1975)
32. J.P. Van Bendegem (ed.), Recent Developments in dialogue logic. Spec. Issue Philosophica **35**, 113–134 (1985)
33. P. Gochet, Belgique, in *La philosophie en Europe*, ed. by R. Klibansky, D. Pears (Gallimard, Paris, 1993), pp. 101–123
34. L. Apostel, J. Van Der Veken, *Wereldbeelden (World Views)* (DNB/Pelckmans, Antwerp, 1991)
35. L. Apostel, D. Aerts, B. De Moor, *Cirkelen om de wereld: concrete invullingen van het wereldbeeldenproject (Circling Around the World: Concrete Proposals for the World View Project)* (DNB/Pelckmans, Antwerp, 1994)
36. L. Apostel, D. Aerts, B. De Moor, S. Hellemans, E. Maex, H. Van Belle, J. Van Der Veken, *Worldviews: From Fragmentation to Integration* (VUBPRESS, Brussels, 1994)
37. S.J. Gould, *Rocks of Ages: Science and Religion in the Fullness of Life* (Ballantine, New York, 1999)
38. W. Nöth, *Handbook of Semiotics* (Indiana University Press, Bloomington & Indianapolis, 1990)
39. L. Apostel, Logique et Preuve. Methodos **5**, 279–321 (1953)
40. F. Van Eeden, *Logische Grundlage der Verständigung – Redekunstige grondslag van verstandhouding* (Wilhelm H. Vieregge, H. Walter Schmitz and Jan Noordegraaf, Hrsgs.) (Franz Steiner Verlag, Stuttgart, 2005)
41. G. Mannoury, A concise history of significs. Methodol. Sci. **2**, 171–180 (1969)
42. L.E.J. Brouwer, Synopsis of the signific movement in the Netherlands. Prospects of the signific movement. Synthese **5**(4/5), 201–208 (1946)
43. L. Bergmans, Gerrit Mannoury and his fellow significians on mathematics and mysticism, in *Mathematics and the divine. A historical study*, ed. by T. Koetsier, L. Bergmans (Elsevier, Amsterdam, 2005), pp. 549–568
44. W.H. Schmitz, *De Hollandse Significa. Een reconstructie van de geschiedenis van 1892 tot 1926 (Dutch Significs. A Reconstruction of its History from 1892 to 1926)* (Van Gorcum, Assen, 1990)
45. W. H. Schmitz (ed.), V. Welby, *Significs and Language* (John Benjamins, Amsterdam, 1985)
46. H. Von Wright, *Norm and Action. A Logical Enquiry* (Routledge & Kegan Paul, London, 1963)
47. L. Apostel, *Communication et Action* (Communication & Cognition, Ghent, 1979)

48. G. Alberts, *Twee geesten van de wiskunde: biografie van David van Dantzig (Two Spirits of Mathematics: Biography of David van Dantzig)* (CWI, Amsterdam, 2000)
49. L. Apostel, F. Vandamme, *Formele Logika. Deel II: Niet-Klassieke Systemen en Toegepaste Logika* (3rd and extended edition, 1986) (De Sikkel, Kappelen, 1980)
50. L. Wittgenstein, *Tractatus Logico-Philosophicus* (Kegan Paul, London, 1922)

J.P. Van Bendegem (✉)
Vrije Universiteit Brussel, Brussels, Belgium

Center for Logic and Philosophy of Science,
Universiteit Gent, Ghent, Belgium
e-mail: jpvbende@vub.ac.be

Tarski's Recantation: Reading the Postscript to "Wahrheitsbegriff"

Philippe de Rouilhan

Abstract Tarski's postscript (Ps) to the second, German edition of his famous work on the concept of truth (Wb, for "Wahrheitsbegriff"), began by a recantation about the theory of semantical categories. In Wb he had unreservedly supported this theory, in Ps he rejected it. But there is strong evidence that something deeper had happened, beyond this avowed change of opinion and irreducible to it, something unavowed, perhaps unthought. I identify this deeper change as an abandonment of the universalism inherited from the founding fathers of modern logic, to which Tarski had previously remained faithful. I argue that this abandonment was, beyond all mathematical proof, a matter for philosophy.

Keywords History of logic · Logical universalism · Philosophy of logic · Tarski · Truth

Mathematics Subject Classification 00A30 · 01A60 · 03A05

Tarski's Profession of Faith About the Theory of Semantical Categories

In his 1933 masterpiece on "The concept of truth in languages of deductive sciences" [25], Tarski limited his investigations to languages satisfying the basic principles of the "theory of semantical categories".[1]

The theory dates back to Husserl, who, in the Fourth Logical Investigation, was the first to have written of "semantical categories (*Bedeutungskategorien*)":

> [On them] a large number of *a priori* laws of meaning rest, laws which abstract from the objective
> validity [...] of such meanings. The laws, which govern the sphere of complex meanings, and
> whose role it is to divide sense from nonsense, are not yet the so-called laws of logic in the pregnant

---

The present paper is a revised edition of [20]. The general line is the same as before, but some substantial changes appeared to be needed for the sake of clarity and persuasiveness. Thank you to Serge Bozon and François Rivenc for their stimulating comments. Thank you also to Claire O. Hill for her linguistic help at an earlier stage of the work. All awkward turns of phrase are of my own invention.

[1] The French edition of Tarski's work, in [28], vol. I, provides a seemingly questionable translation of the original, Polish edition [25], but usefully points out the changes introduced in the subsequent, German [26] and English [27] editions. There is also the second, revised edition [29] of [27], with the same pagination and a wonderful analytical index.

© Springer International Publishing Switzerland 2016   119
F.F. Abeles, M.E. Fuller (eds.), *Modern Logic 1850-1950, East and West*, Studies in Universal Logic, DOI 10.1007/978-3-319-24756-4_5

sense of this term [...] and the name of "laws of pure logical grammar"[2] can be justifiably given [to them]. [10, 2d edn, vol. II, pp. 294–295], [11, vol. II, pp. 49–50]

In his article, regarding the priority I mentioned, Tarski did give Husserl's name[3] without providing any further details, but in his eyes that reference could only be of historical value, the acknowledgment of theoretical debt being directed rather to Lesniewski [13, 14],[4] who had introduced the notion of semantical category into investigations of the foundations of the deductive sciences, and to Russell [22],[5] the inventor of the theory of types. Tarski noted that, from the formal point of view, the theory of semantical categories could be considered to have developed out of the "simple" version of the theory of types given by Chwistek [4] and Carnap [3].[6] Historically, however, unlike the theory of types, the theory of semantical categories owed nothing to the question of paradoxes.

From the theory of semantical categories, Tarski retained the fact that the expressions of a scientific language are split up into mutually exclusive and collectively exhaustive categories. Two expressions are of the same category if, and only if, they can be substituted for each other in every (open or closed) sentence ("sentential function") *salva congruitate*. This relation of substitutivity is an equivalence relation and the categories in question are the corresponding equivalence classes. It was therefore a matter, strictly speaking, of syntactic categories (or categories of expression) rather than semantical categories (or categories of meaning).

According to this theory, to verify that two expressions are of the same category, it suffices to verify that they can be substituted for each other in *some* sentence *salva congruitate*. This is what Tarski called "the first principle of the theory of semantical categories".[7] This principle says, in short, that, if two expressions can be substituted for each other *salva congruitate* in *some* sentence, then they can in *all* sentences. In particular, the category of a variable occurring in a sentence is not only included, by definition, in the class of all the expressions that can be substituted for it in this sentence *salva congruitate*, but even more, according to the first principle, it is identical to it.

Tarski stated that, "from an intuitive standpoint, [this first principle] is indubitable".[8] More generally, about the theory of semantical categories, one page before, he had

---

[2]I correct Finlay's "logico-grammatical laws", which fail to allow for the "rein" of "reinlogisch grammatische [Gesetze]".

[3]No change in subsequent editions, see [26, p. 335]; [27 and 29, p. 215].

[4]See in particular what Tarski said again of the system of Lesniewski in [26, p. 328, n. 56] (but deleted in [27 and 29, p. 210, n. 2]) and [26, p. 338, n. 65] (slightly corrected in [27 and 29, p. 218, n. 2], to allow for the aforesaid deletion).

[5]No change in subsequent editions, see in particular [26, p. 285, n. 18, and p. 289, n. 20], [27 and 29, p. 170, n. 1, and p. 173, n. 2].

[6]No change in subsequent editions, see [26, p. 335, n. 60], [27 and 29, p. 215, n. 1].

[7]No change in subsequent editions, see [26, p. 336], [27 and 29, p. 216].

[8]See [25] as translated in [28, vol. I, p. 216]: "Du point de vue intuitif la réponse est indubitable". In the subsequent editions, the formulation would be more cautious, see [26, p. 336], [27 and 29, p. 216]: "From the standpoint of the ordinary usage of language, [the first principle] seems much more natural [than its negation]."

adamantly stated:

> [T]he theory of semantic categories penetrates so deeply into our fundamental intuitions regarding the meaningfulness of expressions, that it is scarcely possible to imagine a scientific language in which the sentences have a clear intuitive meaning but the structure of which cannot be brought into harmony with the above theory.[9]

Tarski's Recantation

In 1935, Tarski published a revised version of his article in German, entitled "The concept of truth in formalized languages" ("*Der Wahrheitsbegriff in den formalisierten Sprachen*") [26], and added an important postscript to it. In the latter, he first recalled the position that he had espoused in the original, Polish version and he maintained in the German version (postscript not included), regarding the theory of semantical categories:

> In writing the present article I had in mind only formalized languages possessing a structure which is in harmony with the theory of semantical categories and especially with its basic principles. This fact has exercised an influence on the construction of the whole work and on the formulation of its final results. [26, p. 393], [27 and 29, p. 268]

Then, after having cited *in extenso* the profession of faith related above, Tarski made this astonishing recantation:

> Today I can no longer defend decisively the view I then took of this question. In connexion with this it now seems to me interesting and important to inquire what the consequences would be for the basic problems of the present work if we included in the field under consideration formalized languages for which the fundamental principles of the theory of semantical categories no longer hold. (*ibid.*)

At first sight, the postscript is a dozen page completion of this program. But even more astonishing than the recantation with which it begins are Tarski's results. For, upon close inspection, those results are not the ones he should have arrived at owing only to his avowed change of opinion regarding the theory of semantical categories. Something else must have happened, something unavowed, perhaps unthought. But what? I contend that it was a matter of a change in the idea of logic itself.

One needs to compare two texts: one, imaginary, the postscript that the stated change of position alone would have led Tarski to write; the other, real, the one Tarski actually did write. It will then be possible to identify the change in question. It corresponds to what I consider to be the essential turning point in the history of modern logic, which I believe I can date from the 1930s: an abandonment of the universalism of the founding fathers, Frege and Russell, to which Tarski had remained faithful in the Polish, and again in the German, version of the paper (minus the postscript) and the adoption of a new idea of logic, the dawning of a new era, soon ready to fall, beginning in the 1950s, for better or for worse, under the influence of what has been called the "model-theoretical spirit".

---

[9]No change in subsequent editions, see [26, p. 335], [27 and 29, p. 215].

What Languages Complying with the Theory of Semantical Categories Are Taken into Account in "Wahrheitsbegriff" (Wb), and Examples of Language Not Complying with This Theory that Tarski Could Have Given in the Postcript (Ps)

In "Wahrheitsbegriff" itself (henceforth Wb), as languages complying with the basic principles of the theory of semantical categories, Tarski essentially took into consideration only languages grounded on the extensional, simple theory of types, that is languages obtained from the language of this theory or from a fragment[10] of it by adding extra-logical constants, or, more precisely, individual names or predicates ("sentence forming functors") or functors ("name forming functors") whose semantical category is already represented by variables. In his explicit examples, there are no variables representing functors, and there is indeed no essential loss in excluding them once and for all from the considered theory of types itself. So I will limit myself to the theory so obtained (henceforth type theory). In the postscript (henceforth Ps) he limits himself to languages containing only the usual connectives and quantifiers, simple singular terms ("individual names") and variables representing them, predicates ("sentence-forming functors") and variables representing them.[11] [26, p. 394], [27 and 29, p. 268].

As in Wb, certain signs of the languages actually taken into consideration by Tarski in Ps have an *order*, but in a new, albeit related, sense. For individual names and the variables representing them, it was 1 in Wb, it is 0 in Ps. Up to that point, the difference is, insignificant. For the predicates and the variables representing them, however, the order is to be determined in an essentially new way. In Wb, the order of any argument of a unary predicate, for example, was determined by that predicate. Now, that is no longer the case. Correlatively, the order of that predicate was determined by that of any of its arguments. Now, that is no longer the case. To determine the order of this predicate, one must now take into account the orders of all its possible arguments. This is the least ordinal strictly higher than the orders of all those arguments. And the same is so, *mutatis mutandis*, for all predicates in general—and variables representing them.

One consequence of this new policy is that there is no longer anything keeping the order of a predicate, for example, from being infinite. It suffices that it admits arguments of finite, but not bounded, order. And the same is so for a variable representing predicates. At a certain point in Wb, Tarski brought up the idea of "expression of 'infinite order'", only to dismiss it straight away:

> Yet neither the metalanguage which forms the basis of the present investigations, nor any other of existing languages, contains such expressions. It is in fact not at all clear what intuitive meaning could be given to such expressions. [26, pp. 366–367], [27 and 29, p. 244]

---

[10]I borrow this term ("*Bruchstück*") from Tarski (see below, § 6, first quotation), who, if I am not mistaken, never explained what, exactly, he meant by it. In the present case, I think that a *fragment* of the language of the theory of types is a *(sub)language* obtained by removing all variables of finitely or infinitely many types.

[11]Whether in Wb or in Ps, the constraint put by Tarski on extra-logical constants, that their semantical category be represented by variables, seems uselessly restrictive in the case of predicates and functors. It could be replaced by the same constraint put only on their arguments.

Now, on the contrary, an expression (more precisely a predicate or a variable representing it) can well have an infinite, or transfinite, order.

The order of a language is in turn defined as the least ordinal strictly higher than the orders of all its variables. In Wb, a language could be of finite or infinite order. There was only one possible infinite order, corresponding to the least transfinite ordinal, ω. Now, on the contrary, a language can be of a transfinite order strictly higher than ω.

Tarski does not give any example of language that does not comply with the basic principles of the theory of semantical categories and is nonetheless as clearly understandable as the languages he envisioned in Wb. Let me propose examples as simple and natural as possible.

Let place ourselves in the framework of informal, usual set theory (*à la* Zermelo), and, for the sake of simplicity, suppose that individuals (*Urelemente*) form a set. From this point of view, individuals, classes of individuals, dyadic relations between individuals, etc., classes of classes of individuals, etc., type theory is about are nothing other than individuals (i.e., non-sets), sets of individuals, dyadic relations between individuals, etc., sets of sets of individuals, etc., respectively.[12] From the same point of view, individuals, sets of individuals, set of sets of individuals, etc. are what *monadic* type theory is about. Then, calling T_1 the set of individuals, T_2 the set of sets of individuals, T_3 the set of sets of sets of individuals, etc., $(T_n)_{n\geq1}$ is the *monadic* hierarchy of types (note that the T_n are pairwise disjoint); and, calling C_0 the set T_1, C_1 the set $T_1 \cup T_2$, C_2 the set $T_1 \cup T_2 \cup T_3$, etc., $(C_n)_{n\geq0}$ is the so-called *cumulative hierarchy of finite-order types* (this hierarchy is qualified as *cumulative* because C_n is always included in C_{n+1}; its types are qualified as *finite-order* in contradistinction with transfinite-order types that may be added to them, see below). According to the usual interpretation of usual set theory the two hierarchies are quite different.[13]

Now, consider the following language. Its primitive constants are usual connectors and quantifiers. Its primitive variables are, first, 'x^0', 'y^0', 'z^0', ... , which range over C_0 and are given order 0; second, 'x^1', 'y^1', 'z^1', ..., which range over C_1 and are given order 1; third, 'x^2', 'y^2', 'z^2', ... , which range over C_2 and are given order 2; etc. Its primitive sentences are obtained by concatenating any two variables of order m, n respectively, with $m > n$. So, 'z^2x^0' and 'z^2y^1' are sentences and so is 'y^1x^0', but it is not the case of 'z^2z^2',

---

[12]One will perhaps feel like protesting that this embedding of the intended universe of the language of type theory into that of the language of set theory ZFC is but a takeover by force, which reduces the infinite multiplicity of empty classes of the former universe to the one and only empty set of the latter. To which I will respond that it is rather the usual description of the intended universe of the language of type theory that distorts the situation when one thinks to have to distinguish empty classes themselves in parallel with type distinctions of expressions that refer to them. (On this kind of illusion, which dates back to Frege, see Rouilhan [18, Chap. 5] and [21].)

[13]The two hierarchies would become identical, if we identified individuals of usual set theory and their singleton. This move, however unusual, would be legitimate, and even advisable, for none of the objections put forward since Frege and Peano against identifying an object to its singleton applies to individuals, and thus, until further notice, Occam'z razor commends doing it. I will perhaps be told that all this is grist to Quine's mill for his well known explication of the notion of individual in set theory as object identical with its singleton (first published in [16]). Well, let me insist that, in contradistinction with Quine's, my explication gets cleared of all artificiality. Quine got his explication all by a trick made to comply with the maxim of *simplicity*; mine is the fruit of a *correction* made to comply with the maxim of *parsimony*.

thus 'z^2' is substitutable for 'y^1' in 'y^1x^0', but not in 'z^2y^1', *salva congruitate*. Hence it appears that this language does not abide by the first principle of the theory of semantical categories. Is there for all that anything incomprehensible in it?

And the same would be so for any language *grounded* on this language (in a sense analogous to the one explicated at the beginning of the present section), provided that the language under consideration contains variables of three different orders. Note also how natural it would be to contemplate adding variables 'x^ω', 'y^ω', 'z^ω', ... ranging over the union, C_ω, of all C_n for $n \geq 0$, and assigning them the infinite order ω; and even to envisage going farther into the transfinite.

Close-Up on Set-Theoretic Languages: How They Appeared in Ps and Why They Did Not Already Show Up in Wb

Languages so far considered in Ps can have variables of transfinite order, strictly higher than in Wb. Clearly, every concept expressible in a language grounded on type theory is also expressible in one of the languages considered in Ps, but the reciprocal does not hold. However, Tarski thinks it necessary to still introduce something new:

> [I]t seems to me [...] almost certain that we cannot restrict ourselves to the use of variables of definite order if we are to obtain languages which are actually superior to the previous languages in the abundance of the concepts which are expressible by their means, and the study of which could throw new light on the problems in which we are interested. [26, p. 396], [27 and 29, pp. 270–271]

Hence the introduction of "variables of indefinite order which, so to speak, 'run through' all possible orders" (*ibid.*). The flexibility involved in using the former variables (from the morphological point of view) would be such that, obviously, measures should furthermore be taken to prevent any possible return of the antinomies.

Immediately after having brought up the problem of antinomies, in a long footnote of the greatest significance, Tarski comes to "languages of another kind which constitute a much more convenient and actually much more frequently applied apparatus for the development of logic and mathematics" and are associated with the names of "Zermelo and his successors"[14] [26, p. 397, n. 106], [27 and 29, p. 271, n. 1]. He says that "it is but a step" from the languages he has just considered in Ps and these new ones (*ibid.*).

For the sake of an example, consider the language of the well-known set theory ZFC such as it appeared completely clearly for the first time in Skolem [23]. If the variables of that language are of "indefinite order", it is according to a notion of "order" other than the homonymous notion, whatever it may be, that is involved in the common view that the language of set theory ZFC is first-order and that so are its variables.[15] This other notion of order dates back to Mirimanoff [12], and Zermelo, who certainly had some intuition of it as early as 1908 [31], would only expressly use it beginning in 1930 [32]. Let me

---

[14]I.e. Skolem [24], mentioned by Tarski, but also Frænkel, Neumann, etc.

[15]ZFC may taken to be of first order *relatively to its intended universe* and there is nothing wrong about that. It is often taken to be of first order *in the sense of first-order logic*, but it is a mistake, for its intended universe should then be a set, while actually it is a proper class.

recall what this is a matter of. In the intended interpretation of set theory ZFC, the objects (individuals and sets) can be given their place in a term of a certain series, $(C_\alpha)_{\Omega(\alpha)}$, called the "cumulative hierarchy of types[16]", indexed on the class Ω of ordinals and definable within the theory itself by transfinite recursion in the following manner:

$C_0 =$ set of individuals,[17]
$C_\alpha = C_\beta \cup \mathfrak{P}(C_\beta)$, if α is a successor ordinal, of the form $\beta + 1$,
$C_\alpha = \cup_{\beta<\alpha} C_\beta$, if α is a limit ordinal.

In the intended interpretation of ZFC, for every x, there are ordinals α such that $x \in C_\alpha$. By definition, the order of x is the smallest of them. A variable can then well be said to be of "indefinite order", inasmuch as its domain of values is the whole hierarchy (or, more exactly, the union of cumulative types, the class of objects).

As a matter of fact, the language of set theory ZFC evidently complies with basic principles of the theory of semantical categories. As Tarski puts it more generally and more precisely in the footnote we are interested in, about formalized languages satisfying those principles and all of whose variables are of indefinite order, and by resorting to a classification he introduced in section 4 of Wb:

> From the formal point of view these languages are of a very simple structure; according to the terminology laid down in § 4 they must be counted among the languages of the first kind, since all their variables belong to one and the same semantical category.[18] [26, p. 397, n. 106], [27 and 29, p. 271, n. 1]

But, if the language of set theory ZFC, for the Tarski of Ps, satisfies so evidently the basic principles of the theory of semantical categories that this fact is not even worth outlining any argument to prove it, then this language is retrospectively conspicuous by its absence in Wb. For the Tarski of Wb, the inventory of languages satisfying the basic principles of the theory of semantical categories essentially retained languages grounded on type theory. The language of set theory ZFC could not occur in Ps,[19] and Tarski could not be unaware of this. If he did not conclude that his inventory was essentially incomplete, it is probably because he thought that, in spite of its nice syntactical simplicity, the so-called language of set theory ZFC was properly incomprehensible, that, in reality, it was only a "formal" language, a fake language, a non-language.[20]

---

[16]This notion of type must not be confused here with that of one or another theory of types. Set theory is not what one calls a "theory of types".

[17]It is supposed, here, that individuals are intended to form a set.

[18]Tarski distinguished four kinds of language, the first one being precisely defined as that of "languages in which all the variables belong to one and the same semantical category", [26, p. 340–341], [27 and 29, p. 220].

[19]Admittedly, the language of ZFC could be *reinterpreted* so as to appear as a first-order language grounded on type theory, but this reinterpretation would *not* be the *originally intended* interpretation, whose relation to type theory was exposed in section 4.

[20]One will think here of the influence of Kotarbinski and Lesniewski on the young Tarski, but also of the astonishing profession of faith cheerfully made by Tarski twenty years later in a cenacle of logicians: "I happen to be, you know, a much more extreme anti-Platonist. [...] However, I represent this very rude, naïve kind of anti-Platonism, one thing which I could describe as materialism, or nominalism with some materialistic taint, and it is very difficult for a man to live his whole life with this philosophical attitude, especially if he is a mathematician, especially if for some reasons he has a hobby which is called

In Ps, Tarski does not think it any longer. So, besides explicit recantation concerning the universal character of basic principles of the theory of semantical categories, there must have been another recantation, without a word, about the inventory of languages satisfying these principles. However, it is not this extra recantation that I had in mind above (see § 2), when I wrote that "[s]omething else must have happened, something unavowed, perhaps unthought", it is not so modest a clarification that is the aim of the present paper.

Universality of Logic in Wb and the Three Main Theses Proved There

In Wb, type theory constituted the impassable logical framework of the totality of languages envisaged. In other words, the language of this theory played the role of logically *universal* language that logical universalism assigns to the language of logic. In support of this interpretation, let me take this long quotation from a long footnote (yet another one!):

> [T]he language of some one complete[21] system of mathematical logic [...] can be regarded as a universal language in the sense that all other formalized languages – apart from 'calligraphical' differences – are either fragments of it, or can be obtained from it or from its fragments by adding certain constants, provided that the semantical categories of these constants are already represented by certain expressions of the given language.[22] [...] The only complete system of mathematical logic whose formalization – in contradistinction with the system of Whitehead and Russell, for instance – leaves nothing to be desired as for its preciseness is the system constructed by S. Lesniewski [...]. Unfortunately, this system, because of certain of its specific properties, constitutes, to my opinion, a too thankless object of study. [...] Formally speaking, it would even be difficult to subsume this system under the general characterization of formalized deductive sciences admitted [at the beginning of § 2]. [25], [26, p. 328, n. 56].

In the English edition, all of the passage concerning Lesniewski's system was replaced by the following:

> As such a language [viz., a universal one] we could choose the language of the general theory of sets which will be discussed in § 5, and which might be enriched by means of variables representing the names of two- and many-termed relations (of arbitrary semantical categories). [27 and 29, p. 210, n. 2]

---

set theory." (Transcript of remarks, ASL meeting, Chicago, Ill., April 29, 1965, Bancroft Library, p. 3, quoted by Mancosu [15])

[21]The term ("*vollständig*") is obviously not to be understood in the *model-theoretical* sense that it took on in Hilbert and Ackermann in 1928, then in Gödel in 1929 [5] and 1930 [6]. One could at the very most understand it in the "experimental" sense, as Herbrand said, that it could have from the point of view of the founding fathers. But, as the rest of the quotation shows clearly, it was first and above all here a matter of completeness of the expressive power, not of the proof-theoretical one [7, 8].

[22]In other words, which I am used to using, the given language is *logically* universal in the sense that everything expressible in any of the formalized languages envisaged is expressible, not necessarily within the given language *itself*, but within *the framework of* it, meaning here (to follow Tarski, but see above, n. 11) in some extension of it obtained by adding extra-logical constants whose semantical category is already represented by certain expressions of it.

The so-called "general theory of sets", which would appear in section 5 of Wb under the name of "general theory of classes" ("*allgemeine Klassentheorie*" in [26]), was actually nothing other than the monadic type theory.[23]

A few pages later, Tarski again wrote:

> The language of a complete system of logic should contain – actually or potentially – all possible semantical categories which occur in the language of the deductive sciences. Just this fact gives to the language mentioned a certain 'universal' character, and it is one of the factors to which logic owes its fundamental importance for the whole of deductive knowledge. [25], [26, p. 340], [27 and 29, p. 220]

These are essentially remarks that logical universalists like Frege or Russell could not have disowned.

For languages grounded on type theory, Tarski arrived at numerous results, the principal of which he summarized in the last section of Wb[24]:

> A. *For every formalized language of finite order a formally correct and materially adequate definition of true sentence can be constructed [. . .].*
> B. *For formalized languages of infinite order the construction of such a definition is impossible.*
> C. *On the other hand, even with respect to formalized languages of infinite order, the consistent and correct use of the concept of truth is rendered possible by including this concept in the system of primitive concepts of the metalanguage and determining its fundamental properties by means of the axiomatic method [. . .].* [26, pp. 390–391], [27 and 29, pp. 265–266].

In terms of what concerns us here, thesis B is the most interesting one. A language of infinite order has the same order and, in this sense, the same *essential logical richness* as the language, taken to be logically *universal*, of type theory, and this richness is *impassable*. A definition of truth for the language in question would be possible only in languages logically essentially richer than it, but there are *none*. It is for this reason, and for this reason only, that a definition of truth for it could be declared, as it was actually by Tarski in this thesis, to be purely and simply *impossible*.

An Imaginary Ps, with the Language of ZFC in the Role of Logically Universal Language and Three Main Theses Analogous to that of Wb

Let us go back to Ps. At the end of our favorite note, Tarski states:

> Our further exposition also applies without restriction to the languages which have just been discussed [namely the languages of Zermelo and his successors] [26, p. 39], [27 and 29, p. 271]

As for me, I shall keep to *these* languages (those of Zermelo and his successors), of which I have been using that of set theory ZFC as a paradigmatic example (see § 4).

---

[23]In my favorite terminology, to which, here as everywhere else, I strive to remain faithful, in no way is this theory a theory of *sets*, it is a theory of *classes*. I discussed the relationship between the theory of types and set theory and the proper use of set-theoretical terminology at length in [18, chap. VI]; see also [19, chap. II, sect. B].

[24]Without number in [26], numbered as § 6 in [27] and [29].

Imagine in the role of logically universal language, no longer the language of type theory, but that of set theory ZFC, with its variables of indefinite order; and, correlatively, in the role of object-languages, languages obtained from the language of set theory ZFC through (1) possibly, restriction of the domain of variables to some type, C_α, of the cumulative hierarchy, or, more generally, to some strict subclass of the class of objects, and (2) possibly, addition of extra-logical constants whose semantic category is represented by variables, that is constants of object, interpreted in the new domain of variables. Except for this change of logical framework and of object-languages, *mutatis mutandis*, partially repeat Wb.

Let me recall the following (highly ineffective) criterion of sethood, due to Mirimanoff [12], applicable to set theory ZFC: a class is a set if, and only if, the order of its members has an upper bound in the class of ordinals, or, as he put it, a *Cantorian bound*. I shall call the object-languages whose domain of variables is a set, and therefore whose possible values of variable have a Cantorian bound, languages of *relatively finite* order; I shall call the languages whose domain of variables is not a set, but a proper class, and therefore whose possible values have no Cantorian bound, languages of *absolutely infinite* order.[25]

The results at which one arrives (at the end of an unhindered voyage, following step by step, *mutatis mutandis*, that of Tarski in Wb, and which I shall forgo detailing here) are then the following, perfectly analogous to results A, B and C of Wb:

> A*. *For every formalized language of relatively finite order, a formally correct and materially adequate definition of true sentence can be constructed [. . .].*
> B*. *For formalized languages of absolutely infinite order the construction of such a definition is impossible.*
> C*. *On the other hand, even with respect to formalized languages of absolutely infinite order, the consistent and correct use of the concept of truth is rendered possible by including this concept in the system of primitive concepts of the metalanguage and determining its fundamental properties by means of the axiomatic method [. . .].*

As much as Thesis B above (see § 5), here Thesis B* is deserving of attention. A language of absolutely infinite order is logically essentially as rich as the language of set theory ZFC, taken to be logically *universal*. Like it, it therefore has an *impassable* essential logical richness. There is *no* language whose essential logical richness would be sufficient to enable one to define truth for the language in question in it. It is exactly for this reason that a definition of truth for such a language can be declared, as it actually is in this thesis, purely and simply *impossible*.

The Real Ps, with Its Two Main Theses Instead of Three

Interpreted in the way I have said (or even taken literally), the opening recantation in Ps leaves room for hope that, through a method analogous to the one he used in Wb within the logical framework of the type theory, Tarski might now establish results A*, B* and C* above, or some results of the same kind, within the new logical framework of set theory ZFC, or of some logical theory transgressing the basic prohibitions of type theory (or even of the theory of semantic categories). However, this is not at all what he does,

---

[25]Cantor, *1899*, called certain "multiplicities", such as that of ordinals, "absolutely infinite" [1].

or rather, since his remarks remain allusive and programmatic, it is not at all what he says one could do. The results he announces are *completely different* in style (I quote):

A. *For every formalized language a formally correct and materially adequate definition of true sentence can be constructed in the metalanguage [. . .] under the condition that the metalanguage possesses a higher order than the language which is the object of investigation.*
B. *If the order of the metalanguage is at most equal to that of the language itself, such a definition cannot be constructed.* [26, p. 399], [27 and 29, p. 273]

The decisive difference between these results and those one might expect to obtain is not evident at first sight in what they say, but it is in what they do not say, in what they no longer say. There is no longer anything in these results corresponding to thesis B of Wb, or to thesis B* of the imagined Ps, nor is there anything corresponding to thesis C of Wb, or thesis C* of the imagined Ps.

But one must go back to the considerations preparing the statement of results A and B in the real Ps. Among these, one finds this:

[I]t is *always* [italics mine, PR] possible to construct the metalanguage in such a way that it contains variables of higher order than all the variables of the language studied. The metalanguage then becomes a language of higher order and thus one which is essentially richer in grammatical forms than the language we are investigating. This is a fact of the greatest importance from the point of view of the problems in which we are interested. For with this the distinction between languages of finite and infinite orders disappears – a distinction which was so prominent §§ 4 and 5 [of Wb] and was strongly expressed in the theses A and B formulated in the Summary [at the end of Wb]. [26, pp. 397–398], [27 and 29, pp. 271–272]

The first sentence, in which I have italicized "always", is decisive. It shows the sense in which it is appropriate to understand thesis A of Ps. The first part of the thesis must be understood categorically, unconditionally. For *any* formalized language, one can construct a formally correct, materially adequate definition of the notion of a true sentence. The second part does not restrict the universality of the first part. It simply specifies what condition regarding the metalanguage is sufficient for such a construction to be possible there, but henceforth a metalanguage satisfying that condition *always* exists. Thesis B stipulates then that this condition is also necessary. Thus, it is not only in what the results announced in Ps do not say, but already in what they do say, that the rupture occurs.

As for what they do not say, it is appropriate to distinguish between theses B* and C* of the imagined Ps. Thesis C*, or some thesis of the same kind, is missing in the real Ps, but it is only through lack of interest. It remains quite true that one could legitimately and coherently introduce the concept of truth as an indefinable in a metalanguage of the same order as the object-language in keeping with the axiomatic method, but it is henceforth *always* possible to do better by defining this concept in some higher-order metalanguage. Thesis B*, on the contrary, or some thesis of the same kind, is absolutely ruled out. There is no longer any logically universal language. There is no longer any language whose essential logical richness is impassable. There is no longer any language for which undertaking to define the concept of truth runs up against pure and simple impossibility.

Objection Against Logical Universalism and Reply

In the reflections preparing the presentation of results A and B of the real Ps, Tarski makes the decisive statement that "it is *always* possible to construct a metalanguage [etc.]" in the false light of mathematical self-evidence. Let us look at the text again starting one sentence earlier:

> There is obviously no obstacle to the introduction of variables of transfinite order not only into the language which is the object investigated, but also into the metalanguage in which the investigation is carried out. In particular, it is always possible to construct the metalanguage in such a way that it contains variables of higher order than all the variables of the language studied. [26, p. 397], [27 and 29, pp. 271–272]

As if the introducing of variables of transfinite order beyond the variables of finite order, the only ones to be admitted in Wb, mathematically ensured the possibility of the construction in question, and one needed only observe that! But that is not the case. And as if with this introducing "the distinction between languages of finite and infinite orders disappear[ed]", as Tarski claims a little bit further on, and that no other distinction, if I understand him well, could play the same role in the new situation, and one needed only take note of this! But, once again, that is not the case. The imagined Ps shows that rather clearly. The distinction between the languages of relatively finite order and the languages of absolutely infinite order plays precisely this role. Variables of transfinite order are admitted there, and the definition of the concept of truth for a language of absolutely infinite order is not made possible for all that.

An objection may be made to that, which it is appropriate to examine now. I do not know to what extent Tarski would have been ready to endorse it. Given set theory ZFC, there is nothing to keep you from adopting the point of view of a more powerful set theory from which the language of the first theory would appear relative to a certain cumulative type, C_α. For that, it suffices that the axioms of both theories imply no difference between their sets of individuals, and that the axioms of the first theory imply exactly the existence of sets of all the orders lower than α, while those of the second theory imply, in addition to that, the existence of at least some set of order α. Thus, from the framework of your so-called logically universal, absolutely-infinite-order language of the first theory, you will have moved to a framework in which the language of this theory turns out to be of relatively finite order α, as you would put it, and obviously no longer has anything logically universal about it, and it *is* possible to define truth for it. Such is the objection.

But one should not play with words. The notion of order has surreptitiously changed. According to the definition of this new notion of order, given without further comment by Tarski at the end of our favorite note, about the languages of "Zermelo and of his successors", "the order of the language is the smallest ordinal number which exceeds the order of all the sets whose existence follows from the axioms adopted in the language". Strictly speaking, it is no longer a matter, as it was before, of the order of a *language*, independently of any axioms and rules of inference, but of the order of a language relative to such and such axioms and rules of inference, in other words, of the order of a *system*. Admittedly, if the first theory is of order α in the new sense, it can well be *reinterpreted* as a theory relative to C_α, but C_α is not its *originally intended* universe for all that. So the objection is off the point.

Objection

Let me play devil's advocate. The objection of the preceding section deserves to be taken more seriously, if its *underlying thesis* is that, *when speaking about individuals and sets in the framework of an axiomatic extension of ZFC, we are speaking about individuals and the only sets whose existence is implied by the axioms of this extension.* This thesis, which, by the way, puts to work no notion of order at all, is reminiscent of an application of Quine's criterium ontological commitment of a theory,[26] even though it does not exactly result from such an application. Tarski may have had something like it in mind when defining his new notion of order. Let me resume the discussion giving it a more accurate form and allowing for the underlying thesis in the presentation of the objection.

My view of things. — Let $\mathcal{L}$ be the language of ZFC, our logically universal language; U, its universe, definable *from within* as the class of objects by the predicate of self-identity; $\mathcal{T}_1$, set theory ZFC itself; $\mathcal{T}_2$, the theory obtained from set theory ZFC by adding the so-called axiom of inaccessibility, which affirms the existence of an inaccessible cardinal (in U). $\mathcal{L}$ is the common language of $\mathcal{T}_1$ and $\mathcal{T}_2$, and, in the transition from $\mathcal{T}_1$ to $\mathcal{T}_2$, its universe U does not change. The definition of truth for $\mathcal{L}$ (as for any language of absolutely infinite order) is impossible.

The objection. $\mathcal{L}$ Actually, there is no logically universal language at all. In the transition from $\mathcal{T}_1$ to $\mathcal{T}_2$, except for some inconsistency, the universe U of $\mathcal{L}$ changed. Before the transition, U is the set of individuals and the only sets whose existence is provable in $\mathcal{T}_1$, viz., C_{κ_1}, where κ_1 is the first (the smallest) inaccessible cardinal, and $\mathcal{L}$ is of order κ_1; after the transition, U is the larger set of the same individuals and the only sets whose existence is provable in $\mathcal{T}_2$, viz., C_{κ_2}, where κ_2 is the second (the smallest after the first one) inaccessible cardinal, and $\mathcal{L}$ is of order κ_2. Thus the definition of truth for $\mathcal{L}$ as it is before the transition is quite possible in the framework of $\mathcal{L}$ as it is after the transition. And the same would be so with any larger, distinct, inaccessible cardinals in place of κ_1 and κ_2.

My reply (until the end of the section). — The putative fact that, in the transition from $\mathcal{T}_1$ to $\mathcal{T}_2$, U *changes* is but a *corollary* of the underlying thesis. I have two counter-objections against this thesis, here is a hint for each of them.

The first one is directed against its *corollary*. U does *not* change, because its members remain the same whatever may be asked, conjectured, affirmed, denied, postulated, proved, refuted, etc. about *them*. (End of the hint) So, in the transition from $\mathcal{T}_1$ to $\mathcal{T}_2$, $\mathcal{L}$ remains what it has been ever since the beginning, viz., our logically universal language, of absolutely infinite order, and for which the definition of the notion of truth is impossible. And the same would be so with any other distinct, inaccessible cardinal (in the invariant U) in place of κ_1.

The second counter-objection is directed against the *underlying thesis* itself. I should like to ask the objector a question. When you are speaking about "the only sets etc.", your statement *presupposes* there to be a reserve of sets beforehand, from which the "only sets" in question are drawn. What about this reserve? Where are you speaking from? (End of the hint)

---

[26]"[A] theory is committed to those and only those entities to which the bound variables of the theory must be capable of referring in order that the affirmations of the theory be true." [17, p. 33]

I will not go any farther into the controversy between the two views of logic in the heart of the enterprise of knowledge. The one carries the higher ambition, that of a certain variant of classical universalism; according to the other, any such ambition must just be abandoned. I need not heap here counter-objections to the latter, be it sufficient to have given the former a new chance.[27]

A *Non Sequitur*: From Tarski's Recantation About the Theory of Semantical Categories to His Renouncement to Logical Universalism

It may well be that abandonment of the basic principles of the theory of semantical categories (or even only of the basic prohibitions that led to the theory of logical types) induced Tarski to abandon logical universalism, but it could not in any way justify it. In Wb, the language of the theory of types naturally appeared to be a good candidate for the role of logically universal language. One could be a logical universalist effortlessly. One could believe in the logical universalist thesis understood as a thesis of first philosophy and laid down with certitude independently of any other more ordinary kind of knowledge. Given what the world, thought, and languages were taken to be, there was a logically universal language (in this case, the language of the theory of types). That was classical logical universalism.

When, beyond languages of finite or infinite order grounded on the theory of types, one envisions languages of relatively finite or absolutely infinite order grounded on set theory (e.g., on set theory ZFC), when, therefore, one is about to have the language of set theory play the role of logically universal language, the supposed logical universality of this new language seems to be vulnerable to a subtle objection that did not affect the supposed logical universality of the old one.

There was something rigid about the framework of the theory of types that the framework of set theory seems not to have. The idea of a framework of higher order than that of the theory of types, but which one might still call "theory of types" seemed to be completely inconceivable.[28] Now, on the contrary, it seems that one could always supplant the putative logical universality of a set-theoretical framework by that of another, essentially richer, framework of the same kind, and the possibility of a logically universal language loses all of its likelihood.

Then one abandons logical universalism to convert to a form of logical anti-universalism, whose thesis is that there is no logically universal language. I do not approve this conversion, but I must say that I have already failed to endorse the first view and am only ready to subscribe to a revised version of classical logical universalism (see above, n. 27).

---

[27] For a brief description and a defense of the version in question, see [21].

[28] In Wb, about "expressions of 'infinite order'" Tarski could not but say: "[N]either the metalangage which forms the basis of the present investigations, nor any other of the existing languages, contains such expressions. It is in fact not at all clear what intuitive meaning could be given to such expressions."

Be that as it may, between the "Wahrheitsbegriff" and its postscript, Tarski abandoned logical universalism, and that abandonment was beyond all mathematical proof, it was a matter for philosophy. And that is all that I really wanted to prove.

References

1. G. Cantor, Letters to Dedekind dated July 28 and August 31, 1899. In: [2], pp. 447–448 (1932)
2. G. Cantor, in *Gesammelte Abhandlungen mathematischen und philosophischen Inhalts*, ed. by E. Zermelo (Springer, Berlin, 1932)
3. R. Carnap, *Abriss der Logistik* (Springer, Vienna, 1929)
4. L. Chwistek, The theory of constructive types. Annales de la Société Polonaise de Mathématiques **2**, 9–48 (1924)
5. K. Gödel, *Über die Vollständigkeit des Logikkalküls*, Dissertation, University of Vienna, 1929. In: [7], vol. I, pp. 60–101 (1986)
6. K. Gödel, Die Vollständigkeit der Axiome des logischen Funktionenkalküls. Monatshefte für Mathematitik und Physik **37**, 349–360 (1930)
7. K. Gödel, in *Collected Works*, 5 vols, ed. by S. Feferman et al. (Oxford University Press and Clarendon Press, New York and Oxford, 1986–2003)
8. D. Hilbert, W. Ackermann, *Grundzüge der theoretischen Logik* (Springer, Berlin, 1928); 2d edition 1938; 3d, English, edn 1950, see [9]
9. D. Hilbert, W. Ackermann, *Principles of Mathematical Logic* (revised and annotated edn by Luce, R. E.; translated by Hammond, L. M., Leckie, G. G., Steinhardt F.) (Chelsea, New York, 1950)
10. E. Husserl, *Logische Untersuchungen*, 2 vols (M. Niemeyer, Halle, 1900–1901); 2d edition, 3 vols, 1913–1921
11. E. Husserl, *Logical Investigations*, two vols (2d edn of [10], translated by Finlay, J. N.) (Routledge, London and New York, 1970)
12. D. Mirimanoff, Les antinomies de Russell et de Burali-Forti et le problème fondamental de la théorie des ensembles. L'Enseignement Mathématique **19**, 37–52 (1917)
13. S. Lesniewski, Grundzüge eines neuen Systems der Grundlagen der Mathematik. Fundamenta Mathematicae **14**, 1–81 (1929)
14. S. Lesniewski, Ueber die Grundlagen der Ontologie. Comptes Rendus des Séances de la Société des Sciences et des Lettres de Varsovie, Classe III **23**, 111–1332 (1930)
15. P. Mancosu, Harvard 1940–41: Tarski, Carnap and Quine on a finitistic language of mathematics for science. Hist. Philos. Logic **26**, 327–357 (2005)
16. W.V.O. Quine, New foundations for mathematical logic. Am. Math. Mon. **44**, 70–80 (1970)
17. W.V.O. Quine, On what there is. Rev. Metaphys. **2**, 21–38 (1948)
18. Ph. de Rouilhan, *Frege. Les paradoxes de la representation* (Editions de Minuit, Paris, 1988)
19. Ph. de Rouilhan, *Russell et le cercle des paradoxes* (Presses Universitaires de France, Paris, 1996)
20. Ph. de Rouilhan, Tarski et l'universalité de la logique, in *Le formalisme en question. Le tournant des années 30* (Vrin, Paris, 1998), pp. 85–102
21. Ph. de Rouilhan, In defense of logical universalism. Taking issue with Jean van Heijenoort. Logica Universalis **6**, 553–586 (2012)
22. B. Russell, see [30], vol. I, 2d edn (1925)
23. Th. Skolem, *Einige Bemerkungen zur axiomatischen Begründung der Mengenlehre, Mathematikerkongressen i Helsingfors den 4–7 Juli 1922, Den femte skandinaviska matematikerkongressen, Redogörelse* (Akademiska Bokhandeln, Helsinki, 1923), pp. 217–232
24. T. Skolem, Über einige Grundlagenfragen der Mathematik, Sckrifter utgitt av Det Norske Videnskaps-Akademi i Oslo, I. Mat.-nat. Klasse **4**, 1–49 (1929)
25. A. Tarski, Projecie prawdy w jezykach nauk dedukcyjnych (The concept of truth in the languages of deductive sciences). Varsaw (1933)
26. A. Tarski, Der Wahrheitsbegriff in den formalisierten Sprachen. Studia Philosophica **1**, 261–405 (1936)

27. A. Tarski, *Logic, Semantics, Metamathematics – Papers from 1923 to 1938* (edited and translated by Woodger, J.) (Clarendon Press, Oxford, 1956)
28. A. Tarski, *Logique, sémantique, métamathématique 1923–1944*, 2 vols (edited and translated by Granger, G.-G., et al.) (A. Colin, Paris, 1972–1974)
29. A. Tarski, revised edition of [27], ed. by J. Corcoran (Hackett Publishing Company, 1983)
30. A.N. Whitehead, B. Russell, *Principia Mathematica*, 3 vols. (Cambridge University Press, Cambridge, 1910–1913); 2d edn, 3 vols (1925–1927)
31. E. Zermelo, Untersuchungen über die Grundlagen der Mengenlehre, I. Mathematische Annalen **59**, 261–281 (1908)
32. E. Zermelo, Ueber Grenzzahlen und Mengenbereiche – Neue Untersuchungen über die Grundlagen der Mengenlehre. Fundamenta Mathematicae **16**, 29–47 (1930)

P. de Rouilhan (✉)

Institut d'histoire des sciences et des techniques (CNRS, Université Paris 1 - Panthéon-Sorbonne, ENS), Paris, France

e-mail: rouilhan@orange.fr

Paradox of Analyticity and Related Issues

Jan Woleński

Abstract This paper formulates the paradox of analyticity similar to the Liar Antinomy. The proposed analysis shows that **T**-equivalences play a crucial role in generating both logical puzzles. The main lesson derived from the paradox of analyticity suggests that the concept of analytic sentences should be defined in a metalanguage, at least if it is understood in the semantic manner.

Keywords Semantics · The Liar Antinomy · Truth

Mathematics Subject Classification Primary 03A05 · Secondary 03A99

We begin by choosing a random sentence. (According to Polish tradition, I prefer to speak about sentences as bearers of logical properties.) By definition, sentences are syntactically well-formed as well as equipped with meaning and thereby intelligible as far as their understanding is taken into account: A is analytic or is an analytical ($A \in \mathbf{AN}$) under any substantial definition of analyticity (by the adjective 'substantial' I mean here: one of the definitions of the concept of analyticity occurring in standard philosophical or/and logical discussions on this topic). Generally speaking, such a definition falls under the scheme

$$(*)\, A \ \in \ \mathbf{AN} \text{ if and only } \mathbf{C}(A),$$

where the letter $\mathbf{C}$ refers to a condition to be satisfied by an analytical. For instance, $\mathbf{C}$ can state 'is true in virtue of meanings', 'is a tautology', 'is true in all possible worlds', 'it is true on the base of rules of a given language', 'is derivable solely on the basis of logic and definitions', 'its negation is contradictory', etc. (see [5] for an extensive survey of various attempts to define the concept of an analytic sentence from Kant to the end of the twentieth century). However, no particular understanding of analyticity is pre-supposed in my further discussions, except claims captured by the conditions (1) and (2) formulated in the next section.

In order to make things more explicit, I assume two statements, namely:

(1) A is analytic and true if and only if $\neg A$ is analytic and false; symbolically:

$$A \in \mathbf{AN} \wedge A \in \mathbf{VER} \iff \neg A \in \mathbf{AN} \wedge \neg A \in \mathbf{FLS};$$

(2) $A \in \mathbf{VER} \iff A.$

© Springer International Publishing Switzerland 2016

F.F. Abeles, M.E. Fuller (eds.), *Modern Logic 1850-1950, East and West*, Studies in Universal Logic, DOI 10.1007/978-3-319-24756-4_6

Intuitively, (1) and (2) assert the following presumably non-controversial facts (a) the concept of analyticity is actually a semantic one (analytic sentences are either true or false; I do not enter the problem of analyticity in many-valued semantics); (b) analyticity is closed under the operation of negation, that is, denials of analytic sentences are analytic as well, and (c) the concept of truth obeys the condition regulated by so-called **T**-equivalences. Although (1) seems fairly trivial, its content has some relevance from the historical point of view. In particular, Kant defined analyticity only for analytic truths (clearly, if '*S* is *P*' is analytic the content of *P* cannot be a part of the content of *S*), but early Wittgenstein and logical empiricists (in the radically syntactic version of their philosophy) did not attribute truth or falsity to analytical (the issue is more complex, but I skip details). Accordingly, (1) should be explicitly or, at least implicitly, noted in any discussion on analyticals. As far as the issue concerns (2), this assumption corresponds to Tarski's basic requirement concerning the material correctness of the semantic definition of truth.

Consider the sentence

(S) (S) is not analytic.

The sentence (S) asserts (about itself) that it is not analytic. Thus, (S) uses the predicate '(not) analytic' self-referentially.

Suppose that (S) is true. By (2), this assumption entails (S). Thus, (S) is not analytic. On the other hand, ¬(S) is false and, thereby, analytic. By (1), (S) is analytic as well. To sum up, if (S) is true, it is analytic and not analytic. Contradiction! Let us suppose now that (S) is false. Thus, it is analytic. By (1), its negation, that is, the sentence ¬(S), is true as well as an analytical. This implies (by the theorem of sentential calculus: $(A \wedge B) \iff (A \iff B)$) that ¬(S) is analytic if and only if it is false. Consequently, (S) is analytic if and only if it is true. However, the last assertion entails that (S) is not analytic if and only if it is false. This conclusion is at odds with (1). Contradiction! This completes the argumentation that (S) leads to a paradox, analogical to the Liar Antinomy. (I use the terms 'antinomy' and 'paradox' interchangeably.)

According to the Leśniewski–Tarski diagnosis (see [3, 4]), the Liar paradox has its source in three facts:

(A) Classical (bivalent) logic is applied;
(B) **T**-equivalences characterize the concept of truth;
(C) Self-referential sentences are admitted.

These points suggest the following ways out:

(A1) Changing logic, for instance, to qualify self-referential sentences as having other logical values than truth or falsity or being truth-value gaps;
(B1) Rejecting **T**-equivalences as inherently related to the concept of truth;
(C1) Making formal semantic considerations free of self-referential sentences (of course, except reasoning showing their paradoxical consequences).

Tarski decided to adopt (C1) as the most natural solution, because changing logic or rejecting the condition imposed by **T**-equivalences would cost much more than excluding self-referentiality from semantics. The transfer of semantic concepts to metalanguages connected with given object-languages becomes the outcome of this decision. Thus, if **L** is an object-language and the concept of truth is to be adequately defined for **L**, it must be done in its suitably selected **ML** (since requirements of suitability are not relevant here, I omit this question in the further discussion).

Tarski's solution invoked some criticism. It was argued (see [1]) that self-referential sentences have a correct grammatical form and, thereby, they are admissible in natural language as well as its formal idealizations. Moreover, for the stratification of the concept of truth (we need several truth-predicates relatively to the hierarchy L_0, L_1 ($=ML_0$). L_2 ($=ML_1$), ..., L_k ($=ML_{k-1}$, ...). Finally, not all self-referential sentences are paradoxical. Consider, for instance,

(Q) (Q) does not consist of ten words.

Clearly, (Q) does not cause any trouble. It is evidently false (seven words occur in (Q)) and, its negation, that is, the sentence 'it is not true that (Q) consists of ten words' is evidently true. No way of reasoning leads to the conclusion that (Q) is simultaneously true and false. This observation justifies the statement that the sentence (Q) does not lead to a paradox, although it is perfectly self-referential. In fact, safe self-referential utterances occur in colloquial languages much more frequently than their paradoxical cousins.

I will not discuss all the objections to Tarski's solution. Let me observe, at first, that we do not need to qualify self-referential sentences as meaningless. Even if some of Tarski's remarks suggest that the liar sentence and similar utterances are incorrectly formed from the syntactic point of view, this proposal appears artificial. In fact, if we infer something from the liar formula or (S), it is difficult to maintain that our fundamental premise has no meaning. Polish logicians, including Tarski himself (although he was not always transparent in his remarks on (in)correctness of the liar sentence), considered self-referential sentences rather as items which could not be values of sentential variables. Not because they are not grammatically defective, but for their creative role in generating semantic antinomies. Using perhaps a somehow exaggerated comparison, one could say that similarly, we should not divide by 0 or that the absolute vacuum is physically impossible and cannot be measured in any reasonable system of magnitudes. However, we perfectly understand what to divide by 0 means or how to measure an absolute vacuum. Mathematical or physical impossibilities do not necessarily imply logical meaninglessness.

My main focus in the coming remarks is to investigate or at least to indicate the role of **T**-equivalences in producing semantic troubles in which such equivalences are involved. If we look at the steps leading to the Liar Antinomy or the analyticity paradox, we immediately observe that **T**-equivalences play an absolutely crucial role in this business. Roughly speaking, **T**-equivalences compress the left-hand part of the formula $TA \iff A$ to its right-hand part. The analyticity paradox and the Liar Antinomy cannot be derived without such a compression. On the other hand, the sentence (Q) has no such connection with the related **T**-equivalence, that is, **T**(Q) if and only if Q, which produces a semantic antinomy in the case of the liar sentence or (S). It seems (it is an empirical claim) that **T**-equivalences lead to semantic troubles only in the case that they compress semantic concepts. This assertion is very nicely supported by the derivation of the Liar Paradox in formalized Peano arithmetic, when we use the fix point (or diagonal) lemma. Let the symbol **PA** refer to formalized (first-order) Peano arithmetic (see [2], 102–105). The diagonal lemma says that **PA** proves a formula $F(A^*) \iff A$, where A^* is the Gödel number of the formula A and F is an arbitrary property expressible in arithmetic. In particular, **PA** proves **T**-equivalences, if F is interpreted as 'is true'. Since the diagonalization function $f(A^*, A)$ is (strongly) definable in **PA**, the fix point lemma guarantees that **T**-equivalences for self-referential sentences are available as theorems of **PA**. However, since the same procedure leads to their negations, we immediately obtain the Liar Paradox. Thus, in order to avoid the paradox we should either exclude some **T**-equivalences or to admit that the truth-predicate is only partially defined. If we exclude self-referentiality, the predicate **T** becomes coherently definable, but we must go to **ML** in order to provide a consistent definition. This conclusion follows from Tarski's undefinability theorem: if a theory **Th** is sufficient enough to cover **PA**, its truth-predicate is not definable in **Th**. Note that the formal construction using the diagonal lemma can be fairly considered as at least partially parallel to Tarski's analysis of the Liar paradox in natural language. This observation supports the view that formal constructions do not depart very dramatically from ordinary usages.

Coming back to analyticity, the paradox concerning (S) combines analyticity and truth via **T**-equivalence **T**(S) if and only if (S). In fact, **T**-sentences compress analyticity to truth or falsity. Since we assume that analytic sentences have definite logical values, that is, they are true or false, it is not strange that both compressions produce antinomies. What is a solution? Since the diagnosis of the sources of the trouble with (S) closely follows the points (A)–(C), possible ways out are indicated by the recommendations (A1)–(C1). Consequently, if we intend to keep **T**-equivalences as determining the use of the predicate 'is true' and classical logic as our logical environment, the simplest proposal consists in claiming that in order to define analyticity for **L**, we must use resources available in **ML**. This suggestion directly follows not only from the assumption (1), but is also supported by formal constructions related to the very nature of semantic concepts. If classical logical values, that is, truth or falsity, are attributed to analyticals, no other way looks reasonable, unless we resign from considering them as genuine semantic items. Anyway, the real strength of Tarskian (model-theoretic) semantics is sufficient enough to produce a general uniform frame for synthetic as well analytic sentences.

References

1. S. Kripke, An outline of a theory of truth. J. Philos. **72**, 690–716 (1975)
2. R.M. Smullyan, *Gödel's Incompleteness Theorems* (Oxford University Press, Oxford, 1992)
3. A. Tarski, *Pojęcie prawdy w językach nauk dedukcyjnych*, Towarzystwo Naukowe, Warszawskie, Warszawa 1933; Eng. translation, *The Concept of Truth in Formalized Languages* in A. Tarski, *Logic, Semantics, Metamathematics. Papers from 1923 to 1939*, Clarendon Press, Oxford 1956, 152–278
4. A. Tarski, The semantic conception of truth and the foundations of semantics. Philos. Phenomenol. Res. **4**, 341–375 (1944); reprinted in A. Tarski, *Collected Papers,* v. II: 1935–1944, Birkhäuser, 1986, 661–699
5. J. Woleński, Analytic vs. synthetic and a priori vs. a posteriori, in *Handbook of Epistemology*, ed. by I. Niniiluoto, M. Sintonen, J. Woleński (Kluwer, Dordrecht, 2004), pp. 781–839

J. Woleński (✉)
University of Information, Technology and Management, Rzeszow, Poland
e-mail: wolenski@if.uj.edu.pl

Naturalizing Natural Deduction

David DeVidi and Herbert Korté

Abstract A simplified and improved system of natural deduction for classical predicate logic is presented. The inference rules of existential instantiation EI, existential elimination ($\exists$E), and universal generalization UG ($\forall$I) are not employed in this system. Instead, a new proof-theoretic role for the quantifiers is introduced, which we call *commonizing quantifiers*. We argue that the commonizing quantifiers exemplify one of Frege's basic insights by making explicit, in the context of natural deduction, the fundamental difference between the semantics of *proper names* and *expressions of single or multiple generality. Arbitrary-object semantics*, which is the best-developed philosophical justification of $\forall$I and $\exists$E and offers an explanation of the cumbersome restrictions placed on these inference rules, is discussed and criticized. It is argued that by looking at natural deduction in an intuitionistic rather than classical setting, it can be seen that the notion of an arbitrary object is not merely a harmless manner of speaking or notational variant, but a philosophical commitment that has consequences that are disguised if the discussion is framed only in terms of classical logic. In the system on offer, the notion of an *arbitrary object* is entirely dispensed with; in addition to considerations of simplicity, elegance, and parsimony, philosophical considerations decidedly favour the naturalized natural deduction system presented here.

Keywords Arbitrary object · Epsilon operator · Francis Jeffry Pelletier · Frege · Gentzen · Jeffrey C. King · Kit Fine · Logic · Natural deduction · Quantifier

Mathematics Subject Classification (2000). Primary 03A05 · Secondary 03B10

For a philosopher teaching a first logic class to undergraduates, natural deduction systems for the standard, extensional sentential connectives are a beautiful thing. The standard introduction and elimination rules either represent patterns of inference of a familiar sort (as, for instance, *modus ponens* and conditional proof undoubtedly do), or are easily justified in terms of the "intended meaning of the connective" (as with "or introduction"). The rules are readily mastered by most students willing to put in a modicum of effort. The few simple rules are readily shown to be surprisingly powerful. And while the rules

DeVidi wishes to acknowledge the support of the Social Sciences and Humanities Research Council of Canada during the preparation of this paper.

F.F. Abeles, M.E. Fuller (eds.), *Modern Logic 1850-1950, East and West*, Studies in Universal Logic, DOI 10.1007/978-3-319-24756-4_7

are usually introduced in a classical version alongside some version of two-valued truth functional semantics that purports to provide the meanings of these connectives, the introduction/elimination pairs make it easy to introduce alternative theories that spell out the meaning of logical particles in terms of their use, i.e., their inferential role.

Things are less happy when discussion turns to predicate logic. Partly, of course, this results from the increasing power of the system, which is paid for with increased complexity and the loss of certain nice properties (decidability, for instance). Some of the problems, though, do not have the same payoff. Standard versions of natural deduction extend the pattern of introduction/elimination pairs from the sentential operators to the quantifiers. While two of the resulting rules share the appealing features of familiarity and easy useability, two others involve complicated restrictions that make them difficult to use and to teach. Moreover, the benefits of preserving the pattern of intro/elim pairs for each operator are outweighed by substantial philosophical difficulties this approach involves.

Or so we shall suggest. In this paper we will introduce a modified system of natural deduction for classical predicate logic.[1] We offer this system for consideration and claim two advantages for it. While the system includes the familiar rules for existential introduction and universal elimination, the problematic rules of universal introduction and existential elimination are replaced. The first advantage of the system is that it is easier to use and to teach, at least based on the experience of one author over many years teaching the system to undergraduates after many prior years teaching standard natural deduction systems. The second advantage is philosophical. For one thing, giving up the easy intro/elim parallelism avoids the philosophical difficulties that come with it, and so restores the *naturalness* of the system by better reflecting unformalized but rigorous reasoning. For another, the usual explanations for the standard rules (but not our replacements for them) is philosophically dubious, requiring appeal to philosophically dubious *arbitrary objects*.

We will proceed as follows. First, we briefly indicate the aspects of standard natural deduction systems that create philosophical complications, describe some approaches offered as solutions to the problems, and describe some reasons for being unsatisfied by the solutions. We then describe some of the key features of the modified system, and give a few examples that illustrate its naturalness and ease of use. We conclude by indicating how the modified system "solves" the philosophical difficulties—by simply failing to raise them.

Problems of Standard Systems

In 1999, Jeffry Pelletier could quite plausibly say that natural deduction "is by now the most common way to teach elementary logic by far, and indeed is perhaps all that is known in any detail about logic by a number of philosophers (especially in North America)" [17, p. 1]. While in the intervening years, especially in texts that present a range of non-classical logical systems, trees, semantic tableaux and sequent calculi have

---

[1]The system itself is the product of joint work by Korté and the late Robert Coleman.

perhaps become more common, natural deduction systems remain the first encounter with logic for many philosophy students. The virtues of natural deduction systems, at least for the classical propositional and predicate systems that are the centerpieces of most introductory formal logic courses for philosophers, are many. The basic rules of inference are, indeed, "natural" ones that are readily seen to reflect argument patterns in ordinary discourse; especially for systems that stick to Gentzen's ideal of systems where the basic deductive behaviour of each logical operator is governed by introduction/elimination rules, students are well set up to understand subsequent discussions in philosophy of logic, while other common inference patterns are readily seen to be correct "derived rules"; and the systems are easy for students to learn, and easy for instructors to motivate.

Of course, these virtues are all only comparative, and involve trade-offs. It is easier for students to learn to do proofs in natural deduction than in axiomatic systems, but proving meta-theorems tends to be messier; compared to proof trees or semantic tableaux, natural deduction proofs are harder—though they look more like ordinary language arguments than trees do—while the meta-proofs are trickier. What we want to compare is not the usual presentations of natural deduction and other proof systems, but the usual presentations of natural deduction and our alternative version of natural deduction.

The teachability and transparent philosophical virtues of standard natural deduction systems decline markedly when one moves from propositional to predicate logic, as most who have taught the material will attest. The Universal Elimination ($\forall$E) and Existential Introduction ($\exists$I) rules share the virtues—seemingly obvious correctness, ease of application—possessed by most of the standard propositional rules.[2] The same cannot be said for the partner rules. The restrictions and conditions for the application of universal introduction and existential elimination in the systems one finds in logic textbooks are unnatural and complicated, making them difficult to teach.

The complications are perhaps most evident in the existential case. While there are many differences of detail in the systems of natural deduction presented in various textbooks, differences small enough for us to ignore here, one must distinguish two quite different styles of treating existential elimination. We shall distinguish the styles by assigning different names to each.

The first approach, which following Pelletier we shall call "Quine systems," uses the rule we shall call *existential instantiation* (EI, not to be confused with $\exists$I). When applying this rule, one infers $\varphi(a)$ from $\exists x \varphi(x)$, provided certain restrictions, including that a be suitably "new" at that point of the derivation (e.g., that it does not occur in any assumption "active" at that point in the proof), are satisfied:

$$\text{EI} \qquad \frac{\exists x \varphi(x)}{\therefore \ \varphi(a)} \qquad (1)$$

---

[2]For the most part we follow the convention, which goes back to Gentzen, of referring to rules as introduction and elimination rules, though many authors opt for other names, especially where there are traditional or explanatory names available, including ones that often figure in philosophical discussions outside the logic classroom (*modus ponens, conditional proof, reductio, disjunctive syllogism,* etc.). For reasons that will soon be clear, we make an exception for existential instantiation.

The first difficulty here is that the rule of inference is not actually valid. More precisely, the other rules of natural deduction systems satisfy the property that what is inferred by the application of a rule must be true in any interpretation that makes all the "active assumptions" at that point in the derivation true. Since a is new, we have no such guarantee. One of the supposed virtues of natural deduction is in danger here—lack of logical correctness makes a claim of "obviously correctness" hard to sustain. Quine's approach is to adopt a method of *flagging* variables introduced by application of EI and declaring derivations "unfinished" if any such variable is free in the last line of the proof or any premise of that last line. Pelletier detects "a note of desperation" in Quine's description of his approach [17, p. 13]. Desperation or not, the easy parallel to ordinary patterns of reasoning seems broken.

In the second approach, employed by what Pelletier calls "Fitch systems," the term that is introduced when the existential quantifier is eliminated is introduced *in an assumption* that starts a new sub-proof. The rule requires that, under certain conditions designed to ensure that a is "new" and that it does not figure in unacceptable ways in ψ,

$$
\begin{array}{|l}
\exists x \varphi(x) \\
\quad \begin{array}{|l}
\varphi(a/x) \\
\hline
\psi
\end{array} \\
\psi
\end{array}
\tag{2}
$$

This rule perhaps matches ordinary language reasoning more closely. One readily imagines someone saying "Somebody has the property φ; let's suppose it's Bob ...," and so long as the conclusion drawn isn't *about* Bob, we will, other steps in the argument being correct, accept the conclusion. Relatedly, this pattern of inference is *valid* in the sense in which EI was said above to be invalid.

Nevertheless, this approach has a further problematic feature, one shared with ∀I and reflected in the complicated conditions for correct application of these rules. The role played by the term that is introduced in the existential elimination or replaced in a universal introduction is philosophically puzzling. Depending on the details of the system, the term is a variable or a name and the conditions ensure that it is "arbitrarily chosen." As the normal semantic role of a variable or a name is to refer to an individual (under an assignment, in the case of a variable), the semantic role of an "arbitrarily chosen" term, especially when described, as it often is, as an *arbitrary object*, is problematic. The idea is that, one way or another, when a represents an arbitrary object in its occurrence in $\varphi(a)$, its role is not to designate a particular individual but to somehow stand in for the whole set of individuals with the property φ, which in the case where ∀I is applicable, turns out to be the entire universe of discourse.

That there are philosophical problems in this neighbourhood should be of no surprise, as some of major figures in the history of philosophy have considered the matters carefully, and some have detected serious problems. One argument against arbitrary objects goes back at least to [4]. Berkeley quotes Locke's description of the general idea of a triangle, which "must be neither oblique nor rectangle, neither equilateral, equicrural, nor scalenon, but *all and none* of these at once. In effect, it is something imperfect that cannot exist, an idea wherein some parts of several different and inconsistent ideas are put together," pointedly adding the emphasis and asking the reader whether she can really imagine it. His point seems well captured with the slightly simpler example of an arbitrary integer:

1. Consider an arbitrary integer x.
2. Then $E(x) \vee O(x)$, since each integer is either even or odd.
3. Since x is arbitrary, x cannot be even, $\neg E(x)$, since some integer is not even.
4. Since x is arbitrary, x cannot be odd, $\neg O(x)$, since some integer is not odd.

$\therefore$ $(E(x) \vee O(x)) \wedge \neg(E(x) \vee O(x))$

In short, the notion of an arbitrary object, at least *prima facie*, leads to contradiction.

Moving forward a couple of centuries, Frege, for instance in [13], rejected the view that the variables of mathematics and logic have *indefinite* objects as their meanings analogous to the way proper names have *definite* objects as their meanings.

> But are there indefinite numbers? Are numbers to be divided into definite and indefinite numbers? Are there indefinite men? Must not every object be definite? But is not the number n indefinite? I do not know the number n. 'n' is not the proper name of some number, definite or indefinite. And yet one sometimes says 'the number n.' How is that possible? Such an expression must be considered in context. ... One writes the letter 'n,' in order to achieve generality. ...
>
> Naturally one may speak of indefiniteness here; but here the word 'indefinite' is not an adjective of 'number,' but an adverb of, for example, to 'signify.'

Frege wishes to sharply distinguish the semantic relation involved in such expressions as

$$5 + 7 = 7 + 5, \tag{3}$$

from the semantic relation involved in expressions such as

$$x + y = y + x. \tag{4}$$

Expression (3) does not express a universally general claim because the numeral expressions are names of their corresponding objects. The universal generality achieved in (4) is not, however, achieved by replacing definite objects with indefinite ones, that is, by regarding the variables 'x', 'y' as naming indefinite or arbitrary numbers. The semantic relation of expression (3) is one of "naming," that of expression (4) is one of *indefinitely signifying, indicating or referring*. Frege suggests that there is a fundamental difference between the semantics of standard referring terms and the semantic account one must give of sentences containing multiple expressions of generality. The latter, but not the former, requires the notion of a *domain of individuals* over which an expression of generality ranges and the *scope* of such an expression of generality. Generality cannot be achieved, according to Frege, simply by an extension of the semantics of standard referring terms through an introduction of indefinite or arbitrary objects that are referred to by the variables of an expression.

Many subsequent philosopher-logicians have also argued against arbitrary-object semantics: for instance, [26, pp. 90–91], [28, pp. 4–5], [5, p. 13], [24, pp. 69–71], [25, pp. 134–137].[3]

Of course, arbitrary objects are not entirely without defenders. Kit Fine has presented a detailed account of an arbitrary object semantics, and provided a robust philosophical defense for it in [8–10]. According to Fine, arbitrary-object semantics is motivated by two things: First, it provides an explanation and justification of the various restrictions that must be imposed on the application of ∀I and ∃E. Without an explicit explanation and justification, Fine says, we would be left with a purely instrumental justification of these restrictions. Secondly, arbitrary-object semantics provides the underlying rationale for these problematic rules by providing us with an account of why they are truth preserving. At its very simplest, the idea is as expressed above: if a represents an arbitrary object in $\varphi(a)$, then it is associated with, i.e., refers to, the entire set of individuals that have a, rather than with any particular individual.

So construed, instantial terms appear not to function as individual terms at all, but instead seem to be expressions of generality. Echoing Frege, in his critique of Fine's arbitrary–object semantics, King [18, pp. 254–255] observes:

> In introducing new objects for quantifier expressions to refer to one assimilates the semantics of expressions of generality to that of standard names. ... Surely, if anything, the introduction of referents for quantifiers obscures the differences between genuine names and expressions of generality by construing both sorts of expressions as referring terms.

However, the role these terms play in usual presentations of natural deduction leave them with certain deficiencies as quantifiers. If the term in question appears in the formula $\varphi(a)$, we are to suppose that a refers to something about which we know only that it is φ. But while the syntax of first-order languages clearly delineates, in a context-independent manner, the scope and force of ordinary quantifiers, the formulas containing instantial terms involved in applications of ∀I and ∃E do not provide any information, syntactic or otherwise, about the relative scope and force of these quantifier-like expressions of generality. Because of this, King refers to them as *context-dependent quantifiers*.

What is on offer here is a reformulation of natural deduction that learns the lesson of King's Fregean critique of arbitrary-object semantics.[4] That is, what this paper tries to achieve is a reformulation of natural deduction which makes explicit Frege's basic insight by using rules which highlight the difference between the semantics of singular terms and expressions of generality. No instantial terms of the problematic sort, that is, no quantifier-like expressions in King's sense, ever appear in a derivation. Instead, what are used are variables bound by what we call *commonizing quantifiers* in their stead. The result is that the scope and force of expressions of generality within a proof or subproof are syntactically transparent, thus answering King's concern about the context dependence of the instantial terms of standard systems.

---

[3] Other relevant discussion can be found in [12], [15], [16], [19–23] and [29].

[4] King's own Fregean interpretation of the semantics of instantial terms presupposes the framework of standard natural deduction techniques.

It may appear that we are simply accepting that King (and Frege) are right to reject Fine's arbitrary object semantics, without a detailed investigation. However, with the system on offer, Fine's case for arbitrary objects is undercut—neither of the rules that requires an arbitrary-object semantics for its legitimation occurs in the system. What motivates the claim of necessity for arbitrary-object semantics in the first place is the presence of the rules of ∀I and ∃E. Since the rules are not required for a satisfactory natural deduction system, and given the philosophically problematic nature of arbitrary objects, arbitrary object semantics lacks motivation.

The System

We focus here on the distinctive features of the system on offer. While there are differences of detail between systems in the literature that might make some of our proofs look odd to some, we doubt this will prove an obstacle for most readers. When displaying proofs, we will employ the method developed by Fitch [11], which is used in many popular texts and which we presume will be familiar to most readers.

Some of what we have to say applies equally to both the existential and universal quantifiers. To prevent unnecessary repetition, we shall use the symbol ⊔ when a statement or rule holds for both the universal and existential quantifier.

The terms that are taken to designate arbitrary objects in standard presentations of natural deduction will here be replaced by *bound variables*. What allows a bound variable to play the role required is that they will not be bound by quantifiers *that are part of the formulas in the proof*. Rather, they will be bound by what we shall call *commonizing quantifiers*. Like the normal quantifiers, a commonizing quantifier has a well-determined scope. Unlike the normal quantifiers, the scope is not a formula, but a proof (or subproof).

To indicate the scope of a commonizing quantifier, we employ the familiar *Fitch bar*, which is standardly used to indicate that an assumption is made and how long the assumption is in force (or *open*). A commonizing quantifier ⊔v, can be placed where an assumption would normally occur in a Fitch-style proof—i.e., alongside a newly drawn vertical line and above a horizontal line. Of course, since what is being placed is merely a quantifier and not a formula, the Fitch bar is clearly being used for different purposes. We note the following:

1. The length of the vertical bar indicates the *scope* of the commonizing quantifier.
2. While it is a "legal" move to assume any formula at any time, there are some general conditions constraining when a commonizing quantifier can be introduced.

Graphically, the diagram below indicates that the scope of the commonizing quantifier ⊔v includes (the displayed occurrences of) $\varphi_1, \varphi_2, \ldots, \varphi_n$. The intended meaning is that, under the assumptions open at that point in the derivation, $\varphi_1, \varphi_2, \ldots, \varphi_n$ all hold for all/some $v \in UD$, respectively, if ⊔ is ∀/∃.

$$
\begin{array}{c|l}
& \vdots \\
& \text{Ll}v \\
& \overline{\varphi_1} \\
& \vdots \\
& \varphi_n \\
& \vdots
\end{array}
\tag{5}
$$

We say that the proofs in this system are such that *no variable occurs freely*, but this needs to be understood correctly. In general there will be *open formulas* occurring at various lines in the proof. But in all cases, all the free occurrences of variables on that line must be bound by a commonizing quantifier. It is useful to make a *Local/Global Distinction*.

- All variables in a proof are in the scope of either an *ordinary* quantifier or a *commonizing* quantifier.
- If the variable v lies in the scope of an *ordinary* quantifier $\text{Ll}v$, then it is said to be *locally* bound by $\text{Ll}v$, whose scope is a formula or subformula in the usual way. We may correspondingly say that the ordinary quantifier $\text{Ll}v$ is a *local* quantifier.
- If a locally free occurrence of v occurs in the scope of a *commonizing* quantifier $\text{Ll}v$, then it is said to be *globally* bound by $\text{Ll}v$, the scope of which is the proof or subproof as indicated above. We may correspondingly say that the *commonizing* quantifier $\text{Ll}v$ is a *global* quantifier.
- The system is designed so that in a correct proof if a variable v is *locally free*, that is, v is not captured by a *local quantifier* $\text{Ll}v$, then v is *globally bound*, while if an occurrence of v is *locally bound* by $\text{Ll}v$ then it cannot be *globally bound* by a *commonizing* quantifier.

This has implications for the statement of the universal instantiation rule:

Universal Instantiation (UI)

We allow for the instantiation of any term t (constant, variable, or function term), provided that any free variable in t lies within the scope of some commonizing quantifier Ll. (As a result, genuine universal quantifier elimination occurs only in the case of universal instantiation with an individual constant.) Thus, we present the rule UI as follows, where t is a term:

$$
\begin{array}{rl|l}
& \vdots \\
i & \forall v \varphi(v, v_1, \ldots, v_k) \\
& \vdots \\
j & \varphi(t, v_1, \ldots, v_k) & \quad \text{UI } i \\
& \vdots
\end{array}
\tag{6}
$$

provided that:

Universal Instantiation Conditions

1. If t has variables in it then all of the variables in t must be substitutable for v in $\varphi(v, v_1, \ldots, v_k)$, and
2. Each variable in t must be in the scope of some commonizing quantifier ⊔ (to which reference should be made in the justification for universal instantiation).

It is worth taking a bit of care to make clear the conditions under which a commonizing quantifier may be introduced, and incidentally where the name "commonizing" comes from. Let's begin with the universal quantifier. The idea of the commonizing quantifier is that it will pick out the class of objects in the domain which satisfy *all the assumptions open at the point in the proof at which that quantifier is introduced*. That is, any member of this class will have all the properties which the open assumptions require *every* object to have, but otherwise they can differ as much as you please. The general constraint for introducing a universal commonizing quantifier therefore is as follows:

Introducing a Universal Commonizer (UC)

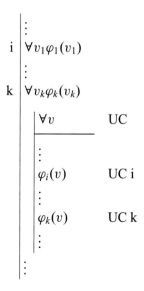

$$(7)$$

We must of course prevent unwanted variable capture in the course of our proofs, and so adopt the following

Universal Commonization Conditions

1. v must not occur freely in $\forall v_i \varphi_i(v_i)$,   for all $i \in \{1 \ldots k\}$.
2. v is substitutable for v_i in $\varphi_i(v_i)$,   for all $i \in \{1 \ldots k\}$.

The idea here is, we hope, clear. We have ensured that v is available for "relabeling" the bound variable in any universal claim that has been shown to follow from the open assumptions at the point the universal commonizer is introduced. Hence in the scope of the commonizing quantifier we can make any of the universal claims *of v*. (Strictly speaking, the inclusion of the lines justified by UC_n are merely a convenience, since $\varphi_n(v)$ could be derived at that point using reiteration and universal instantiation.)

Note that if $v_1 = \cdots = v_k = v$ then conditions (1) and (2) are automatically satisfied. Moreover, there need not be any universal claims $\forall v_i \varphi_i(v_i)$ (i.e., k may be zero), in which case we may commonize as follows:

$$
\begin{array}{ll}
\vdots & \\
\forall v \quad & UC \\
\rule{2cm}{0.4pt} & \\
\vdots & \\
\vdots &
\end{array}
\tag{8}
$$

Given this description of the conditions governing the introduction of a commonizing universal quantifier, the conditions for existential commonization will be no surprise.

Introducing an Existential Commonizer (EC)

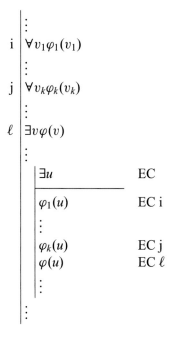

$$
\begin{array}{lll}
& \vdots & \\
i & \forall v_1 \varphi_1(v_1) & \\
& \vdots & \\
j & \forall v_k \varphi_k(v_k) & \\
& \vdots & \\
\ell & \exists v \varphi(v) & \\
& \vdots & \\
& \quad \exists u & EC \\
& \quad \rule{2cm}{0.4pt} & \\
& \quad \varphi_1(u) & EC\ i \\
& \quad \vdots & \\
& \quad \varphi_k(u) & EC\ j \\
& \quad \varphi(u) & EC\ \ell \\
& \quad \vdots & \\
& \vdots &
\end{array}
\tag{9}
$$

Existential Commonization Conditions

1. u must not occur freely in $\forall v_i \varphi_i(v_i)$ (for all $i \in \{1 \ldots k\}$) nor $\exists v \varphi(v)$.
2. u is substitutable for v_i in $\varphi_i(v_i)$, for all $i \in \{1 \ldots k\}$, and substitutable for v in $\varphi(v)$.
3. One cannot existentially commonize on u within the scope of a existential commonizer $\exists u$—roughly, one cannot existentially quantify on the same variable twice.

Once again, if $v_1 = \cdots = v_k = v$ then conditions (1) and (2) are automatically satisfied, in which case one may existentially commonize on v.

As in the case of universal commonization, it is of course possible that there not be any $\varphi_i(v_i)$ (that is, $k = 0$) or $\varphi(v)$, so the following is an allowable pattern of commonization.

$$
\begin{array}{l}
\vdots \\
\exists v \qquad \text{EC} \\
\rule{1.5cm}{0.4pt} \\
\vdots \\
\vdots
\end{array}
$$

$$\text{(10)}$$

Of course, just as rules that begin with assumptions also must specify the conditions under which the assumption is discharged, our rules must specify the conditions under which the commonized variable is "discharged." We call this step "decommomization." We state the general form of the rule, then note the special case that very often is the one applied in practice.

Universal Decommonization (UD) and Existential Decommonization (ED)

$$
\begin{array}{ll}
i \quad \big| \quad \forall v \qquad \text{UC} & \qquad i \quad \big| \quad \exists v \qquad \text{EC} \\
\quad \big| \quad \rule{1cm}{0.4pt} & \qquad \quad \big| \quad \rule{1cm}{0.4pt} \\
\quad \big| \quad \vdots & \qquad \quad \big| \quad \vdots \\
j \quad \big| \quad \psi(v) & \qquad j \quad \big| \quad \psi(v) \\
k \quad \big| \quad \forall v \psi(v) \quad \text{UD } i\text{–}j & \qquad k \quad \big| \quad \exists v \psi(v) \quad \text{ED } i\text{–}j \\
\quad \big| \quad \vdots & \qquad \quad \big| \quad \vdots
\end{array}
$$

$$\text{(11)}$$

We display this proof with a single conclusion, but the system allows more than one formula to be inferred by decommonization from a single subproof.

More importantly, it may be that v does not in fact occur freely in ψ, in which case the quantifier attached by commonization is vacuous. In such cases, we allow as a special case of the decommonization rules that ψ may be inferred without attaching the vacuous universal/existential quantifier.[5]

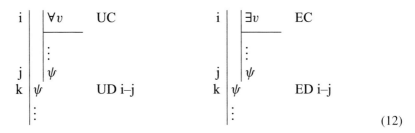

$$(12)$$

Instantiating a Universal Commonizing Quantifier

It would clearly be problematic if the following proof were correct.

$$
\begin{array}{lll}
1 & \forall v & \text{UC} \\
2 & \varphi(v) & \text{A} \\
3 & \varphi(a) & 2\ \forall E \\
4 & \varphi(v) \Rightarrow \varphi(a) & \Rightarrow I \\
5 & \forall v(\varphi(v) \Rightarrow \varphi(a)) & 1\text{--}4\ \text{UD}
\end{array}
$$

$$(13)$$

We clearly don't want a system in which one can derive, for example, "For any integer n, if n is even, then 3 is even."

The reason such proofs are not correct in this system have to do with the scope of commonizing quantifiers, and the requirement of uniform substitution. Just as we may not infer from $\forall x(P(x) \Rightarrow Q(x))$ that $P(a) \Rightarrow Q(x)$, but instead must uniformly replace x by a when applying $\forall E$, similarly when we instantiate on a universal commonizer we must uniformly replace all occurrences of the variable bound by the commonizer. In general, since the scope of a commonizing quantifier is a proof or subproof rather than a formula, the result of a uniform substitution will be a proof or subproof and not a sentence.

In practice, universally instantiating on a commonizing quantifier will not often be of much use. However, we can illustrate its application with a simple example[6] in which we

---

[5]Of course, it would also be possible to simply add rules of vacuous quantifier elimination to the system. If the special case described here offends the reader's sense of how general the statement of a rule should be, we encourage her to think of the system as including the rules of vacuous universal de-quantification (VUDQ) and vacuous existential de-quantification (VEDQ), and the special case as a derived rule of inference.

[6]Though even so, they are more complicated than necessary if all we cared about were actually proving the conclusion.

may prove $\varphi(a_2) \Rightarrow \varphi(a_2)$ in either of the following two ways:

$$
\begin{array}{lll}
1 & \forall v & \text{UC} \\
2 & \quad \varphi(v) & \text{A} \\
3 & \quad \varphi(v) & \\
4 & \quad \varphi(v) \Rightarrow \varphi(v) & \Rightarrow\text{I} \\
5 & \forall v(\varphi(v) \Rightarrow \varphi(v)) & \text{1–4 UD} \\
6 & (\varphi(a_2) \Rightarrow \varphi(a_2)) & \text{5 } \forall\text{E}
\end{array}
\qquad (14)
$$

$$
\begin{array}{lll}
1 & \forall v & \text{UC} \\
2 & \quad \varphi(v) & \text{A} \\
3 & \quad \varphi(v) & \\
4 & \quad \varphi(v) \Rightarrow \varphi(v) & \Rightarrow\text{I} \\
5 & \quad \varphi(a_2) & \\
6 & \quad \varphi(a_2) & \\
7 & \quad \varphi(a_2) \Rightarrow \varphi(a_2) & \\
8 & \varphi(a_2) \Rightarrow \varphi(a_2) &
\end{array}
\qquad (15)
$$

In (15) we have used a universal instantiation for the commonizing quantifier $\forall v$; paying attention to the requirement of uniform substitution, the subproof 5–7 is an instance of 2–4. We then can infer line 8 by UD.

Discussion of Examples

We turn now to presenting a few examples that will allow us to illustrate some features of the system on offer, and contrast it to other systems. Let us start with a very simple example. In most textbooks the proof of

$$
\{\forall x(F(x) \Rightarrow G(x)), \exists x F(x)\} \vdash \exists x G(x) \qquad (16)
$$

would look something like the left side of (17), in which the third line introduces an auxiliary assumption which says that an arbitrarily selected object a has F. On the other hand, thanks to the introduction of a commonizing quantifier on line 3, the derivation on the right of (17), does not invoke the notion of an arbitrary object.

$$
\begin{array}{lll}
1 & \forall x(F(x) \Rightarrow G(x)) & A \\
2 & \exists x(F(x)) & A \\
\hline
3 & \quad\big|\; F(a) & A \\
\hline
4 & \quad\big|\; (F(a) \Rightarrow G(a)) & 1, \forall E \\
5 & \quad\big|\; G(a) & 3,4, \Rightarrow E \\
6 & \quad\big|\; \exists x G(x) & 5, \exists I \\
7 & \exists x G(x) & 2, 3\text{–}6, \exists E
\end{array}
\qquad
\begin{array}{lll}
1 & \forall x(F(x) \Rightarrow G(x)) & A \\
2 & \exists x(F(x)) & A \\
\hline
3 & \quad\big|\; \exists x & EC \\
\hline
4 & \quad\big|\; (F(x) \Rightarrow G(x)) & 1, EC \\
5 & \quad\big|\; F(x) & 2, EC \\
6 & \quad\big|\; G(x) & 4, 5, \Rightarrow E \\
7 & \exists x G(x) & 3\text{–}6, ED
\end{array}
$$

$$\tag{17}$$

We next look at a proof of

$$\neg\forall x\neg P(x) \vdash \exists x P(x) \tag{18}$$

The proof in [1, p. 358] appears to share similarities to ours, but the philosophical story behind it is problematic. They place $[c]$ on the Fitch bar, and interpret it as follows: "Let c denote an arbitrary object satisfying $P(c)$".

$$
\begin{array}{lll}
1 & \neg\forall x\neg P(x)) & A /\therefore \exists x P(x) \\
\hline
2 & \quad\big|\; \neg\exists x P(x) & A /\therefore \perp \\
\hline
3 & \quad\big|\;\big|\; [c] \\
\hline
4 & \quad\big|\;\big|\;\big|\; P(c) & A \\
\hline
5 & \quad\big|\;\big|\;\big|\; \exists x P(x) & 4, \exists I \\
6 & \quad\big|\;\big|\;\big|\; \perp & 2, 5, \perp I \\
7 & \quad\big|\;\big|\; \neg P(c) & 4\text{–}6, \neg I \\
8 & \quad\big|\; \forall x\neg P(x) & 3\text{–}7, \forall I \\
9 & \quad\big|\; \perp & 1, 8, \perp I \\
10 & \exists x P(x) & 2\text{–}9, \neg E
\end{array}
$$

$$\tag{19}$$

Given what we have said above, it will not surprise anyone if we say we find this problematic. We find the words of Rescher [25, p. 137] apt:

> When one introduces random individuals, one can do so meaningfully only subject to the self-denying ordinance represented by the convention that: *Nothing is to be said about a random individual that is not intended about ALL of the individuals of the domain at issue.* A random individual is therefore not a thing but a linguistic principle, a shorthand device for presenting universal statements.[7]

---

[7]The emphasis is Rescher's.

We agree. The universal commonizing quantifier is just the tool to make this intention clear, and for the case in question it does so as follows:

$$
\begin{array}{lll}
1 & \neg\forall x\neg P(x) & \text{A} /\therefore \exists x P(x) \\
2 & \quad \neg\exists x P(x) & \text{A} /\therefore \bot \\
3 & \quad\quad \forall x & \text{UC} \\
4 & \quad\quad\quad P(x) & \text{A} \\
5 & \quad\quad\quad \exists x P(x) & 4, \exists\text{I} \\
6 & \quad\quad\quad \bot & 2,5, \bot\text{I} \\
7 & \quad\quad \neg P(x) & 4\text{–}6, \neg\text{I} \\
8 & \quad \forall x\neg P(x) & 3\text{–}7, \text{UD} \\
9 & \quad \bot & 1,8, \bot\text{I} \\
10 & \exists x P(x) & 2\text{–}9, \neg\text{E}
\end{array}
\tag{20}
$$

One sometimes hears logicians defend or explain the role of an appeal to arbitrary objects in natural deduction reasoning by reference to the frequency with which one will hear mathematicians say things like "since x was arbitrarily chosen, ..." in the course of a proof. Whatever one thinks of the tendency to defer to mathematical practice in other domains such as the philosophy of mathematics, we find it unpersuasive as a general rule in logical matters. It doesn't take much familiarity with some parts of the mathematical literature to realize that many mathematicians routinely ignore the *use/mention* distinction. Not many logicians would use this as a reason to embed such disregard in logic. It's therefore unclear why we should embed arbitrary object semantics into a system of natural deduction based on a mathematical manner of speaking, especially when examples like the one above show that it is quite avoidable.

As a final example, we will show a proof in this system of the theorem that Pelletier focuses on in [17], his informative historical study of natural deduction. It is a useful example because it involves appeal to "arbitrary objects" in both their universal and existential forms in a single short proof. It thus provides an illuminating contrast of the various attempts to implement arbitrary object reasoning in a number of systems, including the "flagging" methods introduced by Quine and Suppes, Fitch's and Copi's systems, and so on.

$$
\emptyset \vdash \exists x\forall y R(x, y) \Rightarrow \forall y\exists x R(x, y).
\tag{21}
$$

$$
\begin{array}{lll}
1 & \quad \exists x \forall y R(x,y) & \text{A } /\therefore \ \forall y \exists x R(x,y) \\
2 & \quad \quad \forall y & \text{UC} \\
3 & \quad \quad \quad \exists x & \text{EC} \\
4 & \quad \quad \quad \forall y R(x,y) & \text{1, EC} \\
5 & \quad \quad \quad R(x,y) & \text{2,4, UI} \\
6 & \quad \quad \exists x R(x,y) & \text{3–5, ED} \\
7 & \quad \forall y \exists x R(x,y) & \text{2–6, UD} \\
8 & \exists x \forall y R(x,y) \Rightarrow \forall y \exists x R(x,y) & \text{1–7, } \Rightarrow\text{I}
\end{array}
$$

(22)

Our inclination is to find this proof perspicuous compared to those in the other systems, and we invite readers to compare it with those for other systems presented in Pelletier's paper.

Hilbert's Epsilon Operator and Ambiguous Names

We find the case made above, based on arguments offered by Frege, Rescher, King et al., persuasive. It suggests that one should adopt the naturalized version of natural deduction described here, even if it were harder to teach. Coupled with the need for complicated restrictions in order for natural deduction systems involving arbitrary objects to be correct, which seems to make the present system easier to teach and to learn, there is much to recommend the system we offer. We turn now to a different sort of objection to natural deduction systems that rest on the idea of arbitrary objects.

The objections to arbitrary objects given above can be broadly regarded as metaphysi-cal. The present objection strikes us as perhaps even more important in a logical context. The objection appeals to some results, well known in the specialist literature but not as widely known to non-specialists as they should be, which show that, even apart from philosophical justification, using bound variables and invoking arbitrary objects are not mere notational variants of one another—there are *differences in logical potency* between the two. Discussing the difference against the background of classical logic disguises this fact, as the differences of potency are masked by the extra logical power provided by indirect proofs (or, equivalently, accepting double negation elimination or excluded middle). Instead, we will look at intuitionistic logic.

As a preamble to our argument, we again refer to Pelletier's useful history of natural deduction. Pelletier notes that Patrick Suppes, in his presentation of natural deduction, employs what he calls *ambiguous names* as a "technical device" to aid in the statement of the quantifier rules. While Suppes takes the notion of ambiguous names to be "apparently new" in the literature on first-order logic, he says "the central idea of this approach is related to Hilbert's ε symbol." The basic idea of the epsilon symbol was that for any predicate $F(x)$, " '$\varepsilon x F(x)$' was to designate 'an arbitrary or indefinite F' about which nothing was known other than that it was an F" [17, p. 22]. Rather than introducing

Hilbert's term-forming, variable binding operator ε, Suppes introduces orthographically distinct "names"[8] to play this role.

Hilbert's symbol was introduced in an axiomatic presentation of logic. Suppressing some details, the fundamental axiom governing its behaviour is:

$$\exists x F(x) \Rightarrow F(\varepsilon x F(x))$$

The epsilon term $\varepsilon x F(x)$ has no free variables, so this axiom scheme picks out a class of sentences, one for each formula $F(x)$ with at most one free variable. Suppes' orthographically distinct "ambiguous names" are introduced when an existential quantifier is eliminated, and a name new to the proof is selected each time. Now, the syntax does not make explicit that α is an arbitrary F in the way the syntax of $\varepsilon x F(x)$ does. However, in the context of the proof, when α appears in $F(\alpha)$ due to the application of EI to $\exists x F(x)$, it is clear that α functions no less as an arbitrary F than $\varepsilon x F(x)$ would.

Next, we want to say a few words about why pointing to the effects of including arbitrary objects in *intuitionistic logic* is not unnatural in a discussion of natural deduction. As [17] notes, if one looks back to one of the two pioneering works of natural deduction, [14], one sees that the basic system Gentzen develops is the one he calls LJ, i.e., a natural system for *intuitionistic* predicate calculus. To get classical logic from LJ Gentzen considers two approaches, each of which requires giving up some nice feature of LJ— either its axiom-free nature, or the uniform intro/elim pairs for each logical operator. Gentzen opts for the former, adding the law of excluded middle as an axiom scheme.[9] Right from the inception of natural deduction, then, one can see good reasons for regarding intuitionistic logic as the most natural version of natural deduction. Arguments to this effect do not go away in philosophy. One sees, for example, kindred claims in [7] that the rules of intuitionistic logic are *harmonious* in a way that classical rules are not. We do not want to wade into those deep waters here. We merely claim that considering the effects of arbitrary objects in intuitionistic logic is not "changing the subject."

Thirdly, as is well known, while the addition of Hilbert's ε operator to classical logic is *conservative*—that is, it does not allow us to prove anything that was not already provable, other than formulas that include epsilon terms—the addition of epsilon terms is a *non-conservative* extension of intuitionistic logic. It obviously allows the derivation, for instance, of $\exists y(\exists x F(x) \Rightarrow F(y))$. Starting with some interesting papers written by J.L. Bell, some less obvious and more interesting non-conservativeness results have been proved for the addition of ε in intuitionistic logic. For instance, given certain modest assumptions, the intuitionistically invalid form of De Morgan's law is provable if epsilon is present, but not when it is absent. For discussion of such results, see [2, 3, 6].

Putting some pieces together, the non-conservativeness does not depend on the specific presence of epsilon terms. For instance, $\exists y(\exists x F(x) \Rightarrow F(y))$ is clearly also provable if we adopt "ambiguous names," as described by Suppes, in an intuitionistic system: assuming

---

[8]That is, lower case Greek letters, while constants and variables are Roman letters from opposite ends of the alphabet.

[9]Though he mentions that the resulting system is equivalent to the one that results by adding a rule of double negation elimination. This approach introduces a rule that allows one to eliminate two operators at once, and so adds a rule of a different character from the others in his system.

$\exists x F(x)$, one may infer $F(\alpha)$, so by $\Rightarrow$I arrive at $\exists x F(x) \Rightarrow F(\alpha)$. By $\exists$I, we have the result. Indeed, the result is likely to be provable in any Quine system of natural deduction, i.e., any system that incorporates a version of EI, barring the addition of significant additional constraints on the use of the rule.

The lesson here, we think, is that the metaphysical language in which talk of arbitrary objects is couched is not logically inert, and so is no mere philosophical *façon de parler*. The addition of excluded middle to intuitionistic logic greatly strengthens the logic— it turns it into classical logic. The extra power conferred by the addition of excluded middle *masks* the extra-power conferred by the addition of arbitrary objects. Moreover, what is hidden is something of considerable philosophical significance—by assuming the existence of arbitrary objects, we are making an assumption that increases logical power. Talk of arbitrary objects does not explain correct quantificational reasoning, because it implies more than correct quantificational reasoning alone implies. We think that this is good reason, when teaching logic to new philosophers, to dispense with this dispensable and misleading way of speaking.

Of course, our system is equivalent to what Pelletier says "could be called 'Gentzen systems,' " [17, p. 12], i.e., those systems which include $\exists$E instead of EI. Gentzen systems, starting with LJ, yield intuitionistic logic, and so no such non-conservativeness results are in the offing. The crucial feature that makes this so is that the terms that seem to refer to arbitrary objects occur only in assumptions and in formulas within the scope of those assumptions. Such systems, at least with respect to the existential quantifier, seem closer in spirit to what we suggest here. When we know $\exists x F x$, we can interpret the rule as merely allowing us to "give a name" to one of the Fs as a convenience. It seems, then, that the commitment to the existence of "an arbitrary F" is less robust than in a Quine system. For the universal quantifier, though, terms seeming to refer to arbitrary objects are not always in the scope of an assumption in this way, even in Gentzen systems, so let us turn to consideration of the universal quantifier.

It is probably a familiar thought among those who have taught standard natural deduction systems that the instantial terms in $\exists$E and $\forall$I behave quite differently, and so it may not mean the same thing when we call each sort of object "arbitrary." We think that this suspicion is quite right, and think that considering Hilbert's epsilon operator together with another operator sometimes discussed by Hilbert and his collaborators, viz., the τ-operator, in intuitionistic logic helps make the point clear. The τ operator is governed by the axiom dual to epsilon,

$$F(\tau x.F(x)) \Rightarrow \forall x.F(x).$$

Once again, the extra strength of classical logic paints over an important difference. In classical logic, the two operators are interdefinable. For instance, one can set

$$\tau x F(x) = \varepsilon x \neg F(x).$$

In intuitionistic logic, this is no longer so. Not only are the two operators no longer interdefinable, but while they are both non-conservative over intuitionistic logic, they give rise to different logics. For a simple example, [2] shows that τ makes provable the "infinite De Morgan's law"

$$\neg\forall x P(x) \Rightarrow \exists x \neg P(x),$$

while epsilon does not. We see, then, that the assumption of "arbitrary objects" of the relevant sort for the two quantifiers are both non-conservative assumptions over intuitionistic logic, but each in different ways. This justifies the suspicion that there is an important difference between the two types of arbitrary object.

Once again, of course, the Gentzen rules for the existential quantifier must be weaker than τ, since after all LJ is intuitionistic logic. The fact that commitment to arbitrary objects as seen in Quine systems is non-conservative in intuitionistic logic seems to us a strong objection to such systems. How, though, do the considerations above add up to a justification for adopting the natural deduction system on offer here, rather than the Gentzen systems equivalent to it that include the usual $\exists$E and $\forall$I rules? If ε and τ are indeed reasonable mechanisms for indicating "an arbitrary F" (in two different senses), then the uses of instantial terms in Gentzen style $\exists$E and $\forall$I do not accurately reflect the idea of arbitrary objects. Since the only reasonable story on offer for the use of those terms in natural deduction systems, whether classical or intuitionistic, is in terms of arbitrary objects, we conclude that there is no good philosophical story to justify using a system with such terms.

Conclusion

We have presented a modified version of natural deduction which we suggest has a number of virtues. One of us has used the system for years in introductory logic classes, and finds it easier for students to learn than standard natural deduction systems for classical logic. It is, we think, a more accurate reflection of rigorous but informal reasoning, and in this sense restores some of the naturalness that is supposed to be one of the selling points of natural deduction in the first place. Finally, it is a system that does not require a philosophical story that conflates the semantics of singular terms and general expressions, nor does it require appeals to the dubious entities sometimes called "arbitrary objects." The arbitrary object justification is problematic, among other reasons, because seriously supposing the existence of such objects increases the number of valid inferences beyond standard quantifier rules (though this effect is disguised in the presence of excluded middle), so they are not a suitable tool to justify standard quantificational logic.

References

1. J. Barwise, J. Etchemendy, *Language Proof and Logic* (Seven Bridges Press, New York, 2000)
2. J.L. Bell, Hilbert's ε-operator and classical logic. J. Philos. Log. **22**, 1–18 (1993)
3. J.L. Bell, Hilbert's ε-operator in intuitionistic type theories. Math. Log. Q. **39**, 323–337 (1993)

4. G. Berkeley, *A Treatise Concerning the Principles of Human Knowledge* (Open Court, Chicago, 1904) [Originally published 1710]
5. A. Church, *Introduction to Mathematical Logic* (Princeton University Press, Princeton, 1956)
6. D. DeVidi, Intuitionistic ε– and τ–calculi. Math. Log. Q. **41**, 523–56 (1995)
7. M. Dummett, *The Logical Basis of Metaphysics* (Harvard University Press, Cambridge, 1991)
8. K. Fine, A defence of arbitrary objects. Aristot. Soc. Supplementary Volume **LVII**, 55–77 (1983)
9. K. Fine, Natural deduction and arbitrary objects. J. Philos. Log. **14**, 57–107 (1985)
10. K. Fine, *Reasoning with Arbitrary Objects*, vol. 3 of Aristotelian Society Series (Basil Blackwell, Oxford, 1985)
11. F.B. Fitch, *Symbolic Logic* (The Ronald Press Company, New York, 1952)
12. G. Forbes, But *a* was arbitrary .... Philos. Top. **21**(2), 21–34 (1993)
13. G. Frege, Was ist eine Funktion? in *Funktion, Begriff, Bedeutung*, ed. by G. Patzig, 6th edn. (Vandenhoeck & Ruprecht, Göttingen, 1986), pp. 81–90 [First published in a Festschrift dedicated to Ludwig Boltzmann on his 60th birthday, 20 February, 1904]
14. G. Gentzen, Investigations into logical deduction, in *Collected Works of Gerhard Gentzen*, ed. by M. Szabo (North-Holland, Amsterdam, 1969), pp. 68–131 [Originally published in German in 1934]
15. H.N. Gupta, On the rule of existential specification in systems of natural deduction. Mind **LXXVII**, 96–103 (1968)
16. A. Hazen, Natural deduction and Hilbert's ε-operator. J. Philos. Log. **16**, 411–421 (1987)
17. F.J. Pelletier, A brief history of natural deduction. Hist. Philos. Log. **20**, 1–31 (1999)
18. J.C. King, Instantial terms, anaphora and arbitrary objects. Philos. Stud. **61**(1–2), 239–265 (1991)
19. E.J. Lemmon, Quantifier rules and natural deduction. Mind **LXX**, 235–238 (1961)
20. E.J. Lemmon, A further note on natural deduction. Mind **LXXIV**, 594–597 (1965)
21. E.J. Lemmon, *Beginning Logic* (Hackett Publishing Company, Indianapolis, 1978) [First published in Great Britain 1965]
22. K.A. Peacock, Definite and indefinite articles in elementary predicate logic. Unpublished MS, presented at the WCPA, Saskatoon (2007)
23. D. Prawitz, A note on existential instantiation. J. Symb. Log. **32**(1), 81–82 (1967)
24. W.V.O. Quine, *Mathematical Logic*, revised edn. (Harvard University Press, Cambridge, 1965)
25. N. Rescher, Can there be random individuals? in *Topics in Philosophical Logic*. Syntheses Library (Reidel Publishing Company, Dordrecht-Holland, 1968), pp. 134–137
26. B. Russell, *The Principles of Mathematics* (Cambridge University Press, Cambridge, 1903)
27. P. Suppes, *Introduction to Logic*. The University Series in Undergraduate Mathematics (D. Van Nostrand Company, Inc., Princeton, 1957)
28. A. Tarski, *Introduction to Logic*, 2nd edn. (Oxford University Press, New York, 1951)
29. N. Tennant, A defence of arbitrary objects. Aristot. Soc. Supplementary Volume **LVII**, 79–89 (1983)

D. DeVidi (✉)
Department of Philosophy, University of Waterloo, Waterloo, ON, Canada N2J 2P3
e-mail: David.Devidi@uwaterloo.ca

H. Korté
Department of Philosophy, University of British Columbia, 1866 Main Mall, Buchanan E370, Vancouver, BC, Canada V6T 1Z1
e-mail: Herbert.Korte@ubc.ca

Category Theory and the Search for Universals: A Very Short Guide for Philosophers

Alberto Peruzzi

Abstract The aim of the paper is to present the categorical notion of an adjoint functor as a key to formally capturing the philosophical notion of "universal" especially as it figures in relation to semantics and epistemology. In the first part (first section to seventh section) the relevance of category theory for the main topics of analytic philosophy is suggested, in opposition to a widespread conservative attitude towards the entrenched conjunction of logic and ∈-based set theory. In the second part (8th section to 16th section) the concept of an adjunction is introduced and shown to provide the framework of some fundamental examples of universality. The paper is of an introductory character, because it is addressed to a broad philosophical audience, in particular to philosophically-oriented logicians and logically-educated philosophers with no previous knowledge of category theory.

Keywords Adjoint functor · Analytic philosophy · Category theory · Logic · Universality

Mathematics Subject Classification Primary 97E20 · Secondary 18C10 · 03G30

Motivations and Questions

Asked to characterise twentieth century philosophy, the list of features to be considered could be the subject of long debate. But among them there is undoubtedly an unprecedented attention to language and structure: an attention which also results in the mutual use of these notions within the study of each. As for the structure of language, it is widely acknowledged that its expressive power, its role in controlling the range of possible inferences, and finally the internal limitations related to self-reference have all been specified in logical terms. As for the language best suited to the description of structure in general, it was generally believed in the community of logicians and philosophers working in the analytic tradition that this was supplied by ∈-based set theory and that no structure could be defined without appealing to the notion of membership. Thus, logic + set theory in this view was taken to offer the main, and most comprehensive, framework for any philosophical investigation aiming at conceptual clarity and inferential rigour in the analysis of meaning and knowledge.

In fact, philosophy has always addressed issues which call for more than just thoughtful attention to meaning, rigour and the formal articulation of ontological inquiry. The twentieth century is no exception. But it is disputable whether the goals that philosophy

© Springer International Publishing Switzerland 2016

F.F. Abeles, M.E. Fuller (eds.), *Modern Logic 1850-1950, East and West*, Studies in Universal Logic, DOI 10.1007/978-3-319-24756-4_8

aims at beyond this can be pursued without such an underlying framework; and even the claim that the results of such investigations are untranslatable into such and such a framework already presupposes some competence in logic and set theory. Given the variety of formal languages, logical systems and axiomatic set theories we face today, any thesis as to the manner in which to pursue general philosophical inquiry on language and structure is at risk of vicious regress or being burdened with idiosyncratic technicalities.[1]

The search for underlying principles which can provide a unified view of the variety of languages, logics and structure-kinds (with their corresponding "signatures") brings philosophy back to the longstanding debate on "universals" in dealing with the structure of concepts. "Universals" have been taken in different senses, each deserving careful analysis. But such analysis can be expected to have consequences for our view of ontological architecture and the architecture of knowledge as well. If there is a mathematical clarification of the concept of "universality", it affects the search for "universal" principles on which to rely in order to supply a foundation of knowledge in any domain, once that knowledge is expressible in mathematical form. In the foundations of mathematics, such a clarification already exists and deeply affects the way we look at them.

Category theory identifies the principles that govern the architecture of mathematical structures in their mathematically significant relationships, and expresses these in the form of mutual constraints satisfying certain *universal* properties, which have consequences for the characterisation of each specific structure involved whilst at the same time providing cross-domain principles of characterisation.

The resulting conception of mathematical structure is independent, in general, of the internal composition of the objects in question—more specifically it does not depend on their being treated as *sets*—but this altered conception does not involve some vague appeal to the no less vague totality of relationships in which something stands to something else. Thus, as a conceptual framework, category theory is neither "substantialist" nor "relationist". What this theory aims at is rather the detection of a highly specific kind of relationship between structures, one that takes the form of special maps which we can define as "universal". Moreover, the explanatory power of such universality lies in its being sufficient to recover the known properties of the structures standing in such relationships, and in making it possible to discover new properties of those structures.

To what extent does the general notion of structure, as commonly used in mathematics, linguistics, physics and other sciences, fit with such a categorical framework? Can the principles governing such a framework lead to a satisfactory foundation of mathematics? How do they affect our understanding of logic? One principal achievement of category theory is that the basic notions of logic can be defined in terms of universal properties as expressed in the theory. But what is that specific kind of universality?

---

[1] For any given formalised philosophical thesis, by its "idiosyncratic" character I mean one *essentially* relative to a chosen formal "base", thus one that cannot remain stable under change of base.

Topics of Philosophical Interest

These cursory remarks, in much need of further clarification, and this list of questions suggest reasons why category theory deserves the attention both of philosophers and mathematicians interested in foundational issues. Further reasons for its study by philosophers are related to the way categorical universality affects the treatment of the following topics:

(a) MEANING. The concepts of category theory make possible a semantics that is more general, but no less rigorous, than the extensional semantics which forms the basis of classical model theory and the semantics for modal idioms that makes use of quantification over possible worlds. This gain in generality and flexibility allows a more abstract, but also more finely articulated, analysis of the traditional opposition between extensional and intensional contexts.

(b) THEORY AND TRUTH. Category theory strongly constrains and reformulates the concepts of "theory", "interpretation" and "model" in an original and fruitful way, so that the resulting framework contains Tarski's notion of truth as a special case, and at the same time also provides a sophisticated analysis of predication in natural languages.

(c) FORM OF PRINCIPLES. The special connection between these two features, (a) and (b), already reflects a striking and novel feature of categorical semantics. Of further philosophical relevance is the *equational* character of the principles expressing "universal" properties. This character, together with the emphasis on the search for what is "invariant", suggests new ways of connecting the foundations of physics with the foundations of mathematics.

(d) LOGIC. As already mentioned, category theory raises the possibility of identifying the basic notions of logic (such as connectives and quantifiers, but also those of variables and constants) in terms of universal properties and hence offers a profound re-orientation in the understanding of those notions. For instance, the choice between classical logic and intuitionistic logic turns out to be neither a pragmatic matter (as selecting between two different conventions) nor due to any commitment to either a realistic or a constructive view of mathematical entities. Rather, it is determined by the algebra of sub-objects of any object in a given category (one equipped with a suitable structure to deal with logical notions). Just how far-reaching are the philosophical consequences of this re-orientation and the new meaning of the "universality" of the constructions involved will be a central theme in the following exposition.[2]

---

[2]There is already an extensive literature on each of the aspects (a)–(d), but it presupposes familiarity with categorical language. Those who wish to learn more about how the notions introduced here (along with others) have been put to work could begin by consulting the following papers: [3, 10, 13, 24]; in particular, for categorical logic see [5, 6, 31].

Which Organon for Philosophy?

Until now, the community of philosophers who studied and made use of category theory has been narrow and largely confined to those concerned with the philosophy of mathematics, see [21, 22, 30]. However, the substantial impact of the theory is now evident in various fields of research outside mathematics itself: the notions of category theory have proved to be relevant in theoretical computer science, linguistics, physics, systems theory, theoretical biology and even economics.

This is not the place to analyse the reasons for the delayed and circumscribed impact of these notions on topics of major importance for philosophers of language, science, logic and mathematics. Those who recognise the centrality of the analysis of language to philosophy should also recognise the scope of the re-orientation within that analysis which category theory produces—particularly for (a)–(d). The very idea that one has to formulate any philosophical question carefully and unambiguously, in order to check the relevant arguments, intrinsically poses the demand for appropriate tools. It seems that members of the philosophical community who follow the line of development initiated in the linguistic turn still think the tools devised a century ago for this aim cannot be superseded.

Ever since Aristotle, analysis of the structure of propositions and inferences has been an essential tool of philosophy. No-one lacking familiarity with this tool could expect to make progress in many central areas of philosophy.

Such lack of familiarity implies lack of awareness of the patterns of inference constitutive of rationality itself and hence the likelihood of errors in reasoning, given that most knowledge-claims are the result of inference and truth, is not transparent. While logic as the instrumental hygiene of all inquiry may not be sufficient to guarantee truth, it is a *sine qua non* of the *search* for truth. That lesson should never be forgotten. But the specific form the ideals of clarity and rigour have taken in history should not be transformed into a dogmatic specification of the form they must take.

If our intellectual spectacles distort or darken our vision, even our most modest cognitive claims may be compromised. Hence it is necessary to have the right equipment to ensure the spectacles are kept clean; and syllogistics was intended as part of that equipment. This was no longer so for early modern philosophers such as Bacon and Descartes, who made harsh criticisms of syllogistics. Descartes assigned the place previously held by logic to algebra. Bacon and the empiricists of the seventeenth and eighteenth centuries insisted that the warrant for knowledge-claims was unattainable by purely rational methods and the advance of knowledge was hindered by a logical attitude, which presumed knowledge to be expressed in subject-predicate form. Moreover, since Galileo the relational character of physical laws was recognised and emphasised as a characteristic feature of the new physics. Whereas, even at the period of its origin, the theory of the syllogism in the Organon was already ill-adapted to the relational principles and patterns of inference found in Euclidean geometry, the language of which was used by Galileo before the discovery of the Calculus and indeed by Newton afterwards.

The same limitation is far more apparent in relation to the principles and patterns of inference of post-Newtonian science, but already in ordinary discourse there are many arguments which stand in need of a richer formal framework. It is commonly thought that the lesson was learned and that the tools of formal logic fit much better with contemporary scientific theories, insofar as they are expressed in mathematical language. Mathematical logic was developed over a century ago precisely to identify and clarify the structure of thought, in particular as manifested in mathematics.

It was even held by some to have achieved such a final clarification, in the form of the claim that there is One True Universal Logic and that all mathematical notions are definable within that Logic. Such a claim seems very far-fetched today, when the manifold formal systems at the disposal of logicians are so numerous that the task of finding the one appropriate to a given universe of discourse (or of manufacturing the required system *ex novo* by means of suitable tinkering with logical axioms and rules) is the everyday activity of logicians. The presumption that a universal logic able to serve as an umbrella for all such systems, is the pre-ordained and timeless underlying format of thought has been of little help in guiding philosophical inquiry, except in pushing some philosophers towards the conclusion that there are contents of thought which escape the possibility of formalisation.

Ordinary language, in which most philosophical problems are still expressed and considered, is every bit as complex as mathematical language. Though modern formal logic has expanded into many other areas of philosophical inquiry, it has there faced further challenges posing such questions as to the limits of any axiomatic formal system.

That attention to language which has been so central in shaping "analytic philosophy" could not have existed without the formal treatment of connectives and quantifiers for a language with $n + 1$-ary relations; in turn, that treatment is argued for in terms of some underlying mathematical structure, thought of as set-theoretic in nature. As already noted, not only are the specific resources of such a formalism assumed by arguments aimed at an ideal language in which the structure of thought is (allegedly) finally rendered transparent, but the very same resources are assumed in arguments supposed to demonstrate the inadequacy of formal methods to capture the expressive subtleties of natural language. Indeed, no serious account of such notions as truth, meaning, knowledge, explanation, objectivity or even free will can dispense with mastery of logical tools, though such mastery can by no means be a sufficient condition for such an account.

It is true that there have been philosophers, even amongst those inspired by Aristotle, whose work showed little familiarity or engagement with the theory of the syllogism. It is also true that those who do not care whether the lenses of their intellectual spectacles are clean or not, typically end up not being able to see clearly or very far. What is sometimes overlooked is that those who are so focused on the state of their spectacle lenses as to lose interest in the objects in their field of view also end up in the same way. Hence the ironic quality of Kant's remarks on "the gymnastics of the learned".

In the twentieth century, many analytic philosophers identified logical investigation of language as the core of First Philosophy. Differences as to which fragments of logic were of the deepest philosophical consequence were often at the heart of their arguments. For example, classical first order logic was customarily treated as the central tool and paradigmatic idiom of ontological inquiry—sometimes outfitted with modal or epistemic operators, sometimes modified to take account of other constraints and desiderata, such as constructivity. Irrespective of whether that choice of organon has always been fruitful and whether the categorical approach suggests a viable alternative and in reality a more comprehensive perspective, it is certainly a matter of historical record that this choice of logical tools regulated the way the scope and character of systematic ontological inquiry were conceived in analytic philosophy.

The Bias for Sets

The analytic "turn" has remained essentially tied, albeit often implicitly, to the assumptions embodied in semantics *à la* Tarski, which find expression in the language of set theory. The adoption of non-classical logical systems in treating certain problems or areas of inquiry did not fundamentally alter this situation. Set theory, conceived as the ontological hinterland of quantification theory, provided the fundamental framework. That some authors opted for an axiomatisation of set theory differing from classical ZF through a desire to reduce ontological commitments (e.g. to higher infinities) as a result of constructivist leanings did not essentially change matters.

It is true that more conceptually far-reaching alternative ontological frameworks have been proposed, such as mereology, prompted by philosophical motivations distinct from those underpinning the Tarskian option. But they have not established themselves as serious rivals, since there has been little convincing evidence either for (1) their relevance to mathematical practice, (2) their import for foundations, (3) their transferability to domains structured in terms of basic notions other than those they were originally intended to model, (4) their effectiveness in providing a general idiom of systematic scientific inquiry or (5) their help in clarifying our ontological commitments.

In all five of these respects, set theory continues to provide an inescapable point of reference: one that permits further and flexible specification tailored to the syntactic and semantic structure of a wide range of formal and natural languages and to the expression of the ontological commitments of a wide variety of bodies of knowledge.

No one should ignore the power of set theory, nor underestimate its adaptability. How else would its vocabulary have penetrated so many areas of mathematics? To this must be added the ease of grasping its basic notions—a feature which has made possible its introduction even into the primary school curriculum

Finally, regardless of the axiomatisation adopted, the theory of sets has allowed us to do an extraordinary thing, namely to give a unified foundation for the whole of mathematics: within specific branches specific axioms select certain sets rather than others, but the fact remains that all the subject matters of which mathematicians speak in different regions of their discipline can *au fond* always be regarded as variously structured sets.

In the early twentieth century, the efforts to axiomatise set theory were closely related to positions in the philosophy of mathematics. But later, as the landscape of alternative axiomatic systems for set theory developed into an intellectual mirror of the Amazonian rain forest, the technicalities involved in detecting and comparing the ontological commitments of one proposed system relative to another became more and more demanding (mainly in connection with issues about the size of large cardinals). As a result, philosophy of mathematics retreated into concerns about the nature of mathematical entities, and to a concern with the limitations on both provability and definability, stemming from Gödel's Second Incompleteness Theorem. This was paralleled by a concern with the consequences of Tarski's Theorem for the internally "ineffable" character of truth, with the Löwenheim-Skolem Theorem (leading to the Skolem Paradox), and with a special attention to the consequences deriving from the non-computable character of notions relevant for the philosophy of mind.

While mathematics has been an area of minority concern among linguists and philosophers of language, it is nonetheless striking that those philosophers for whom the analysis of language occupies a primary place in philosophy, should seem to think that such a strategy can stop short of trying to analyse the sources of the expressive power of the language of set theory in understanding the structure of any universe of discourse. In particular, analytic philosophers frequently discuss the notion of truth as if the problems posed for the notion of truth about the set-theoretic hierarchy could be by-passed.

The intuitive familiarity of the notion of membership seems tailor-made to illustrate a doubt, encountered in many areas of inquiry, about the status of definitions; a doubt which plays a crucial theoretical role—namely, whether the definiens is really less problematic than the definiendum. For, in set-theoretic terms, the truth (in a fixed valuation v) of a presumedly atomic statement such as "Mike is blond" is equivalent to the fact that the individual referred to by the name "Mike" (interpreted by v) belongs to the set of elements (of the given domain) which the valuation assigns to the predicate, i.e., $v(\text{Mike}) \in v(\text{Blond})$. But notice: there are two far-reaching presuppositions involved in such a claim, namely, (1) that the reference of an individual constant could bypass the structure which makes the stability of individual entities possible and (2) that the use of any predicate could bypass the question of how a (purportedly timeless) collection is accessed.

Set theory, however, is not the last word, or at least was never proved to be so. Since the mid-twentieth century, category theory has developed to occupy a position in which it offers an alternative to $\in$-based set theory, since it provides a language no less expressive and, most importantly, aims to provide a more adequate framework for both the foundations and the architecture of mathematics (the two being no longer treated as separate issues). Moreover, it is one endowed with greater simplicity and ease of use, given suitable training. But above all it is one which provides a superior understanding of the relationship between different areas of mathematics and better reflects the evolving character of that relationship.[3]

To go back to the previous example, every philosopher is familiar with the distinction between type and token, but only a few care for investigating the structure of a collection of types. At its birth, the idea of a hierarchy of types with mutual relations (of logical significance) could be thought of simply as an antidote to paradoxes. But after the development of a typed lambda-calculus, type theory became a practical resource (think of "data-types"), proved to offer a general framework for computer science and, through the Curry-Howard correspondence (proofs-as-programs, propositions-as-types), for a new perspective on the whole field of logic itself. Over the last decades, various kinds of constructive type theory have been introduced, suitably reformulated and powered with respect to its original non-constructive version (whether the hierarchy is simple or ramified) and have been argued to provide as appropriate foundational systems, from Martin-Löf type theories to the recent "univalent program" focused on homotopy type theory. Such foundational projects have deep implications for the way philosophical questions about foundations of mathematics are formulated.[4]

Analytic philosophers have thus far given little attention to this new generation of type theories. However, they contain at least two aspects which should be of interest to them. The first is that, in view of Cantor's Theorem, the standard universe of sets contains no

---

[3]The phenomenology of these relationship is extensively described in [18].

[4]As for the relationships between intuitionistic type theory and topos theory, [4, 11] are essential references. At a more basic level, [1] is a clear presentation of the categorical approach to computability.

X (other than a singleton) such that $X^X \subset X$, whereas there are domains of theoretical relevance which call for such an X and these can be described in categorical terms. The second is that semantics for higher-order lambda-calculi makes an essential appeal to category theory. Conversely, there are type-theoretical translations of the language of categories, but such translations rely on assumptions which, however well motivated, are philosophically demanding: in fact, they involve a commitment to specific properties for the categories which support such translatability.

Passing to Categories

Many continue to believe that the language of sets has conceptual priority in some manner with respect to the primitive notions of category theory. Though this issue could be discussed endlessly, this introductory paper is not the place for such discussion. But to provide just one (biased) example of a parallel issue in intellectual history, the replacement in physics, of a language which centered around static qualities with one formulated in terms of dynamical quantities met with a similar resistance at the period of Galileo and his struggle with the Aristotelians of his time.

Further: (1) since the notion of collection is exploited in defining a category and (2) since the very notion of membership is itself defined within the categorical approach, rather than being taken as primitive, a certain puzzlement on the part of the mathematician, logician or philosopher trained to think in terms of set theory is to be expected.

For many years a wide-ranging debate has been conducted on whether and in what sense category theory can be claimed to provide a foundation which is really an alternative to the theory of sets, and as to the very sense—or multiple senses—which the expression "is a foundation for" can take with respect to mathematical notions and about the ultimate meaning of the methods of definition and construction of category theory. Beyond that foundational debate, a further controversy concerns the philosophical significance of category theory.

For example, there are those who have seen in category theory the revenge of Heraclitus over Parmenides in mathematics while others have seen the theory as the most precise formulation of the position which among philosophers goes under the label of structuralism in mathematics (and in some versions also in physics and in metaphysics). From the Erlangen Program to Bourbaki's *structures mères*, there is indeed a "structuralist" tradition in mathematics which might support such a reading. But even setting aside the historical circumstance that Bourbaki mingled aspects of that tradition with a poorly designed set theory, the issue of "structuralism" deserves careful re-examination, especially on the part of dialectically-minded philosophers. Such examination is needed, on the one hand because previous attempts to develop a dialectical logic did not produce a system able to provide a foundation for the whole of mathematics; on the other hand such re-examination of the notion of structure is suggested by the fact that category theory does not aim to embody some exotic non-standard logic, but rather emphasises the intrinsic "emergence" of logical notions themselves out of mathematical structure.

The resulting changes in viewpoint on logic resulting from category-theoretic developments have been the subject of extensive discussion, although the task of putting these developments to work in the analysis of natural language has thus far been less explored, apart from some aspects associated with "categorial grammar".[5] Needless to say, such analysis can be of great importance in advancing research in this field. But thus far, recognition of the pertinence of category-theoretic notions for the logical analysis of natural language has been noticeably lacking in this area of philosophy.

Just as happened in the early years of analytic philosophy with respect to what was then the *novissimum organum*, i.e., mathematical logic, so with category theory: developments and changes in viewpoint in philosophy resting on the use of a new language and a new set of tools have met with widespread resistance and hostility.

Like it or not, the terms of the new debates which have opened about the foundations of mathematics, and to a lesser extent and in a more preliminary form, in semantics and the theory of meaning, with the advent of category theory are intelligible only to those with some knowledge of that theory. In this there is nothing strange or suspect, since to undertake philosophical reflection on a topic as it appears by means of a tool, one must first know something not only about the subject but also about the use and purpose of that tool. Just as to judge a musical instrument one must know how to play it, as well as knowing something about music.

After one century of analytic philosophy, it would be unthinkable that a graduate student wanting to do research on any topic, certainly on a "recognised" topic within the norms of the analytic community, and one which might lead to an academic career, could totally ignore logic. However, if that same student is told that, from a categorical standpoint, the investigation of logical notions involves some familiarity with topology and geometry, the reaction is often one of bewilderment or resistance. Yet the task of making explicit the knowledge which is presupposed or hidden in our use of language was at the core of the project initiated by the founding fathers of the "linguistic turn". The claim that some implicit knowledge of topological or geometric concepts is necessary for any understanding either of the way we talk about the common sense world or the realm of science should thus strike such a student as not only a profoundly consequential suggestion from the standpoint of general philosophical inquiry, but one firmly located within the agenda or syllabus which has shaped the methods and outlook of the analytic tradition.[6]

Universality Only Becomes Recognisable in an Expanded Universe

Whereas for the mathematically educated reader there are (at least in English) many good introductory texts to category theory (some of them freely accessible on the internet), the same is not true for those educated in philosophy.[7] As an aid for young philosophers

---

[5] Jim Lambek pioneered the investigation of categorial grammar in terms of category theory, see [9]. *En passant*, a purely lexical note: since the adjective "categorical" was already established in model theory as having a very specific meaning, one might opt for "categorial" in application to category theory. But unfortunately, "categorial" has a no less specific meaning within grammar. In the face of such an impasse, standard usage in the community in question is declared the winner.

[6] Readers who know Italian can find these aspects examined in [29].

[7] A starting point is [23]. For a book, see [16].

interested to learn what category theory has to offer, this essay aims to introduce, albeit very briefly, its basic notions and illustrate them with some examples. Therefore, in the following pages you will not find an account of the stages through which category theory has developed and which led to the program of the categorical foundation of mathematics; nor a discussion of the applications of these basic notions to other philosophical topics, such as those mentioned above.

Even cursory exposition of the categorical reformulation of logic, in particular of semantics, and of the motives underlying that reformulation would presuppose familiarity with those basic notions of category theory this work aims to provide. Still less will we be able to study specific applications of category theory to classical problems of philosophy, and analytic philosophy in particular. Such topics include the contrast between "intensional" and "extensional" aspects of meaning, the semantics of mass nouns, the non set-theoretic aspects of predication, whether the objectification of functions provides a good "meter-stick" for registering ontological commitments, and the study of recursion and diagonal arguments (in relation to self-reference)—to list only some relatively well-established uses of categorical methods in philosophy.[8]

But, just to illustrate the relevance of such methods for the analysis of language, we can point to the radical change of setting for the theory of definite descriptions, the oft-cited "paradigm of philosophy", that results from its being re-presented in a form drawing on the conceptual toolkit made available by category theory: the greater subtlety of analysis for singular terms made possible here by the use of these tools, [25], also suggests lines on which other central issues in philosophy may not just benefit from categorical treatment or reformulation, but may actually require it.

The following pages do not aspire to originality. There is, however, one aspect of the presentation here that differentiates it from other introductory material addressed to a philosophical audience. This is the emphasis on the concept of *universality*: one of the most recurrent and debated in philosophy. Not only does this concept receive a precise formulation in category theory: it can be considered its very leitmotif.

Among the tasks of philosophy, some presuppose that others have already been accomplished. The latter can thus be considered as basic. To this class belong the tasks of identifying the fundamental concepts involved in a given subject-matter, clarifying their exact sense (to be selected for theoretical aims) and distinguishing it from other senses manifested in common language. No less basic is the exploration of the precise ways in which such fundamental concepts are combined and in which their mode of combination contributes to the theoretical articulation of the subject in question. To understand the philosophical significance of a theory—in particular its epistemological basis—it is not enough to ask what kinds of things the theory speaks about, the ways in which they are related, how do we access them, and which grounds can support the theory's principles, once it is axiomatised. It is also necessary to provide an account of those patterns of construction that make it possible to use these concepts in unintended or unforeseen contexts and as a consequence of this cross-domain character, their role can be motivated in a non *ad hoc* manner.

In the case of category theory this additional requirement is decisive, (1) because these very patterns are the subject matter of the theory itself. Thus we find ourselves dealing

[8]These topics are covered by essays collected in [19].

with a situation involving an epistemically productive example of self-reference, and (2) because it is a theory that deals equally with objects and maps, entities and processes, sets and functions, data-types and algorithms, structures and their transformations: in sum, whenever it alludes to the existence of something, that existence condition is always characterised as the (unique)[9] result of a *universal* construction. Within category theory, the very notion of "for all x" which at first sight seems free of any reference to maps is shown actually to be the track of an implicit map, relatively to which the notion is defined by means of a universal property. The same applies to the notion of "for some x". Thus, both universal and existential quantification are instances of one and the same notion of universality, associated with the *adjunction* pattern (between functors), which is the core of the theory.

If, at the end of these philosophical considerations, I may be permitted a personal remark: I have written elsewhere that analytic philosophy, having provided the most important chapter of twentieth century philosophy, is dead. Indeed I thought I had contributed to the collection of arguments required for its epitaph. Now I want to add a rider. I should be willing to withdraw this obituary notice if it becomes clear it is possible to re-fashion the program of analytic philosophy in a consistent manner by expanding its underlying logical resources with the aid of category theory, rather than keeping them confined to the universe of sets (whilst not neglecting the role of sets and functions as a limiting instance of that of objects and maps).

To be brought to full fruition, such a re-conceived version of the analytic program likely requires a new generation of philosophers free from the burden of sacred texts and the evocation of standard authorities. The shift of focus over the last 20 years by many philosophers of language and those working on cognition, towards the new frontiers of computational science and neuroscience has, I believe, not so much undermined the neo-scholastic attitude inherent in much current work in analytic philosophy as spread it to further fields. But seminal work of a quite different sort, which cannot be indefinitely ignored, promises to bring about a shift in the overall picture and some of these developments lead me to believe such a re-conceived program is already taking shape. If the conceptual tools presented in the following sections were of help in its development, I would be glad to concede that my former epitaph was wrong.

Of Substance and Structure: Which Ontology?

It requires no appeal to *a priori* principles to recognise (1) that the objects which we refer to (be they individuals or species, particulars or universals, concrete or abstract entities) as well as the systems which they compose are more then just bags of dots; (2) that the features whereby objects and systems can be identified are not reducible to a fully bottom-up or fully top-down representation in terms of discrete quantities; (3) that, rather, the very existence of the objects and systems we refer to is the outcome of processes which are described in terms of continua and with respect to which objects and

---

[9]"Uniqueness" is always intended as *up to isomorphism*, in a sense that will be defined in what follows. The fact that a theory allows us to prove the existence only of those x's that are unique (in satisfying a given condition) marks one of the distinguishing features of the categorical approach.

systems preserve enough of their structure to be re-identified, to the extent needed for the relevant "qualitative" states to be discretely distinguished in language; (4) that, if objects and systems change, their variation must be subject to law-like regularities, otherwise they could not be reliably detected and stably referred to. This is a requirement quite independent of our success (or failure) to understand the principles governing the specific conditions of being and becoming.

The syntactic manipulation of symbols is no exception: if language is not extramundane, its patterns of construction are those of a hosted subsystem, and the existence of a range of interpretations for any set of linguistic expressions is made possible by the properties of the hosting system. In its turn, the structure of the whole system is constrained by the requirement that it jointly hosts the structure of language and the structure of models for any theory expressible in the language. Though the possibility of a common formalisation for both structures was one of the great achievements of set theory, the strict dependence of a theory on its syntactic presentations by means of axioms as well as the absence of constraints on interpretation (for it only has to be a function) left the relation between system and subsystem unprincipled.

The categorical approach, centered on a very general kind of algebra, provides a framework which takes account of (1)–(4) in a way which highlights the geometric structure of variable objects and systems, and better reflects the fact that relationships between language and world themselves have just this kind of principled genesis in the world.[10]

What matters most in this way of thinking is not the search for an ultimate ontology (the questions of what kinds of entities there are, which are entities of a given type and which of another, which possess a given property and which do not, in which relationships objects—and types—stand to one another, which are to be treated as fundamental— the basic "substances"—and which are to be seen as the derived/dependent/supervenient ones). But it does not at all follow from this that such a way of thinking is "structuralist" in nature (in spite of what many believe), nor does it follow that we should content ourselves with a pluralist ontology or a linguistic reformulation of the above substantialist perspective by means of even the most sophisticated type theory.[11]

Indeed, it should be noticed in passing that structuralism, in its most widespread current versions, is itself the limiting case of a particular ontological perspective: if entities are defined by the structure to which they belong, i.e., from the collection of mutual relations in which they stay, Structure itself takes on the role of Substance in former ontological traditions. For if Substance lies in a web of relations rather than in things, then, or so the structuralist seems to be claiming, it is Structure itself that now takes on the role of that-in-virtue-of-which any specific thing is identified.

One might object that structuralism is always confined to a specific language L and to a specific class of structures associated with L as the class of possible "universes-of-

---

[10]There are also "systemic" approaches to semantics, but so far they are inadequate for the management of logical syntax and call for notable modifications as we move from one semantic field to another. Thus they lack the uniformity found in the categorical approach.

[11]The objective conditions for the possibility of exercising any typing activity are usually set aside in the construction of a type theory. Consequently, the entire philosophical discussion on the choice of one type theory rather than another is essentially pre-Kantian.

discourse". But a consistent relationism based on the notion of role-in-a-structured-system lends itself naturally to recursive iteration. Therefore, any single structured system would in turn be defined by the whole set of relations it has with the others and with itself. As a whole, this recursive hierarchy of structured systems is absolutely indeterminate, since no one can have access to it in the limit and the totality of relations which link structured systems to each other is no less indeterminate.

But then: how to avoid a vicious circle or an infinite regress? In order to define each structured system S, some determinate (possibly finite) set R of relations between S and some determinate (possibly finite) set of systems ought to be sufficient. But which R? And why just those relations and those other systems, rather than others? Integral relationism cannot answer such questions without contradicting itself, because in order to answer it must appeal to constraints of a more than generically relational nature.[12] Category theory permits an answer. That is by no means its only claim on the attention of philosophers, but this feature of category theory alone should already be enough to compel such attention.

The Search for Universality at the Core of Category Theory

The old debate between substantialism and relationism took on many new guises in the twentieth century and the two opposing positions attracted new adherents whose articulation and exploration of the issues reflected their points of entry into the debate through concerns specific to disciplines as diverse as linguistics, physics, cognitive science, epistemology, as well as mathematics. For at least as long as the terms of the debate remain locked in the forms sketched here, it will presumably never end.

But whether objects are defined by the elements that compose them or by a network of mutual relations, it has to be the case that we can only identify the objects we are talking about within a given system (and furthermore, can only identify that system itself) if the reference of some expressions of our language to objects external to that system is kept fixed. This is to recognise what I have elsewhere labeled the Principle of Invariant Referential Potential [27]. Otherwise we could not even understand what we intend to uniquely characterise or argue why we are able (or not) to characterise them in one way rather than another.

Here it is crucial to recognise that the uniqueness in question does not refer to the identification of a thing-in-itself (be it a ur-element, an object, or a network of relations) but rather to the existence of universal properties, intended not so much as relative to an isolated system but rather as concerning a special kind of relation between coupled systems, with possibly different structure. It is precisely this requirement that is met by category theory and, more particularly, which is supplied by the concept of adjunction between functors.

In the definition of adjunction four concepts come into play: *naturalness, isomorphism, functor* and *category*. They are not mutually independent. Indeed, each concept assumes the following one. Only when all of these concepts, starting from that of category, have

---

[12]The escape route provided by saying that anything (hence any such added constraint too) can be relationally conceived, though not all at the same time, is still waiting for a consistently relational justification.

been defined is it possible to understand why adjunction allows us to capture a notion as philosophically important as that of "universality", one which spans the whole history of philosophy.

It is not the first time the concept of adjunction has been assigned such a central role. It may be objected that the concept of universality is itself more general than the concept of adjunction and that it retains this wider generality in category theory too. But, in all mathematically significant cases of universality, it is noteworthy in my view that we find ourselves dealing with adjoint functors. This circumstance brings to mind a saying of Saunders Mac Lane, that in mathematics it is correct and fruitful generality, as opposed to maximal generality, that is to be sought. As was noted for the first time by Bill Lawvere, the notion of adjunction allows us to make precise a sense of the notion of the foundations of mathematics which is distinct, in conception aim and method, from the main proposals advanced under that label in the first half of the twentieth century and formally refined in the second half.

To reiterate: this paper is not the first to place the concept of adjunction at the heart of category theory or to recognise it as a tool of great power in the genesis of particular mathematical discoveries and in the overall conceptual organisation of mathematics. Nor is it the first time that someone suggested its overall philosophical importance, [2]. Lawvere was the first, [12], to propose such a conception of foundations. He also argued that the Thesis of the Dialectical Unity of Opposites can be given exact expression in the form of adjoint functors, [15]. This paper tries to provide the minimal toolkit required for understanding such claims and for devising applications to further topics.

The Definition of an Adjunction and Its Prerequisites

The most common definition of an adjunction is as follows: a functor F from a category $\mathbf{C}$ to a category $\mathbf{D}$, which is written $F: \mathbf{C} \to \mathbf{D}$, is a *left adjoint* of G (and G is a *right adjoint* of F) if and only if, for each object A in $\mathbf{C}$ and each object B in $\mathbf{D}$, there is a natural isomorphism between morphisms in $\mathbf{D}$ from FA to B and morphisms in $\mathbf{C}$ from A to GB. The existence of an adjunction thus defined between F and G is denoted by $F \dashv G$.

Three concepts are involved in this definition: *natural isomorphism, functor*, and *category*.

Definition of a Category and Some Examples

We shall begin by saying that a *category* $\mathbf{C}$ is given by a collection of objects $A, B, C, \ldots$ and morphisms (arrows, maps) $f, g, h, \ldots$ (between objects) such that each morphism has a unique object as its domain (source) and a unique object as its codomain (target). So, $f: A \longrightarrow B$ signifies the morphism f is from A to B, where A is the domain and B the codomain of f.[13]

---

[13]The uniqueness condition in this definition can be relaxed if one aims at an even more general framework.

In addition, two consecutive morphisms $f: A \longrightarrow B$ and $g: B \longrightarrow C$ compose, giving rise to a morphism $g \circ f: AcC$. It is assumed that this composition is associative, i.e., that if $h: C \longrightarrow D$ is a third morphism then $h \circ (g \circ f) = (h \circ g) \circ f$. Finally, it is assumed there is an identity morphism id_- (also noted as 1_-) for each object "$-$", i.e., for any objects A and B there are morphisms $id_A: A \longrightarrow A$ and $id_B: B \longrightarrow B$, such that for every $f: A \longrightarrow B$ and any $g: B \longrightarrow C$, $id_B \circ f = f = f \circ id_A$. This special morphism therefore behaves for composition of morphisms as the identity element of a monoid, except that instead of one there are many: one for any given object.

Examples of categories: topological spaces form a category **TOP**, where the morphisms between spaces are continuous functions. Partially ordered sets (posets) with order-preserving ("monotonic") maps form the category **POS**. The category of **GRP** has groups as objects and its morphisms are maps between one group $< G, \odot, e >$ and another $< G', \odot', e' >$ ("homomorphisms") $f: G \to G'$ such that $f(x \odot y) = f(x) \odot' f(y)$ and $f(e) = e'$, i.e., the morphisms in **GRP** coincide with group homomorphisms. Sets form another category, **SET**, which has sets as objects and functions as morphisms.

Each object of these three categories, **POS**, **GRP**, **SET** can in turn be described as a category. Every partially ordered set **P** (such as a tree or a set with linear order) is a category, whose objects are the elements and between an element p and another q there is a (unique) morphism iff $p \leq q$. Any group as usually defined in algebra, that is, as a set with a structure such that between any two elements there is defined a binary operation $\odot$ which is associative, has an identity element e, and is such that every element is invertible, forms a category: it is a one-object category G, in which the only morphisms are endomorphisms $G \longrightarrow G$, corresponding to the elements of the group as usually defined, $\odot$ is composition of morphisms, and the identity element is the identity morphism. Any set can be seen as a category in which the only morphisms are the identities on each element of the set considered as an object. Note, however, that this feature is not general: the category of graphs has graphs as objects and arc-preserving maps as morphisms but a graph, with nodes as objects and arcs as morphisms, is *not* a category.

Functors and Natural Transformations

A functor is a map between two categories that meets certain minimal properties of conservation, such as that of being a category. To be a functor, a map $F: \mathbf{C} \to \mathbf{D}$ must be such that, to each object A of **C**, F associates an object FA of **D** and, to each morphism $f: A \longrightarrow B$ in **C**, F associates a morphism $Ff: FA \longrightarrow FB$ in **D**; in addition, the map F must be such that $F(g \circ f) = (Fg) \circ (Ff)$ and $F(id_A) = id_{FA}$ for each object A of **C**. That is, a functor preserves both the composition of morphisms and identities. Thus, if the objects of **POS**, **GRP**, **SET** are described as categories, the morphisms in these three categories are functors.

A functor defined in this way is said to be "covariant", because it preserves the order in which morphisms in a category are composed. A functor that reverses the composition is said to be "contravariant" and is a functor from $\mathbf{C}^{op}$ to **D**, where $\mathbf{C}^{op}$ is the opposite category of **C**: $\mathbf{C}^{op}$ is obtained by simply reversing the direction of morphisms. So, for any f from A to B, f^{op} goes from B to A).[14] We can also define functors that, in addition to preserving the category-structure from **C** to **D**, also preserve specific properties of C. For example, a functor is said to be "faithful" if it never confuses two distinct morphisms having the same domain and codomain, i.e., for any f and g: $A \longrightarrow B$, with A and B any objects of **C**, if $F(f) = F(g)$ then $f = g$.

A *natural transformation* between two functors F and H, both from **C** to **D** is a map φ such that for any object A to **C**, the component φ_A is a morphism that goes from FA to HA in **D** and it is also the case that, for f: $A \longrightarrow B$ in **C**, $\varphi_B \circ F(f) = G(f) \circ \varphi_A$.

The functors from one category **C** to another **D** in turn form a ("functorial") category $\mathbf{D}^{\mathbf{C}}$: its objects are the functors from **C** to **D** and the morphisms between its objects are the natural transformations. In the particular case of $\mathbf{C}^{\mathbf{C}}$ it will also have the identity functor $1_{\mathbf{C}}$: $\mathbf{C} \longrightarrow \mathbf{C}$. Considering the categories as objects and functors as morphisms, one can easily verify that the functors can be composed and that their composition is associative. We have thus satisfied all the conditions required to define the category **CAT** of all categories.[15]

Isomorphisms and Monomorphisms

Among the morphisms of a category, some have distinguished properties with respect to composition. Recall that our aim here was to make explicit the meaning of the notions which go into the definition of the notion of adjunction. Here is how the last of these are defined.

A morphism f from A to B in **C** is an *isomorphism* (in **C**) if there is (in **C**) a morphism g from B to A such that $g \circ f = id_A$ and $f \circ g = id_B$, that is, if f has both a left inverse and a right inverse. Finally an isomorphism between functors in $\mathbf{D}^{\mathbf{C}}$ is said to be "natural" if it is a natural transformation each component of which is an isomorphism.

There will be different types of isomorphisms as we move from one category to another, but it is remarkable that with such a compact definition, *and without the slightest nod in the direction of the notion of membership* ($\in$), we can characterise the notion of isomorphism in general, regardless of which category we are in. In **SET**, the category of sets and functions, isomorphisms coincide with the bijective correspondences, while an isomorphism in **TOP** between two spaces is a homeomorphism.

---

[14]Passing from a category to its opposite (or "dualising") highlights the fact that many mathematical notions, including those of logical interest in particular, are the mirror images of others. This provides a theoretical economy: if we have proved a proposition which is supposed to hold in general, its dual holds too (by a dual proof).

[15]As in the case of the set of all sets, care is needed with CAT too in order to avoid paradoxes. Such care usually takes the form of principles of size limitation and the most common method consists of dealing with *locally small* categories, i.e., such that for any two objects the collection of morphisms from one to the other is *small* (is a set). One might object that, by such a move, category theory demonstrates its conceptual dependence on set theory in a manner which runs counter to the claim that it can be viewed as an alternative to set theory for foundational purposes. There is more than one response to this objection and both the objection and the various responses to it are the subject of much debate. Since this paper was intended to avoid an excursion into such foundational controversies, I shall here simply refer the reader to the works of Colin McLarty for a detailed treatment. See, e.g., [19].

Among the various kinds of morphisms, there is also another particularly significant one: the monomorphisms. Given a category $\mathbf{C}$ as reference, a $\mathbf{C}$-map $f\colon A \longrightarrow B$ is a monomorphism (abbreviated as "f is monic") if $f \circ h = f \circ k$, for every h and k from an arbitrary X to A, then $h = k$. In this case we write $f\colon A \rightarrowtail B$. The notion of epimorphism is the dual of that of monomorphism, i.e., f is an epimorphism (abbr. "f is epic") in $\mathbf{C}$ if it is a monomorphism in $\mathbf{C}^{op}$). If a category has a terminal object, i.e., an object, labelled 1, such that from any other A there is one and only one morphism $A \rightarrow 1$, it is unique up to isomorphism. All morphisms $1 \rightarrow A$ are monic.

Example. In $\mathbf{SET}$ monics are injective functions and surjective functions are epic. Note that in algebra and topology there are categories in which it is easy to find examples of a monomorphism which is not injective (as a set-function) and of an epimorphism which is not surjective. Furthermore, in $\mathbf{SET}$, elements x of a set A correspond to morphisms from any singleton to A and any singleton acts as a terminal object, thus one can write $a\colon 1 \rightarrow A$ rather than $a \in A$.

This example is useful for two reasons: (1) it illustrates how two fundamental concepts of set theory—surjections and injections—can be defined without reference to $\in$; (2) it underlines that the notions of monic and epic are more general than the manner in which they occur in $\mathbf{SET}$. In fact, In $\mathbf{SET}$ for there to be an isomorphism between two objects it is enough to ensure that there is a monic map between them which also turns out to be epic. This is not true in other categories.

Subobjects

Given $f\colon A \rightarrowtail B$, we consider another monic $f'\colon A' \rightarrowtail B$. If f is factored through f', that is, if there is a $g\colon A \longrightarrow A'$ such that $f = f' \circ g$, and if vice versa f' is factored through f, then there is an isomorphism between A and A' which allows us to define an equivalence between f and f'. The corresponding equivalence class identifies a *sub-object* of B. Just as in $\mathbf{SET}$ every set A has a family of sub-sets (which is in turn a set, $\mathcal{P}A$, i.e., the power-set or set of all subsets of A), so we can associate with each object A of a given category $\mathbf{C}$ the family of its sub-objects and this association is *functorial*: it is expressed by the existence of a functor $Sub\colon \mathbf{C}^{op} \rightarrow \mathbf{SET}$ which behaves in the expected way.

In philosophical terms, the categorical analysis of the relationship between objects and sub-objects allows us to address the relationship between parts and wholes in a manner intrinsic to the universe of discourse (as a category) within which Part-Whole relationships are to be considered. Thus, what can be a part in one case may not be so in another. Part-Whole relations have to respect the constraint of what may be termed ontological homogeneity. This constraint has non-negligible effects when interpreting the predicates of a language in a universe of discourse U; and by induction (in passing from atomic to compound formulae) it affects the semantic evaluation v of the whole language. For instance, take a binary predicate P: if P is interpreted by v on a U in $\mathbf{SET}$ then $v(P)$ can be any subset of the Cartesian product $U \times U$; but if U is also a group and the interpretation is confined to **GRP**, $v(P)$ will have to be a subgroup of the product group $U \times U$.[16]

The same applies to the notion of element. The classic principle of extensionality ensures that the functor *Sub* is determined by elements. Indeed, in **SET**, an object being a collection of elements in extension, each subset A of B is associated with a characteristic function $\chi_A: B \to \{0,1\}$, with $\chi_A(x) = 1$ if and only if $x \in A$; so, by contraposition, two subsets A and A' of B will be different if and only if $\chi_A \neq \chi_{A'}$. This condition holds if and only if there exists a morphism $x: 1 \to B$ such that $\chi_A \circ x \neq \chi_{A'} \circ x$.

Many categories in which one can interpret first order languages or languages of order n, for $n > 1$, do not satisfy the property just defined: the elements ("global" elements) are not sufficient. Since 1 has a minimal internal structure in **SET** (the only proper sub-object just being $\varnothing$), there is no way in that category to talk about "partial" elements. Many categories have terminal objects which are richer in internal structure and so they allow for such "partiality". At the same time morphisms from 1 to another object have to meet further conditions than in the case of **SET** and for this reason are typically more rare.

As an example, take the category **SET^T** of sets linearly ordered with respect to a parameter, which may be here thought of as sets indexed by Time. At any given time t there is the set $A(t)$ of things belonging to A at that time. In this category, which can also be described as a "slice" category **SET/T** of sets over Time, the terminal object is the identity functor I_T from **T** to **T**, so that for there to be a (global) element of such a variable set, it is necessary that there is never a time t when the set is empty. But even though this condition is not satisfied by A, the algebra of its parts can be rich in structure. In category theory the phenomenology of relations between the concepts of "being an element of " and that of "being part of " is extremely wide and provides an indispensable tool for anyone who wishes to analyse the concept of an *individual* entity, taking into account the way entities vary. So it is possible, within one and the same mathematical picture, to progressively approximate the way individuals are identified and referred to in the real world, rather than resigning ourselves to the disconnect between a semantics in the format provided by set theory and the highly constrained language-world interface to which real speakers in physical reality are subject.

Universality by Adjunction

We can now return to the concept of adjunction to pinpoint its most important characteristics. The natural isomorphism φ that must exist between morphisms $f: FA \longrightarrow B$ and morphisms $g: A \longrightarrow GB$ means that, for any B' in **D** and $k: B \longrightarrow B'$, $\varphi(k \circ f) = Gk \circ \varphi f$ and that, for any A' in **C** and $h: A' \longrightarrow A$, $\varphi^{-1}(g \circ h) = \varphi^{-1}(g) \circ Fh$. That is, the association of a unique morphism $\varphi(f)$ with f is stable with respect to the composition of f with other morphisms in **D**, and the same is true for g and $\varphi^{-1}(g)$. Therefore, the association is uniform, in the sense that it does not depend on the choice of the particular objects (A, B) and morphisms (f, g) considered.

---

[16]In order for the interpretation to be *functorial*, the syntax too must be categorically reformulated. This step introduces intrinsic constraints on semantics, but it requires further details than those strictly needed for the aim of this paper and therefore will not be examined here.

Now, if $F \dashv G$, there are two special morphisms, denoted as the *unit* (η) and *co-unit* (ε) of the adjunction, which allow us to grasp the sense of universality which has been repeatedly mentioned. Indeed, in the case where $B = FA$, among the possible morphisms $f: FA \longrightarrow FA$ there will necessarily be id_{FA}, matched by φ with a morphism $\varphi(id_{FA})$: $A \longrightarrow GFA$ and this morphism, labelled η_A, is the *unit* of the adjunction. Thus, the unit is characterised by the fact that, since φ is natural, each morphism $g: A \longrightarrow GB$ is factored through η_A in one and only one way, namely $g = Gf \circ \eta_A$ for a unique $f: FA \longrightarrow B$.

As A varies in **C**, each morphism η_A behaves as a component of a natural transformation from the identity functor $1_\mathbf{C}$ to the functor **GF**. Symmetrically: in the case where $A = GB$, among the possible morphisms $g: GB \longrightarrow GB$ there will necessarily be id_{GB}, which φ^{-1} will put in correspondence with a morphism $FGB \longrightarrow B$ and this morphism, labelled ε_B, is precisely the *co-unit* of the adjunction; and, similarly, this dual notion of the unit is characterised by the fact that, since φ is natural, each morphism $f: FA \longrightarrow B$ is factored in one and only one way through ε_B, i.e., $f = \varepsilon_B \circ F(g)$ for a unique $g: A \longrightarrow GB$.

Note that if a functor has a (left or right) adjoint, this is unique (up to isomorphism). Note also that the left (right) adjoint of a functor may in turn have a further left (right) adjoint. Indeed this type of situation is often iterated, signalling the existence of deep intrinsic bonds between the different kinds of structure of **C** and **D**, [14].

The traditional way of defining a mathematical structure, such as an algebraic system, or an ordering or a space, makes reference to a set (as the underlying domain of the structure) satisfying additional properties, which often have the form of closure under certain operations. This indeed facilitates the procedure of "forgetting" the structure. For example: given a group G, one can consider just the set of its elements. The procedure can be made precise as a "forgetful" functor $U: \mathbf{GRP} \rightarrow \mathbf{SET}$, but in place of **GRP** one can consider any other category **D** of sets with structure.

Here, what is relevant for universality is that such a procedure admits an "inverse" F: $\mathbf{SET} \rightarrow \mathbf{D}$, in a particular sense: namely, the construction of a *free* structure over a given set, the elements of which are tagged as "generators". In the case of $\mathbf{D} = \mathbf{GRP}$, it will be a *free group*. The point is that any free structure X^* of a given kind has a universal property, i.e., given any function f from its set of generators to another set $U(Y)$, f factorizes through the underlying set $U(X^*)$ of X^* along a unique map g of the same kind-of-structure. In the case of a free group over a set X, with $f: X \longrightarrow U(Y)$, $f = U(g) \circ i$, where i is the insertion of generators in the underlying set of X^*. Thus there is a functor F, such that $X^* = F(X)$ and $U \dashv F$.

Now, it may be the case that, given a functor $G: \mathbf{D} \rightarrow \mathbf{C}$, only for a fixed object A in **C** an object B is available in **D** for which there exists a map $e: A \rightarrow GB$ with the analogous property of unique factorisation, i.e., such that there is a unique k with $Gk \circ e = g'$, for all $g': A \longrightarrow GB'$. The pair $<B, e>$ is said to be a *universal morphism* from A to G, in the sense that it is *free* with respect to G, from which one can solve a problem relative to **C** in a canonical way.

In addition, it may be that such a pair also exists for other objects in **C**, rather than for A only, but with characteristics different with respect to either B or e. Finally, this may occur without there being a universal morphism from F to B, dual to that from A to G. These cases lack the uniform character seen in the case of an adjunct pair. If and only if a universal morphism exists for *every* A in **C** and the dual condition holds for *every* B in **D**, do we have that $F \dashv G$.

So there is a notion of universality that is more general, because "sensitive to data" and therefore *weaker* than that captured by the concept of adjunction. However, Mac Lane was right to propose the slogan "Adjoint functors arise everywhere in mathematics",[17] because the basic constructions which arise time and again across different fields of mathematics correspond to that uniformity which is guaranteed by the existence of adjunctions. The fact that a category **C** admits, for example, the formation of products or function spaces between any two objects, is due to the existence of an adjoint functor to a functor from **C** to another category or from another category to **C**, independently of specific objects and maps in **C**, but even independently of their nature: be they sets and functions, spaces and continuous maps, groups and homomorphisms, propositions and proofs, etc.

Examples of Adjunction

From the algebra of natural numbers to that of the complex numbers, numerical systems differ in being closed or not with respect to certain operations. For example, the natural numbers are closed under sums and products but not with respect to the inverse of those operations. The rationals are also closed with respect to both inverses but not with respect to the inverse of the operation of raising by a power. Similarly, there are many types of categories, differentiated by their closure with respect to certain constructions. Closure constructions are examples of a fundamental mathematical concept—that of limit (and, dually, of co-limit[18]), which in its own turn tracks an underlying adjunction.

To define the notion of limit one starts from the consideration of a diagram within a given category **C**. A (finite) diagram $\mathcal{D}$ in **C** is any collection of **C**-objects and **C**-morphisms between them. A cone for $\mathcal{D}$ is given by an object V (the "vertex" of the cone) and a family of morphisms f_i (having domain V) to each object A_i of $\mathcal{D}$ such that, if h: $A_i \longrightarrow A_j$ is a morphism in D, $f_j = h \circ f_i$.

A limit (cone) $\mathcal{L}$, with vertex L, for $\mathcal{D}$ is a *universal* cone for $\mathcal{D}$ in the sense that, in addition to being a cone on $\mathcal{D}$, with a family of morphisms such that $l_j = h \circ l_i$, it has the following property: every other cone $\mathcal{D}'$ factorizes through L, i.e., there exists a (unique) morphism u from L to the vertex V of each other cone, such that $f_i = l_i \circ u$. A co-limit is nothing more than a limit for $\mathbf{C}^{op}$.

For example, if **C** admits a limit for each diagram of the form $A_1 \leftarrow A_0 \longrightarrow A_2$, **C** has finite Cartesian products of each pair of objects $<A_1, A_2>$. In fact, if a category has binary products (for pairs of objects), then it has products for each n-tuple of finite objects. The notion of sum, or co-product, is simply the dual of this, reversing the direction of the two morphisms: $A_1 \rightarrow A_0 \leftarrow A_2$. If **C** has a limit for the empty diagram, **C** has a terminal object denoted by 1. The dual notion is that of the initial object, denoted by 0.

In a category that is a partial order **P**, the existence of finite limits is equivalent to the fact that every finite set of nodes in **P** has a lowest bound ($\cap$) and the existence of colimits (even infinite ones) to the fact that every set (even infinite) of nodes in **P** has a supremum ($\cup$). Note that the usual definition of a topological space says that the category of open

---

[17] In the Preface to the first edition of [17], p. vii.

[18] Usually written "colimit".

subsets of a space X has finite limits and possibly infinite colimits. Indeed, in **TOP** a morphism f, that is a continuous function, from a space X to a space Y is characterised by the fact that the inverse image, $f^{-1}(-)$, of any open subset—of Y is an open subset of X and also $f^{-1}(U \cap V) = f^{-1}(U) \cap f^{-1}(V)$ and $f^{-1}(\cup U_i) = \cup f^{-1}(U_i)$, for each U, V opens in Y, with $i \in I$, where I may be infinite. That is, f^{-1} preserves finite limits and arbitrary colimits.

In **SET**, the product as defined above coincides with the usual Cartesian product, the terminal object is any singleton, the co-product is a disjoint sum and the initial object is the empty set Ø. In the category of groups, **GRP**, these categorical notions have further properties. In particular, the initial and the terminal object coincide: the trivial group consisting only of the identity is both 0 and 1.

If we now consider as objects propositions $p, q, r \ldots$ and deductions as morphisms, i.e., $p \longrightarrow q$ means that q is deducible from p, we have another category **PROP**, which, once endowed of finite limits and colimits, directly mirrors a fragment of propositional logic. The fact that **PROP** has (finite) products is equivalent to the possibility of forming the conjunction of any (finite) number of propositions. Similarly, co-products corresponds to disjunctions. The initial object is the absurd proposition (False, corresponding to the bottom element $\perp$ of a Heyting or Boolean algebra) and the terminal object is a logical truth (True, corresponding to the top element $\top$ of the given algebra). In **CAT**, the concepts will be those of the corresponding product category $\mathbf{C} \times \mathbf{D}$, the terminal category **1** (formed from a single object with its identity morphism) and so on.

Limits (and colimits) in many other categories can be defined and they turn out to be the outcome of a suitable adjunction. For simplicity, however, we shall confine attention to the examples given so far.

Denoting by $\Delta: \mathbf{C} \longrightarrow \mathbf{C} \times \mathbf{C}$ the diagonal functor which associates to each object A of **C** its duplication <A, A> and every morphism $f: A \rightarrow B$, its duplication <f, f>, the existence of binary products in C is equivalent to the fact that Δ has an additional right adjoint (the "product" functor $\times$). For a fixed object B, the existence of a right-adjoint for the parametric product functor - $\times B: \mathbf{C} \rightarrow \mathbf{C}$ (which associates $A \times B$ to an arbitrary A) is equivalent to the fact that **C** admits "exponentials", i.e., objects C^B. **C** is said to be "Cartesian closed" if there is a natural isomorphism between morphisms $A \times B \rightarrow C$ and morphisms $A \longrightarrow C^B$. In **SET**, C^B becomes the set of functions from B to C. If the objects B and C are spaces, the exponential is the space of all continuous functions from B to C, with the usual topology.

As in the case of products, the existence of sums is equivalent to the fact that Δ has a left adjoint (+ as a functor). In addition, considering the functor !: $\mathbf{C} \rightarrow 1$ (of functors to 1, there exists necessarily only one), if ! has a left adjoint, then **C** has an initial object, and if ! has a right adjoint, **C** has a terminal object.

In the case of **PROP**, in particular, the basic structure corresponding to the logical properties of the conjunction $\wedge$ and the disjunction $\vee$ is definable purely by *equations* between morphisms and their composites and these equations are associated with the existence of adjoints: $\wedge \dashv \Delta \dashv \vee$ and $\perp \dashv ! \dashv \top$. The exponential adjunction in this case gives us the connective of implication as right adjoint to - $\times B$.

The extraordinary progress in this approach to logic was made possible by Lawvere's pioneering research, which showed that such an equational presentation of logic by means of adjoints can also be extended to the quantifiers. In fact, $\exists$ and $\forall$ turn out to be respectively the left and right adjoints of the functor $Sub: \mathbf{C}^{op} \rightarrow \mathbf{SET}$ which associates with each object the set of its subobjects: A *topos* is a Cartesian closed category with a "subobject classifier" (which acts as an intrinsic truth-values-object). Consequently, all the notions required for logic acquire equational form.

It can be shown that a left adjoint preserves colimits and a right adjoint preserves limits. It follows that the existence of a functor which is both a left and a right adjoint tells us that a substantial part of the structure of a category is preserved by the functor. Further constraints due to the form of the axioms of a theory result in a tight control on the variation of its models. A *generic* model of a theory T, roughly speaking, is one with "no junk" and "no noise": formally, it has a property analogous to that seen in the case of free objects, but now holding for the category of models of T, thus endowing it with the corresponding universality.

Even apart from the new aspects of logic which are related to its internalisation in suitable categories (such as a topos),[19] the effects of all this on our understanding of semantics are profound. Not only is the notion of model refined by passing from the universe of sets to different categories (every bit as rich in structure), but also the notion of theory itself, once re-described in categorical terms, leads to a much richer semantic landscape (for example, to a far from trivial semantics for theories that had no models in **SET** and therefore have to be considered inconsistent by "standard" formal semantics).

Concluding Remarks

The concept of universality that has emerged with the notion of adjunction between functors allows us both to rethink and to refine a number of philosophical notions that have been the centre of many controversies, both in semantics and epistemology. Among the most significant are (1) the notion of *prototype* (in cognitive semantics), definable as the unit of an adjunction,[20] and (2) the notion of *transcendental subject* (in epistemology), definable by means of the notion of generic model.

As for (1), once the category **K** of Kinds and the category **I** of Individuals are specified, a subcategory **B** of **K** can be identified as that of *basic* kinds; if the restriction of the typing functor $\sigma: \mathbf{I} \to \mathbf{K}$ to **B** has a right adjoint $\pi: \mathbf{B} \to \mathbf{I}$, it means that any relation between an individual a of basic kind B and another a' of a possibly different kind can be uniquely recovered by factorisation along the unit of the adjunction, i.e., through $\pi\sigma(a)$, which then acts exactly as the prototype of B.

As for (2), once any theory T is described as a category and a suitable collection of theories is associated with each knowing subject S, then the access S can have to a theory T_U relative to a given empirical universe of discourse U takes the form of a functor $F: T_U \longrightarrow S$. In case where F is uniquely factorisable through a generic subject Σ, we are justified in considering Σ as a universal subject which, by coding the conditions of possibility for any knowledge about any domain by any subject, can be taken as "transcendental", i.e., as a generic model of the notion of knowing subject.[21]

---

[19]The most extensive treatise on topos theory is [8].

[20]An adjunction-based approach to the theory of concepts was first explored by Ellerman in [7].

[21]For the details relative to these two constructions, see respectively [26, 28].

The fact that so many different notions related to the search for "universals" are instances of one and same formal pattern, namely that of adjunction, should be of considerable interest to philosophers. Of no less interest should be that behind this unification is a theoretical perspective focusing on the analysis of change-of-structure (of systems emerging within and constrained by the physical world) and the idea that in general this change is not reducible to point-like components, but requires attention to the inner cohesion of any structure under variation, without such a recognition inevitably leading to some loose "structuralist" variant of holism. Previous knowledge of the "change of base" technique in algebraic geometry was indeed one of the sources of inspiration for such a unifying framework; conversely, this generalisation shed light on the relevance of algebraic geometry for logic—something most philosophers, and especially those in the analytic tradition, would consider an alien topic.

Admittedly, what is still lacking is a *systematic* treatment that would clarify the linkages between the form taken within a categorical setting by classical philosophical issues about (1) the nature of mathematical entities and the status of principles supposed to have a foundational role, (2) the nature of meaning, as a main concern for philosophy of language and (3) the architecture of knowledge, as it is investigated in epistemology and philosophy of science. Those starting research in philosophy lack the guidance that could provide a unified understanding of the various faces of universality. Category theory provides the tools for that understanding and for the treatment of those various aspects in a comprehensive and mathematically precise framework.

Only a collaboration between category theorists, logicians, and philosophers will overcome that lack and advance our understanding. But this assumes a motivation and a readiness to co-operate and, to date, there has been little evidence of that on the part of the philosophers. This reflects the fact that, with a few notable exceptions, there has been no recognition on their part of the potential offered by category theory to advance the understanding and resolution of the problems studied particularly in the various sub-disciplines of analytic philosophy. This state of affairs urgently needs to change. It may shortly do so. My aim in this compressed and incomplete presentation of categorical "universality" has been to help that to happen.

References

1. A. Asperti, G. Longo, *Categories, Types and Structures* (MIT Press, Cambridge, MA, 1991)
2. S. Awodey, *Category Theory* (Oxford University Press, Oxford, 2010)
3. J.L. Bell, From absolute to local mathematics. Synthese **69**, 409–426 (1986)
4. J.L. Bell, *Toposes and Local Set Theories: An Introduction* (Clarendon, Oxford, 1988)
5. J.L. Bell, The development of categorical logic, in *Handbook of Philosophical Logic*, ed. by D. Gabbay and F. Guenthner (Springer, New York), vol 12 (2005), pp. 279–361
6. V. De Paiva, A. Rodin, Elements of categorical logic. Log. Univers. **7**, 265–273 (2013)
7. D. Ellerman, Category theory and concrete universals. Erkentnnis **28**, 409–429 (1988)
8. Johnstone, P.: *Skteches of an Elephant: A Topos Theory Compendium*, vol. 1–2 (Oxford University Press, Oxford, 2002–2003)
9. J. Lambek, Categorial and categorical grammar, in *Categorial Structures and Natural Language Structure*, ed. by R.T. Oehrle et al. (Springer, Dordrecht, 1988), pp. 297–317
10. J. Lambek, Are the traditional philosophies of mathematics really incompatible? Math. Intell. **16**, 56–62 (1994)

11. J. Lambek, P. Scott, *Introduction to Higher-Order Categorical Logic* (Cambridge University Press, Cambridge, 1986)
12. F.W. Lawvere, Adjointness in foundations. Dialectica **23**, 281–296 (1969)
13. F.W. Lawvere, Taking categories seriously. Revista Colombiana de Matematicas **20**, 147–178 (1986)
14. F.W. Lawvere, Cohesive toposes and Cantor's "lauter Einsen". Philos. Math. **2**, 5–15 (1994)
15. F.W. Lawvere, Unity and identity of opposites in calculus and physics. Appl. Categ. Struct. **4**, 167–174 (1996)
16. F.W. Lawvere, S. Schanuel, *Conceptual Mathematics*, 2nd edn. (Cambridge University Press, Cambridge, 1997)
17. S. Mac Lane, *Categories for the Working Mathematician* (Springer, Berlin, 1971)
18. S. Mac Lane, *Mathematics: Form and Function* (Springer, New York, 1986)
19. J. Macnamara, G. Reyes (eds.), *Logical Foundations of Cognition* (Oxford University Press, Oxford, 1994)
20. C. McLarty, Defining sets as sets of points of spaces. J. Philos. Log. **17**, 75–90 (1988)
21. C. McLarty, Learning from questions on categorical foundations. Philos. Math. **3**, 44–60 (2005)
22. J.-P. Marquis, *From a Geometrical Point of View: A Study of the History and Philosophy of Category Theory* (Springer, New York, 2008)
23. J.P. Marquis, Category theory, Stanford Encyclopedia of Philosophy (2013) http://plato.stanford.edu/entries/category-theory/
24. A. Peruzzi, Forms of extensionality in topos theory, in *Temi e prospettive della logica e della filosofia della scienza contemporanee*, ed. by M.L. Dalla Chiara, M.C. Galavotti, vol. 1 (CLUEB, Bologna, 1988), pp. 223–226
25. A. Peruzzi, The theory of descriptions revisited. Notre Dame J. Formal Log. **30**, 91–104 (1988)
26. A. Peruzzi, Towards a real phenomenology of logic. Husserl Stud. **6**, 1–24 (1989). errata corrige, 253
27. A. Peruzzi, From Kant to entwined naturalism. Annali del Dipartimento di Filosofia **9**, 225–334 (1993)
28. A. Peruzzi, The geometric roots of semantics, in *Meaning and Cognition*, ed. by L. Albertazzi (John Benjamins, Amsterdam, 2000), pp. 169–211
29. A. Peruzzi, Il lifting categoriale dalla topologia alla logica. Annali del Dipartimento di Filosofia **11**, 51–78 (2005)
30. A. Peruzzi, The meaning of category theory for 21st century's philosophy. Axiomathes **16**, 425–460 (2006)
31. A. Peruzzi, Logic in category theory, in *Logic, Mathematics, Philosophy: Essays in Honor of John Bell*, ed. by M. Hallett, D. Devidi, P. Clark (Springer, New York, 2011), pp. 287–326

A. Peruzzi (✉)

Department of Philosophy, University of Florence, Florence, Italy

e-mail: alberto.peruzzi@unifi.it

On the Way to Modern Logic: The Case of Polish Logic

Roman Murawski

Abstract The aim of this paper is to consider views concerning logic and its philosophy of two Polish philosophers: Henryk Struve and Władysław Biegański. The analysis of their works shows that both belonged rather to the traditional understanding of logic as a science of rational thinking. Their views on the place of logic as well as on its relations with philosophy and mathematics are presented. They are contrasted with views of their contemporaries working in logic (in particular Wł. Kozłowski and Wł. M. Kozłowski) and with views of Jan Łukasiewicz, the leading representative of the Lvov-Warsaw School. Our analysis leads to the conclusion that Struve and Biegański, being representatives of the traditional pre-mathematical approach to logic, stood on the threshold of the new paradigm.

Keywords Biegański · Mathematical logic · Paradigm in logic · Psychologism · Struve · Traditional logic

Mathematics Subject Classification (2010). Primary 03-03 · 01A55 · Secondary 01A60

Polish logicians played a significant role in the development of logic in the twentieth century—the Warsaw School of Logic with its leaders Jan Łukaszewicz, Stanisław Leśniewski, Alfred Tarski and many others are known to any historian of logic. They developed logic in its modern form as mathematical logic. Nevertheless, interest in logic existed in Poland much earlier, though it had been understood and developed in a traditional way. In this paper we shall discuss two scholars: Henryk Struve and Władysław Biegański who can be considered as representatives of the traditional pre-mathematical approach to logic that stood on the threshold of the new paradigm.

Henryk Struve

Henryk Struve was a philosopher who lived at the turn of the nineteenth and the twentieth centuries. He is regarded as one of the most important figures of Polish logic in the nineteenth century (cf. for example Woleński [33, p. 30]—yet he has been almost forgotten. In the interwar period he was frequently referred to but his works were not

© Springer International Publishing Switzerland 2016

F.F. Abeles, M.E. Fuller (eds.), *Modern Logic 1850-1950, East and West*, Studies in Universal Logic, DOI 10.1007/978-3-319-24756-4_9

analysed or reprinted.[1] He was a professor at the Main School [Szkoła Główna] and at the Russian Imperial University of Warsaw, he taught logic. He was the author of numerous textbooks on logic and wrote a history of logic (cf. [28]) as well.

We are first of all interested in Struve's views on logic as a science and his conception of logic, which is important because, in a way, Struve stood on the threshold of the new way of understanding and cultivating logic, combining the old and new paradigms. As Kazimierz Twardowski wrote about him:

> So Struve was a link connecting this new period with the previous one. Between generations of Cieszkowskis, Gołuchowskis, Kremers, Libelts, Trentowskis[2] and the contemporary generation there appears the distinguished figure of this thinker and writer, who saved from the past what was of lasting value. He showed the workers of today's Polish philosophy a prudent direction through rational thinking and was himself devoid of all prejudices and opinions [30, p. 101].[3]

Struve presented his views on logic in his fundamental work *Historya logiki jako teoryi poznania w Polsce* [History of Logic as the Theory of Knowledge in Poland] [28] and in the textbook *Logika elementarna* [Elementary Logic] [27] (cf. also [25]) as well as in various papers. Certain difficulties in reconstructing his views result from the fact that he wanted to create a coherent system of philosophy that would embrace all the traditional branches of philosophy. This caused the limits between particular branches to be flexible and imprecise. The principles of one division influence the foundations of the other and conversely. He balanced between materialism and idealism aiming at the golden mean, which, among other things, was reflected in his understanding of the object of logic. At first he thought that the object of logic was principles and rules of thinking. In his talk given in 1863, inaugurating his lectures at the Main School, he said:

> Gentlemen! Logic is most generally the science of rational thinking, having thinking, its principles and rules as its object ([24], Lecture 1).[4]

However, he added that thinking is one of the powers of the soul; it is "an objective, neutral consideration of this world by the soul" [24, p. 55].[5] Thus he introduces a psychological element, and indirectly—an ontological one. Since the soul is the ideal

---

[1]It is worth adding that the exception was Samuel Dickstein who published posthumously Struve's handwritten sketch dedicated to Hoene-Wroński.

[2]August Count Cieszkowski (1814–1894)—Polish philosopher, economist and social and political activist; Józef Gołuchowski (1797–1858)—Polish philosopher, co-creator of the Polish Romanticist "national philosophy"; Józef Kremer (1806–1875)—Polish historian of art, a philosopher, an aesthetician and a psychologist; Karol Libelt (1807–1875)—Polish philosopher, writer, political and social activist, social worker and liberal, nationalist politician; Bronisław Trentowski (1808–1869)—Polish "Messianist" philosopher, pedagogist, journalist and Freemason, and the chief representative of the Polish Messianist "national philosophy" [my remark—R.M.].

[3]"Był tedy Henryk Struve jakby ogniwem, łączącym ten okres nowy z poprzednim. Między pokoleniem Cieszkowskich, Gołuchowskich, Kremerów, Libeltów, Trentowskich a pokoleniem współczesnem widnieje czcigodna postać myśliciela, nauczyciela, pisarza, który z przeszłości ocalił to, co miało w niej wartość trwałą, a dzisiejszym na polu filozofii polskiej pracownikom wskazał drogę rozważnym i dalekim od wszelkiego uprzedzenia sądem."

[4]"Panowie! Logika jest to w najogólniejszem pojęciu *nauka myślenia* mająca myślenie, jego zasady i prawidła za przedmiot."

[5]"obiektywne, neutralne rozpatrywanie tego świata przez duszę."

embryo of the human being and "the limits of our being are the limits of our correct thinking" [24, p. 35].[6]

In Struve's opinion logic concerns objective reality; nonetheless, it does not concern it directly—the mediator between logic and the world is the thought. However, this does not lead to the thesis that thought reflects the logical structure of the world or to the thesis that the world has some logical structure at all.

Struve's earlier views were even more inclined towards psychologism. Initially, he claimed—as indicated above—that thinking is "an objective, neutral consideration of this world by the soul." He upholds this thesis in *Logika elementarna* [27], but here he separates logic from psychology, writing that logic deals with thinking as "an auxiliary mean to get to know the truth" whereas psychology is interested in emotional and volitional motives of cognition. Logic has both a descriptive and normative character and is to oversee the application of the established norms and thus to evaluate the degree of the truth of cognition.

The foundation of logic is philosophy, but also conversely: philosophy can be developed only on the foundation of logical laws. The title of the main analysed work of Struve, *Historya logiki jako teoryi poznania w Polsce*, may suggest that he identified logic with the theory of knowledge. In the first editions of *Logika elementarna* in Russian[7] he made no clear distinction between these two disciplines, but in the Polish version of the textbook [27] he wrote:

> While it is true that *thinking* is the main co-factor of cognitive activity but not the only one; it unites directly and constantly with the suitable expressions of *emotion* and *will* [27, p. 3].[8]

The examination of emotions and will as well as their relationships with thinking belongs to the sphere of psychology whereas logic deals with thinking merely in one aspect, namely:

> As *an auxiliary mean of getting to know the truth* [...]. Simultaneously, logic is not satisfied with the real course of mental activity but seeks principles, i.e. laws and rules which one *should* follow as norms if one wants to get to know the truth as exactly as possible. This separate view on thinking gives logic the character of an independent science, which is strictly different from *psychology*, namely this part of logic that investigates *thinking* as well [27, p. 5].[9]

In *Historya logiki* we find the following words:

> [...] many separate the theory and criticism of *cognition* from *logic* as science dealing only with *thinking*. Despite that, the connection between the development of correct thinking as well as

---

[6]"granice naszego bytu są granicami naszego myślenia prawidłowego."

[7]"Елементарная логика" [Elemientarnaja logika] was first published in 1874; there were altogether 14 Russian editions. It was the obligatory manual of logic in classical junior high school from the year 1874. Its Polish version appeared in 1907.

[8]"*Myślenie* jest wprawdzie głównym, ale nie jedynym współczynnikiem czynności poznawczej; jednoczy się ono zarazem bezpośrednio i stale z odpowiednimi objawami *uczucia* i *woli*."

[9]"Jako *środek pomocniczy poznania prawdy* [...]. Przytem nie zadowala się logika danym faktycznym przebiegiem czynności myślowej, lecz odszukuje zasady, t.j. prawa i prawidła, któremi w myśleniu jako normami kierować się *należy*, chcąc dojść do możliwie ścisłego poznania prawdy. Ten odrębny pogląd na myślenie nadaje logice charakter samodzielnej nauki, ściśle różniącej się od *psychologii*, a mianowicie tej części jej, która bada również *myślenie*."

arriving at and getting to know the truth is so close from the psychological perspective that these mental activities cannot be separated [28, p. 1].[10]

One can see here some traces of his discussions conducted with Kazimierz Twardowski and the Lvov-Warsaw School. On the one hand, one can notice a certain readiness to recognise the new understanding of logic and on the other hand, a desire to abide by his current understanding of logic.

Speaking of getting to know the truth, Struve differentiates between objective and subjective truth. The former is an ideal that is independent of human cognition and the latter is the reconstruction of the content of being, of what exists in reality, in the mind—done through correct thinking. Logic controls this reconstruction and thus through formal means it reaches the real being. Thus thinking has a reconstructive and not a creative character. In Struve's opinion there are three forms of logic: (1) formal, (2) metaphysical and (3) logic treated as the theory of knowledge. Formal logic considers the principles and laws of thinking regardless of its object. Metaphysical logic (developed by Plato, Neo-Platonists, Spinoza and the German idealists: Fichte, Schelling and Hegel) states that since thinking contains its object directly in itself we get to know the very objective reality knowing the principles and laws of thinking. Struve accepts neither the first nor the second conception of logic. He opts for the third solution, treating it as the golden mean. Thus he understands logic as a method of investigation and cognition of truth. Its task is to discover the principles according to which man reconstructs the structure of the real world in his mind. Naturally, Struve sees the difficulties connected with this view. In *Logika elementarna* he wrote:

The difficulties of examining the relation [. . .] between thinking and the objective world are obvious and can be reduced mainly to the fact that we are not able to compare directly our images and concepts of objects and our views on them with the objects themselves. The question concerning the objective knowledge of truth could be solved in favour of thinking only when it turned out that the laws of our mind, and thus thinking, were fundamentally consistent with the laws of the objective being which is independent of us. [. . .] Nonetheless, showing the accordance between the laws of the mind and the laws of the objective being requires a series of critical investigations concerning the results of scientific studies [27, pp. 6–7].[11]

Struve calls this logic "logic of ideal realism", uses to constitute the framework of a coherent system of philosophy to give a general outlook on the world. We should add that Struve is far from ascribing Messianic tendencies to logic (as Hoene-Wroński did—cf. Murawski [21, 22]). He opts for a balance between the knower and the known, seeing the

---

[10]"[. . .] wielu odróżnia zarówno teorią, jak i krytykę *poznania* od *logiki* jako nauki samego tylko *myślenia*. Pomimo to łączność pomiędzy rozwojem prawidłowego myślenia a dochodzeniem i poznaniem prawdy tak jest ścisła ze stanowiska psychologicznego, że tych czynności umysłowych rozerwać nie podobna."

[11]"Trudności zbadania stosunku [. . .] myślenia do świata przedmiotowego są oczywiste i sprowadzają się głównie do tego, że nie jesteśmy w stanie porównywać bezpośrednio naszych wyobrażeń i pojęć o przedmiotach ani poglądów na nie z samymi przedmiotami. Kwestya przedmiotowego poznania prawdy mogłaby być rozwiązaną na jego korzyść dopiero wtedy, gdyby się okazało, że prawa naszego umysłu, a więc i myślenia, są zasadniczo zgodne z prawami niezależnego od nas bytu przedmiotowego. [. . .] Wykazanie atoli tej zgodności praw umysłu z prawami bytu przedmiotowego wymaga szeregu badań krytycznych nad wynikami dociekań naukowych."

role of emotions and will in cognition. He was interested in Leibniz's view to which he referred many times, the view that logic is abstracted from reality.

Struve begins his lecture on logic by giving images and concepts. Then he introduces judgments. By "image" he means a kind of representation of the object through its characteristics, and "concept" is a set of essential features. Moreover, images result from certain mental processes. It is the object that makes the mind create images.

Struve attached great importance to the teaching of logic. He thought that teaching how to think correctly is much more important than giving students concrete contents. Consequently, he placed a strong emphasis on the teaching of logical culture.[12]

We have already shown that Struve's conception of logic places him between the old and new paradigm (or perhaps even completely in the old paradigm). We have mentioned that he did not value the role and significance of symbolic and mathematized formal logic but he stressed psychological questions. Consequently, one should ask what made him not see the advantages of the new attitude. It seems that one of the reasons was the fact that Struve saw no cognitive value in pure form devoid of content (cf. Trzcieniecka-Schneider [29]). According to his conception it is the object, i.e. the external world, that stimulates our thinking that realises its existence and the characteristics of objects, and then using logical methods it creates notions which in turn it uses, applying logical methods, to formulate judgments. Thus there can be no cognition without content. Another reason may be that he set a low valuation on the role and importance of mathematics. Trzcieniecka-Schneider even claims that Struve "did not understand mathematics, reducing it only to the techniques of operations on numbers" [29, p. 93]. In "Filozofia i wykształcenie filozoficzne" [Philosophy and philosophical education] he wrote:

> Mathematics considers only quantitative factors: [...] But in its activities the human mind is not limited only to quantitative factors but everywhere supplements quantity with quality. [...] One cannot say that truth contains more or fewer thoughts than falsity: [...] These are all qualitative differences and they cannot be defined quantitatively; they cannot be understood rightly and characterised closely from the mathematical point of view but they can only be understood from a more general standpoint, going much beyond the scope of the quantitative factors alone [26, pp. 156–157].[13]

Struve also opposed the introduction of quantifiers. He permitted quantitative elements in logic only in the case of the conversion of judgments and was otherwise squarely in opposition. However, he was not consistent in his views when, analysing the relations between the scopes of concepts using Euler diagrams, he actually used arithmetic notation.

---

[12]The additional activities of Polish logicians Tadeusz Kotarbiński and Kazimierz Ajdukiewicz promoting logical culture fit well with this tendency and can be treated as the continuation of Struve's activities.

[13]"Matematyka rozpatruje wyłącznie czynniki ilościowe: [...] Tymczasem umysł ludzki w działalności swej nie jest bynajmniej ograniczonym samemi tylko pojęciami ilościowymi, lecz uzupełnia wszędzie ilość jakością. [...] Nie można powiedzieć, że prawda zawiera w sobie więcej lub mniej myśli niż fałsz: [...] To wszystko są różnice jakościowe, których ilościowo określić nie można, które należycie zrozumiane i bliżej scharakteryzowane być nie mogą z punktu widzenia matematycznego, lecz pojęte być mogą tylko ze stanowiska ogólniejszego, wynoszącego się wysoko ponad zakres samych tylko czynników ilościowych."

In his opinion formal logic "leads [...] only to the development of one-sided formalism without elevating the essential cognitive value of the relevant forms"[14] [27, p. IX] and "mathematical logic depends on the completely dogmatic transfer of the quantitative and formal principles to the mental area where quality and content are of primary importance" (*ibid.*).[15] He also claims that "reducing judgment to *equation* and basing conclusion on *substance*, i.e. *substituting equivalents*, does not correspond to the real variety of judgments and conclusions"[16] [27, p. X].

Struve's aversion towards mathematics and mathematical methods in logic was connected with his views on the function of language in logic and cognition as well as with his conception of truth. Since if—in accordance with Aristotle—truth is the conformity of thought to the content of propositions, conducting operations on propositions as symbols means losing sight of this property to some extent. In fact, the characteristic of being true does not refer to propositions but to their contents. Symbolically identical propositions can differ with respect to their contents. Thus the operations conducted on the symbols of propositions do not have much in common with establishing their truth. And yet logic leads to the truth about the real world.

Władysław Biegański

By profession Władysław Biegański was a general practitioner, held a medical doctorate and had his own medical practice. He worked in a hospital and was physician for a factory and the railways. His scientific interests included many medical disciplines as well as the philosophy of medicine, and in particular the methodological and ethical issues connected with it (cf. [4], [5]). Biegański represented the Polish school of the philosophy of medicine. However, his true passion was logic. As a student he listened to Henryk Struve's lectures (cf. the previous section). Besides his medical practice, he taught logic in local secondary schools for some time. In 1914 there was even an initiative to appoint Biegański as professor of the Jagiellonian University Chair of Logic.[17] It did not occur because of his poor health condition and the outbreak of the First World War.

Before discussing Biegański's philosophical views on logic we should mention his works on epistemology and the methodology of medicine. As far as the theory of cognition is concerned, Biegański promoted a view which he called "previsionism" (from the Latin *praevidere*—predict, foresee). Thus he referred to A. Comte's motto "savoir c'est prévoir" (to know is to foresee). In Biegański's opinion the main task of science is to foresee phenomena (cf. Biegański [9, 13]). He understood the concept of prevision in a broad sense. He claimed that cognition consists in the prevision of events, their causes and

---

[14]"doprowadza [...] tylko do rozwoju formalizmu jednostronnego bez podniesienia istotnej wartości poznawczej odnośnych form."

[15]"logika matematyczna polega na zupełnie dogmatycznym przeniesieniu zasad ilościowych i formalnych na pole umysłowe, gdzie jakość i treść mają znaczenie pierwszorzędne."

[16]"sprowadzanie sądu do *równania* oraz oparcie wniosku na *substytucyi*, czyli *podstawianiu równoważników*, nie odpowiada rzeczywistej rozmaitości ani sądów, ani wniosków."

[17]Cf. Borzym [15], p. 13.

effects as well as their properties. The aim of cognition is not to reproduce reality—thus he opposed the so-called reproductionism. At the same time he stressed the importance of the teleological point of view, which in his opinion was more effective than causal explanation in many disciplines of science, especially in medicine and biology. Laws formulated in science are not identical with reality—as he wrote "there is no identity of laws with the real order of phenomena" [13]. However he believed that, in our constructions there are always "real elements", and the aim of cognition is not to reproduce reality but to obtain proper orientation in the environment.

In the field of the methodology of medicine, Biegański formulated the first systematic general theory of diagnosis. He dealt with scientific observation, worked on the theory of experiments and was interested in the theory of induction. Of importance are his considerations concerning analogy. He presented them mainly in his work *Wnioskowanie z analogii* [Deduction by Analogy] [8]. He understood analogy in the traditional way as reasoning from details to details. However, he did not agree to reduce analogy to deduction or induction. In his opinion the foundations of analogy always include the suitability of some relations. Analogy is both something different from identity, which is the conformity of all features, and similarity in which we are to deal with the consistency of only some features.

In discussing Biegański's philosophical views on logic, we first should notice that he was neither a formal nor a mathematical logician, but—as Woleński mentions [32]—he was a philosophical logician from the standpoint formulated by Łukasiewicz. The latter characterised philosophical logic in the following way:

> If we use here the term "philosophical logic" we mean the complex of problems included in books written by philosophers, and the logic we were taught in secondary school. Philosophical logic is not a homogenous science; it contains various issues; in particular, it enters the field of psychology when it speaks not only about a proposition in a logical sense but also this psychological phenomenon, which corresponds with a proposition and which is called "judgment" or "conviction." [...] Philosophical logic also embraces some issues from the theory of knowledge, for example, the problem of what truth is or whether any criterion of truth exists [19, pp. 12–13].[18]

At this point it is worth adding that Łukasiewicz himself did not value philosophical logic, thinking that the scope of problems it considers is not homogenous, and also that philosophical logic mixes logic with psychology. Moreover, both fields are different and use different research methods.[19]

How did Biegański understand logic? In *Zasady logiki ogólnej* [Principles of General Logic] he wrote:

> Logic is the science of the ways or norms of true cognition [6, p. 1].[20]

---

[18]"Jeżeli używamy terminu logika filozoficzna, to chodzi nam o ten kompleks zagadnień, które znajdują się w książkach pisanych przez filozofów, o tę logikę, której uczyliśmy się w szkole średniej. Logika filozoficzna nie jest jednolitą nauką, zawiera w sobie zagadnienia rozmaitej treści; w szczególności wkracza w dziedzinę psychologii, gdy mówi nie tylko o zdaniu w sensie logicznym, ale także o tym zjawisku psychicznym, które odpowiada zdaniu, a które nazywa się "sądem" albo "przekonaniem". [...] W logice filozoficznej zawierają się również niektóre zagadnienia z teorii poznania, np. zagadnienie, co to jest prawda lub czy istnieje jakieś kryterium prawdy."

[19]Cf. Woleński [31] or Murawski [23].

[20]"Logika jest to nauka o sposobach albo normach poznania prawdziwego."

In *Podręcznik logiki i metodologii ogólnej* [Manual of Logic and General Methodology] we find the following definition:

> We call logic the science of the norms and rules of true cognition [7, p. 3].[21]

Therefore, the laws of logic concern the relationships of mental phenomena because of their aim, which is true cognition. Consequently, logic aims at investigating cognitive activities of the mind. At the same time Biegański claims that one must separate and distinguish between logic and the theory of knowledge on the one hand, and psychology on the other, since logic is a normative and applied science whereas both the theory of knowledge and psychology are theoretical. However, in practice Biegański—like other authors of his time—did not distinguish strictly between logical and genetic questions, investigating logical constructions both from the precisely logical and psychological points of view. Yet, it should be noted that in *Zasady* [6] Biegański suggests that his conception of logic makes him reject the division into formal and material truth whereas in *Podręcznik* [7] he regards this distinction as correct. He also adds that logic embraces the formal side of cognition.

In his large (638 pages) monograph entitled *Teoria logiki* [Theory of Logic] [11]—an attempt to consider the foundations of logic comprehensively[22]—Biegański writes:

> The main aim of logic is to control argumentation [11, p. 34].[23]

He continues:

> Logic, as the science and art of argumentation, is an *a priori* science, i.e. science that draws its content not from experience and not from the facts given in experience, but from certain *a priori* presumptions and constructions [11, p. 35].[24]

Thus logic appears as a normative science—accordingly, Biegański proposes to use the name "pragmatic logic." He separates logic from psychology, ontology and epistemology. The basis of logic is axioms: "the most general laws which are directly obvious, i.e. requiring no proof" [11, p. 41]. The axioms are the laws of identity, contradiction, excluded middle and sufficient reason.[25] It should be added that Biegański distinguished between the context of discovery and the context of justification.

---

[21]"Logiką nazywamy naukę o normach i prawidłach poznania prawdziwego."

[22]This work presents general problems concerning logic, the study of concepts, the study of judgments, the study of argumentation and the study of induction. Every problem is considered in historical and comparative perspectives on the one hand and a systematic perspective on the other hand. Although Biegański focuses on the views of the representatives of traditional logic, he also analyses the algebra of logic.

[23]"Logika ma na celu głównie kontrolę dowodzenia."

[24]"Logika, jako nauka i sztuka dowodzenia jest nauką aprioryczną, tj. taką, która swoją treść czerpie nie z doświadczenia, nie z faktów w doświadczeniu nam danych, lecz z pewnych naprzód powziętych założeń i konstrukcji."

[25]The last axiom states that nothing is without a causation. It has a variety of expressions. In the case of logic it can be best expressed as: For every proposition P, if P is true, then there is a sufficient explanation for why P is true.

This understanding of logic as the art of argumentation is found in his earlier treatise "Czem jest logika?" [What is Logic?] [10] where he wrote:

[...] logic does not reproduce the processes of thought and it does not aim at doing it at all. Therefore, the definition of logic as the science or art of thinking is actually devoid of any basis. [...] But the origin of logic shows that this ability [i.e. logic – remark is mine, R.M.] is neither a science nor art of thinking, but was created by Plato and Aristotle as the art of argument. Such differences in views cause serious consequences. If logic is a science or even an art of thinking, it is or should be a branch of psychology; on the contrary, if it is only the art of argument, it becomes a separate science that is independent from psychology. Logic as the art of argument does not describe the ordinary course of thoughts, used in argumentation; it does not reproduce it; it does not find laws for it, laws expressing the mutual causal relationship of thoughts, but uses ideal constructions which serve to control the ways of argumentation and in this respect it is explicitly separated from psychology [10, p. 144].[26]

Consequently, logic "must be of normative character" [10, p. 145], which Biegański explains:

The essence of argumentation consists in valuing. Looking for a proof of any proposition we always follow the question about its cognitive value [10, p. 145].[27]

What is meant here is not the meaning of the proposition and its content but its veracity. "Every proof consists in stating the consistency between the content of the proposition and the principles, which we recognise as true, and it is in this consistency that the essence of truth lies" [10, p. 145].[28]

What then is the relation between logic and psychology? Biegański stresses their autonomy:

Any direct [...] dependence here is out of the question. Nonetheless, psychological investigations are not completely meaningless to logic since they constitute an important control for logical constructions [...]. An ideal logical construction would be one that is the closest to the real course of thoughts, that completely guarantees to distinguish truth and is easy to apply. [...] Thus psychological investigations are undoubtedly of great importance for the development of logic because they can contribute to formulating new constructions which are the closest to the natural course of thoughts [10, pp. 147–148].[29]

---

[26]"[...] logika nie odtwarza procesów myśli i nie ma wcale na celu tego zadania. To też określenie logiki jako nauki lub sztuki myślenia jest pozbawione właściwie wszelkiej podstawy. [...] Tymczasem geneza logiki wykazuje, że umiejętność ta [tzn. logika – uwaga moja R.M.] nie jest ani nauką, ani sztuką myślenia, lecz utworzona została przez Platona i Arystotelesa jako sztuka dowodzenia. Takie różnice w zapatrywaniach prowadzą za sobą poważne konsekwencje. Jeżeli logika jest nauką lub nawet sztuką myślenia, to w każdym razie jest lub powinna być działem psychologii, przeciwnie, jeżeli jest tylko sztuką dowodzenia, to staje się nauką odrębną, niezależną od psychologii. Logika jako sztuka dowodzenia nie opisuje zwykłego biegu myśli, stosowanego przy dowodzeniu, nie odtwarza go, nie wynajduje dla niego praw, wyrażających wzajemny związek przyczynowy myśli, lecz posługuje się konstrukcyami idealnymi, które służą dla kontroli sposobów dowodzenia i pod tym względem odgranicza się wyraźnie od psychologii."

[27]"Istota dowodzenia polega na wartościowaniu. Poszukując dowodu dla jakiegokolwiek zdania, kierujemy się zawsze pytaniem o jego wartości poznawczej."

[28]"Każdy dowód polega na stwierdzeniu zgodności treści zdania z zasadami, które uznajemy za prawdziwe i w tej właśnie zgodności tkwi istota prawdy."

[29]"O bezpośredniej [...] zależności nie może tu być mowy. Pomimo to badania psychologiczne nie są zupełnie bez znaczenia dla logiki, stanowią bowiem bardzo ważną kontrolę dla konstrukcji logicznych. [...] Ideałem konstrukcyi logicznej byłaby taka, któraby się najbardziej zbliżała do rzeczywistego biegu

What did Biegański mean by argumentation? In fact, he did not give any clear idea of inference. He neither used the concept of logical deduction nor distinguished between deductive and inductive reasoning. He says that inference is based on the idea of necessity, that the principles of logic refer to the form and not the content of cognition. But these ideas are not fully clear and additionally, they are mixed. Biegański's misconception concerning deduction and its role is confirmed, for example, by the fact that in his work "Sposobność logiczna w świetle algebry logiki" [Logical Modality in the Light of the Algebra of Logic] [12] he speaks about reliable and possible deduction, which is a misunderstanding.

Finally, discussing Biegański's conception of logic we should add that his departure from psychologism was not definitive since in *Podręcznik logiki ogólnej* [Manual of General Logic] [14] one can see his return to psychologism. He writes:

> We call logic the science about the ways of controlling the truth of our cognitive thoughts [14, p. 1].[30]

Biegański's conceptions concerning the foundations and philosophy of logic did not evoke much interest and his work was criticised. In *Ruch Filozoficzny* Łukaszewicz published a review of Biegański's work "Czem jest logika?" [10], stressing his departure from psychologism but noticing that it was not completely consistent. He also emphasised the fact that Biegański's conception of logic was too narrow because he limited it to inference. In Łukaszewicz's opinion the object of logic should be reasoning in general, which should include non-deductive reasoning.[31]

Involved only in the traditional paradigm of logic Biegański could, however, see the advantages of the new approach, in particular the values and advantages of the algebra of logic. In the introduction to his work "Sposobność logiczna w świetle algebry logiki" [12], in which he attempted to (admittedly, with a poor result) apply the algebra of logic to the theory of modal categories, he wrote:

> Although logical calculus, called the algebra of logic or logistics, has not and cannot have a large practical application, considering the logical evaluation of our judgments and conclusions, it has undoubtedly important theoretical significance. [...] Yet, algebraic symbols, which we use in logical calculus, separate clearly the object of investigation from psychological factors and objective relations, and bring to light all the properties of pure logical relations. Therefore, the main value of the algebra of logic consists in the fact that using it we can explain more thoroughly and mark strictly the relations that are explained variously in school logic [12, p. 67].[32]

---

myśli, dawała zupełną gwarancyę w odróżnianiu prawdy i była łatwa do stosowania. [...] Toteż badania psychologiczne mają niewątpliwie duże znaczenie w rozwoju logiki, gdyż mogą się przyczynić do wynalezienia konstrukcyi nowych, najbardziej zbliżonych do naturalnego biegu myśli."

[30]"Logiką nazywamy naukę o sposobach kontrolowania prawdy naszych myśli poznawczych."

[31]For more on Łukasiewicz's philosophy of logic—see Woleński [31], Murawski [23].

[32]"Rachunek logiczny, zwany algebrą logiki lub inaczej jeszcze logistyką, jakkolwiek nie ma i nie może mieć rozległego zastosowania praktycznego przy ocenie wartości logicznej naszych sądów i wniosków, posiada jednak niewątpliwie ważne teoretyczne znaczenie. [...] Tymczasem symbole algebraiczne, jakimi się w rachunku logicznym posługujemy, odrywają wyraźnie przedmiot badania zarówno od czynników psychicznych jako też od stosunków objektywnych i wydobywają na jaw wszystkie właściwości czystych stosunków logicznych. To też główna wartość algebry logiki polega na tem, że przy jej pośrednictwie możemy dokładniej wyjaśnić i ściślej wyznaczyć stosunki, które w logice szkolnej rozmaicie bywają tłomaczone."

Conclusion

The presented analyses show that Struve and Biegański should be included among those traditional logicians whose approaches to logic are "pre-mathematical". Although the first signs of interest in mathematical logic appeared in Poland in the 1880s (suffice it to mention the treatise of Stanisław Piątkiewicz *Algebra w logice* [Algebra in Logic] published in 1888—cf. Batóg [1, 2] and Batóg–Murawski [3]), the logical culture was decisively "pre-mathematical" in Poland at the turn of the nineteenth and the twentieth centuries. This opinion is supported, for example, by the first edition of *Poradnik dla samouków* [A Guide for Autodidacts] which includes Adam Marburg's paper "Logika i teoria poznania" [Logic and the Theory of Knowledge] [20], written in a rather old-fashioned manner. Struve's and Biegański's views on logic were shared by their contemporaries: Władysław Kozłowski (1832–1899) and Władysław Mieczysław Kozłowski (1858–1935). The former wrote in *Logika elementarna* [Elementary Logic] that "logic is the science about mental activities with the aid of which we reach truth and prove it" [16, p. 1].[33] In turn, Władysław M. Kozłowski wrote in *Podstawy logiki* [The Foundations of Logic] that "Logic is the science about the activities of the mind which seeks truth" (1916, p. 8).[34] He calls the first chapter of his work "Thinking as object of logic" [17, p. 22]. He repeats this thought in *Krótki zarys logiki* [A Brief Outline of Logic], claiming that logic is a normative science and its task is "to examine the ways leading the mind to truth" [18, p. 1]. However, he stresses that logic:

> analyses mental operations conducted to reach the truth in a form that is so general that it could be applied to any content. It investigates its form, separating it completely from the content. Logic shares this property with mathematics [. . .]. [. . .] This formal character, common to logic and mathematics, made these sciences close in their attempts, which were less or more developed, and led to the creation of mathematical logic [18, pp. 8–9].[35]

Finally, he states that logic can be defined "as the science about the forms of every ordered field of real or imaginary objects" [18, p. 9].[36]

In Poland the road to the new paradigm in logic, to the new understanding of it, was long and difficult. Only in the next generation of Polish scholars can one see the new mathematical approach to logic.

Acknowledgements   The financial support of the National Center for Science [Narodowe Centrum Nauki] (grant no N N101 136940) is acknowledged. The paper is based on my book *Philosophy of Logic and Mathematics in the 1920s and 1930s in Poland*, Birkhauser Verlag, Basel 2014.

---

[33]"Logika jest nauką o czynnościach umysłowych, za pomocą których dochodzimy prawdy i jej dowodzimy."

[34]"Logika jest nauka o czynnościach umysłu poszukującego prawdy."

[35]"bada operacye umysłowe, wykonywane w celu osiągnięcia prawdy w formie tak ogólnej, iżby mogły zastosować się do jakiejkolwiekbądź treści. Bada je ze stanowiska ich formy, odrywając się zupełnie od treści. Własność tę podziela z logiką matematyka [. . .]. [. . .] Ten formalny charakter, wspólny logice z matematyką, spowodował zbliżenie do siebie obu nauk w próbach mniej lub dalej posuniętych i znalazł wyraz w utworzeniu logiki matematycznej."

[36]"jako naukę o formach każdej uporządkowanej dziedziny przedmiotów rzeczywistych lub urojonych."

References

1. T. Batóg, Stanisław Piątkiewicz – pionier logiki matematycznej w Polsce. Kwartalnik Historii Nauki i Techniki **16**, 553–563 (1971)
2. T. Batóg, Stanisław Piątkiewicz – pionier logiki matematycznej w Polsce. Z Dziejów Kultury i Literatury Ziemi Przemyskiej **II**, 325–330 (1973)
3. T. Batóg, R. Murawski, Stanisław Piątkiewicz and the beginnings of mathematical logic in Poland. Historia Mathematica **23**, 68–73 (1996)
4. Wł. Biegański, *Logika medycyny, czyli Zasady ogólnej metodologii nauk lekarskich*, W. Drukarnia W. Kowalewskiego, Warszawa (1894); second edition: 1908; German translation: *Medizinische Logik: Kritik der ärztlichen Erkenntnis*, C. Kabitzsch, Würzburg 1909
5. Wł. Biegański, *Zagadnienia ogólne z teorii nauk lekarskich* (Drukarnia Towarzystwa S. Orgelbranda Synów, Warszawa, 1897)
6. W. Biegański, *Zasady logiki ogólnej* (Kasa Józefa Mianowskiego, Warszawa, 1903)
7. Wł. Biegański, *Podręcznik logiki i metodologii ogólnej dla szkół średnich i dla samouków*, E. Wende, Warszawa; W.L. Anczyc, Lwów; H. Altenberg, Kraków (1907)
8. W. Biegański, *Wnioskowanie z analogii* (Polskie Towarzystwo Filozoficzne, Lwów, 1909)
9. Wł. Biegański, *Traktat o poznaniu i prawdzie*, Zapomoga Kasy dla Osób Pracujących na Polu Naukowem im. Józefa Mianowskiego, Warszawa (1910)
10. Wł. Biegański, Czem jest logika?, Sprawozdania z Posiedzeń Towarzystwa Naukowego Warszawskiego, Wydział Nauk Antropologicznych, Społecznych, Historii i Filozofii, year **III**, issue **6**, 119–152 (1910)
11. W. Biegański, *Teoria logiki* (E. Wende i S-ka, Warszawa, 1912)
12. Wł. Biegański, Sposobność logiczna w świetle algebry logiki, Sprawozdania z Posiedzeń Towarzystwa Naukowego Warszawskiego, Wydział Nauk Antropologicznych, Społecznych, Historii i Filozofii, year **V**, issue **9**, 67–79 (1912)
13. W. Biegański, *Teoria poznania ze stanowiska zasady celowości* (Towarzystwo Naukowe Warszawskie, Warszawa, 1915)
14. Wł. Biegański, *Podręcznik logiki ogólnej*, E. Wende i spółka (Ludwik Fiszer), Warszawa; H. Altenberg, G. Seyfarth, E. Wende i S-ka, Lwów; M. Niemierkiewicz, Poznań; Ludwik Fiszer, Łódź (1916)
15. S. Borzym, Poglądy epistemologiczne Władysława Biegańskiego. Filozofia Nauki **6**(3–4), 11–17 (1998)
16. W. Kozłowski, *Logika elementarna* (Nakładem Towarzystwa Nauczycieli Szkół Wyższych, Lwów, 1891)
17. Wł.M. Kozłowski, *Podstawy logiki, czyli Zasady nauk: wykład systematyczny dla szkół wyższych oraz dla samouctwa* (Wydawnictwo M. Arcta, Warszawa, 1916)
18. Wł.M. Kozłowski, *Krótki zarys logiki wraz z elementami ideografii logicznej* (Wydawnictwo M. Arcta, Warszawa, 1918)
19. J. Łukasiewicz, *Elementy logiki matematycznej*, ed. by M. Presburger. Komisja Wydawnicza Koła Matematyczno-Fizycznego Słuchaczów Uniwersytetu Warszawskiego, Warszawa (1929). Reprint: Wydawnictwo Naukowe Uniwersytetu im. Adama Mickiewicza, Poznań 2008
20. A. Marburg, Logika i teoria poznania, in *Poradnik dla samouków*, Wydawnictwo A. Heflera i St. Michalskiego, Warszawa (1902)
21. R. Murawski, Genius or Madman? On the life and work of J.M. Hoene-Wroński, in *European Mathematics in the Last Centuries*, ed. by W. Więsław (Stefan Banach International Mathematical Center/Institute of Mathematics, Wrocław University, Wrocław, 2005), pp. 77–86
22. R. Murawski, The philosophy of Hoene-Wroński. Organon **35**, 143–150 (2006)
23. R. Murawski, *Filozofia matematyki i logiki w Polsce międzywojennej*, Monografie Fundacji na rzecz Nauki Polskiej, Wydawnictwo Naukowe Uniwersytetu Mikołaja Kopernika, Toruń (2011)
24. H. Struve, *Logika poprzedzona wstępem psychologicznym. Odczyt Uniwersytecki Dra Henryka Struve profesora Szkoły Głównej w Warszawie 1863*, lithograph (1863)
25. H. Struve, *Wykład systematyczny logiki czyli nauka dochodzenia i poznania prawdy*. Volume 1: *Część wstępna* (*Historya logiki u obcych i w Polsce*), Warszawa (1868–1870)
26. H. Struve, Filozofia i wykształcenie filozoficzne, in *Encyklopedia wychowawcza*, ed. by J.T. Lubomirski, E. Stawiski, J.K. Plebański, vol. IV (J. Drukarnia J. Sikorskiego, Warszawa, 1885)

27. H. Struve, *Logika elementarna. Podręcznik dla szkół i samouków z dodaniem słownika terminów logicznych* (Wydawnictwo M. Arcta, Warszawa, 1907)
28. H. Struve, *Historya logiki jako teoryi poznania w Polsce*, printed by the author, Warszawa (1911)
29. I. Trzcieniecka-Schneider, *Logika Henryka Struvego. U progu nowego paradygmatu* (Wydawnictwo Naukowe Uniwersytetu Pedagogicznego, Kraków, 2010)
30. K. Twardowski, Henryk Struve. Ruch Filozoficzny **2**(6) (1912)
31. J. Woleński, *Logic and Philosophy in the Lvov-Warsaw School* (Kluwer Academic Publishers, Dordrecht, 1989)
32. J. Woleński, Władysław Biegański jako logik. Filozofia Nauki **6**(3–4), 19–26 (1998)
33. J. Woleński, Mathematical logic in Warsaw: 1918–1939, in *Andrzej Mostowski and Foundational Studies*, ed. by A. Ehrenfeucht, V.W. Marek, M. Srebrny (IOS Press, Amsterdam, 2008), pp. 30–46

R. Murawski (✉)

Faculty of Mathematics and Computer Science, Adam Mickiewicz University, ul. Umultowska 87, 61-614 Poznań, Poland

e-mail: rmur@amu.edu.pl

Russian Origins of Non-Classical Logics

Valentin A. Bazhanov

Abstract This paper reviews the conditions under which non-classical logic was historically promoted in Russia. It identifies those Russian scholars who, beginning in the second part of the nineteenth and in the turn of twentieth century, contributed most to the formation and development of non-classical logic along with the socio-cultural milieu crucial to the birth of non-classical (non-Aristotelian) ideas.

Keywords Non-Aristotelian logic · Imaginary logic · Law of excluded middle · Law of non-contradiction · Paraconsistent logic · Relevant logic · Substructural logic

Mathematics Subject Classification (2000). 03–03 · 03A05 · 03B50 · 03G20 · 03A70

The path to modern non-classical logic has always been long and arduous. The starting point of this path may be seen in rather vague ideas about the imperfections of Aristotle's logic, its basic laws, style and the nature of its argumentation. Among the mathematicians who made significant historical contributions to the emergence and formation of the shape of non-classical logic were a number of eminent Russian scientists.

What were the key conditions that promoted the emergence of non-classical logic ideas in Russia in particular? Who among Russian scientists contributed the most to the formation of non-classical logic's birth and development? What factors of the socio-cultural background contributed to the emergence in Russia of important ideas namely in the field of non-classical logic? These are the key questions in our present discussion.

Near and Distant Approaches to Non-Classical Logic

Leaving aside the consideration that the founder of non-classical logic was Aristotle, the real story of non-classical logic should be measured from the middle and second half of the nineteenth century.

Dedicated to Dr Irving Anellis, whose God given talent in the field of history of logic is highly appreciated by his Soviet/Russian colleagues.

© Springer International Publishing Switzerland 2016
F.F. Abeles, M.E. Fuller (eds.), *Modern Logic 1850-1950, East and West*, Studies in Universal Logic, DOI 10.1007/978-3-319-24756-4_10

Since the mid nineteenth century, classical logic developed in two forms: first, so to speak, in traditional Aristotelian and, second, in the form of mathematical logic, in its algebraic and logistics styles of presentation. The basic principles of classical logic—the laws of (non)contradiction and excluded middle—were analyzed more and more thoroughly, mainly through the language of traditional logic. Dialectics also paid close attention to these laws and were both used and castigated.

At the end of the 1880s C.S. Peirce suggested the possibility of "non-Euclidean" logic. A little bit later he began to think about a non-Aristotelian logic, having in mind the idea of three-valued logic.

In 1902, C.S. Pierce pointed out several situations in which, according to his opinion, the laws of (non)contradiction and excluded middle may be violated: "I have long felt that," C.S. Peirce wrote in a letter to W. James in 1909, "there is a serious defect in existing logic ... I do not say that the Principle of Excluded Middle is downright *false*, but I do say that in every field of thought whatsoever there is intermediate ground between *positive assertion* and *negative assertion* which is just as Real as they. Mathematicians always recognize this, and seek for that limit as the presumable lair of powerful concepts; whereas metaphysicians and the old-fashioned logicians—the sheep and goats separators—never recognize this. The recognition does not involve any denial of existing logic, but involves a great addition to it..." [1, p. 116].

In 1910, in a letter to F. C. Russell from Chicago and published by Paul Carus, the Editor of the journal "The Monist", C. S. Peirce presented "sundry" topics of modern logic (to paraphrase P. Carus): "Before I took up the general study of relatives, I made some investigation into the consequences of supposing the laws of logic to be different from what they are. It was a sort of non-Aristotelian logic, in the sense in which we speak of non-Euclidian geometry. Some of the developments were somewhat interesting, but not sufficiently so to induce me to publish them. The general idea was, of course, obvious to anybody of sufficient grasp of logical analysis to see that logic reposes upon certain positive facts, and it is not mere formalism. Another writer afterward suggested such a false logic, as if it were the wildest lunacy, instead of being a plain and natural hypothesis worth looking into [notwithstanding its falsity]" [2, p. 45]. C.S. Peirce explained that he intended to modify the law of transitivity, and did his best to foresee what would happen with logic in this case in the absence of this law [3, pp. 158–159].

Commenting on the letter of C.S. Peirce, P. Carus noticed that the latter's view did not agree with his own interpretation of the nature of non-Aristotelian logic and the proposed method of its construction, because, firstly, this view suggested a different and nonuniform interpretation of the nature of classical and non-classical logic and, secondly, the law of transitivity was never included in the basic laws of logic and therefore its revision could hardly be considered the starting point for creating a non-Aristotelian logic.

In any case, for us it is important to stress that in the late XIX to early XX centuries, Charles S. Pierce and several other thinkers discussed the possibility of creating a non-Aristotelian, non-classical logic and considered such a possibility quite real and worthy of close study.

Intellectual Milieu Just Before the Emergence of Non-Classical Systems

If we try to summarize the facts surrounding the emergence and creation of non-classical systems in science, it is possible to notice an interesting pattern: these systems have had a better chance to appear and establish themselves within research centers in those places (referring to the classical branches of science) where young and well-trained scientists were available, but also where no strong scientific school or tradition had yet arisen and developed. This is quite understandable: a well-established scientific school with its longstanding traditions and the canons of scientific criteria will likely (or even inevitably?) suppress the sprouting of novel—in this case, the non-classical—paradigm, the formation of which requires freedom from the dictates of authority and scope for the development of "crazy" ideas. Probably, this pattern may relate to the mechanisms of the birth of new paradigms in general, not only in the field of logical knowledge.

The history of science at the University of Kazan largely confirms this trend. Almost until the end of the nineteenth century it was the most eastern European university with quite favorable working conditions due to carefully planned and well-supplied libraries, specific academic autonomy and freedom of thought and action, with the possibility of travel, including overseas. At the same time, conditions favorable for the formation of stable scientific schools had not developed (except, maybe, in the case of the school of chemistry).

Indeed, quite a few pioneers in science (and in culture) either worked or were associated with the University of Kazan for long periods of time. These included N.I. Lobachevsky (mathematics), I.A. Baudouin de Courtenay (linguistics), Velimir Khlebnikov (poetry), and N.A. Vasiliev (logic).

Each of the aforementioned thinkers may be considered the forerunner of a non-classical system: N.I. Lobachevsky of "imaginary", non-Euclidian geometry; I.A. Baudouin de Courtenay of sociolinguistics; V. Khlebnikov—"imaginary" poetry; and N.A. Vasiliev—"imaginary", non-Aristotelian logic. Perhaps it is no coincidence that characterizing these three systems as "imaginary" emphasizes the fact that the "reality" needs to be described in terms of, in a certain sense, non-classical concepts.

V.I. Ulyanov-Lenin, who for a short period (3 months) studied at the University of Kazan, can be assigned to this cohort of heretics as well. He had the goal to shake—and indeed shocked—customary foundations.

An important source of non-classical logic may be seen in the article by Ivan E. Orlov, published in 1928. It is the only strictly logical work by the author, who launched the quest for the radically novel—dialectical in essence—logic of science during a period when the dialectical method in the early Soviet Union was seen as the most versatile and powerful method for acquisition of scientific knowledge. I.E. Orlov is often considered the predecessor of relevant, paraconsistent, substructural logics.

Autonomous Development of Non-Classical Logic in Russia

Russian scientists kept close ties with their Western counterparts and thus had excellent knowledge of their achievements. Almost all leading Western journals and books instantaneously arrived at Russia's University libraries and bookshops, and the scientists themselves regularly traveled abroad. Nevertheless, the ideas of A. Meinong (1907), L.E.J. Brouwer (1908), and J. Łukasiewicz (1910), which express non-classical ideas, were unknown in Russia at the moment of their emergence and did not have any effect upon Russian scientists. Russian logicians did their best to follow closely the achievements of their Western counterparts, but developed their ideas to a large extent independent of the latter.

Back in 1901–1902, Professor S. O. Shatunovsky of Novorossiysk University (Odessa) declared that the law of the excluded middle has no implication for infinite sets (see [4, p. 208]).

According to Professor E. A. Sidorenko [5], Father P.A. Florensky, in the book "The Pillar and Ground of the Truth: An Essay in Orthodox Theodicy in Twelve Letters" (1914), in his speculations related to the Holy Spirit, had in mind the idea of paraconsistent and nonmonotonic logic. Professor B.V. Biryukov claimed Russian neo-Kantians A. I. Vvedensky and N. O. Lossky were rather close to the idea of paraconsistency, though they did not express it in explicit form (see [6]).

Expectations Prior to the Emergence of Non-Classical Logic

P. Carus in 1910 wrote that "Aristotelian logic is incomplete and insufficient. It treats only the most simple relations and does not cover the more complicated cases of thinking, but so far as it goes is without fault... [I]t is possible to imagine a fairy-tale world where our scientific conception of cause and effect could be crossed by a causation of miracle... The purely formal logic rules of Aristotelian logic would not be upset thereby. The mill remains the same even if the grist is changed" [2, p. 44].

"... [W]hy should there as well exist a curved logic and a mathematics of curved space? A curved logic would be a very original innovation for which no patent has yet been applied. What a splendid opportunity to acquire Riemann's fame in the domain of logic!", exclaimed the scholar. "The world has seen many new inventions. Over the telephone we can talk at unlimited distances, and some of our contemporaries fly like birds through the air. Radium has been discovered which is often assumed, with a certain show of plausibility to have upset the laws of physics, but the invention of non-Aristotelian logic would cap the climax" [2, pp. 45–46]. Meanwhile, in the same year [2] on May 18, in the trial-lecture N.A. Vasiliev at Kazan University constructed a formal system of non-classical, non-Aristotelian logic which he called "imaginary" (scientific biography of N. A. Vasiliev presented in [7, 8]).

Heuristic Background of Imaginary Logic

When Nikolai Vasiliev was 17 he carefully read C.S. Peirce's then recently published paper related to logic relatives and had written up a resumé. In this article, C.S. Peirce develops a logic of relatives in a certain sense alternative to Aristotelian logic. Apparently, at this moment young Nikolai Vasiliev realized that Aristotelian logic was not absolute by its nature.

Later, when N. A. Vasiliev discussed theoretical sources of imaginary logic he mentioned the following milestones on the way to a new logic:

- The Hegelian dialectic (and all the dialectical tradition in general);
- Inductive logic of J.S. Mill, and his criticism of the Aristotelian syllogism;
- Criticism of Ch. Sigwart of modal judgments classification;
- Development of mathematical logic [9].

Analysis of the life and scientific legacy of N. A. Vasiliev allows us to list probable heuristic assumptions that served as catalysts in the process of creating an imaginary logic. These are:

1. The *logic of relatives* of C.S. Peirce, which convinced N. A. Vasiliev that Aristotelian logic is not unique, and that we can conceive of the idea of a plurality of logical systems.
2. Poetry in the spirit of *symbolism*, to which N. A. Vasiliev paid close attention and had written poems in the same style. Symbolist poetry is characterized by the topic of "other worlds". N.A. Vasiliev, perhaps, was the only symbolist who endowed the worlds of his poetry with contradictory properties;
3. *Psychologism* which intertwined the human psyche, perception, and the possible changing of the nature of negation in imaginary worlds;
4. The ideas of *Charles Darwin*, which, as claimed Ch. Sigwart (N. A. Vasiliev had specially emphasized this point), are of paramount importance for logic and force us to move from a static interpretation of concepts to the analysis of their dynamics—an analysis involving a logic different from Aristotelian logic;
5. The *analogy with non-Euclidean geometry*, which N. A. Vasiliev deliberately used when constructing his imaginary logic. He specifically pointed out that this logic is constructed by the "method of imaginary geometry".

The true meaning of N. A. Vasiliev's innovative work was revealed gradually: generations of scholars have found in it ideas consonant with the current situation in science. For imaginary logic, presentation was informal, and each successive generation that followed disclosed and developed the ideas that it deemed dominant and relevant to the demands of the moment.

For example, Professor N. N. Luzin, founder of the Moscow Mathematical School, in his 1927 review of N. A. Vasiliev's works (written while he was still a corresponding member of the Academy of Sciences of the USSR), draws attention to a rejection of the law of excluded middle, which, in his opinion, anticipated intuitionism and "is in line with efforts in this direction by creating a new mathematical theory". N. N. Luzin never mentioned N.A. Vasiliev's much more radical abandonment of the law of non-contradiction.

The prominent Soviet algebraist A.I. Maltsev emphasized the merit of N. A. Vasiliev in the development of logical thought in Russia. He claimed (by the way, along with N. Rescher, G. Kline, and M. Jammer) that N. A. Vasiliev, due to his rejection of the law of excluded middle, was the founder of many-valued logic. However, the abandonment by N. A. Vasiliev of the law of non-contradiction went virtually unnoticed (see [10]). Thus an intellectually radical move by N. A. Vasiliev was ignored. His works mark the prehistory of paraconsistent logic.

Despite the fact that the works of N. A. Vasiliev were included in the additional listings (1936–1938) of A. Church's famous "A Bibliography of Symbolic Logic, 1666–1935", they were barely noticed by the scientific community. Only A. Fraenkel and Y. Bar-Hillel in their book "Foundations of Set Theory" [11] mentioned one of the works of N. A. Vasiliev—one published in English; R. Feys in his book on modal logic included in the bibliography the same paper of N. A. Vasiliev [12], but ideas of N. A. Vasiliev were developed neither by A. Fraenkel and Y. Bar-Hillel, nor by R. Feys.

The ideas of N. A. Vasiliev that were related to abandonment of the law of non-contradiction were rediscovered decades later—by S. Jaśkowski (1948), D. Nelson (1957), and especially N.C.A. da Costa (1958). The true history of paraconsistent logic is linked to these scholars. While N.C.A. da Costa was visiting Moscow in the summer of 1987, he met the author and told him that, at a time when he was already a quite well-known expert who had developed paraconsistent logic, he accidently found that there was a Russian predecessor who was ahead by 50 years in rejection of the laws of non-contradiction and excluded middle. He urged his pupil A. Arruda to study the works of N.A. Vasiliev. A. Arruda, unfortunately, died early, but she managed to popularize the ideas of N. A. Vasiliev among Western logicians [13].

Non-Classical Logic as a "By-Product" of Dialectical Research

The phenomenon of so-called ideological science prevailed in the Soviet Union in the 1920s–1940s. This phenomenon had a very negative impact on the development of Soviet science, although in some cases it produced very interesting and unexpected "by-products". By the latter, I mean the psychological research by L.S. Vygotsky and A.R. Luria and the logic of propositional consistency by Ivan E. Orlov (see [9, 14]).

I.E. Orlov was vigorously engaged in the quest for a special logic of natural science, which in his opinion should have been dialectical in its essence. He was very actively published and participated in numerous discussions that were quite saturated by ideological pockets and which criticized the works of philosophers and scientists from the rather orthodox standpoint of Marxist dialectics. His work included numerous philosophical papers and would become known (as would the author himself) to a narrow circle of historians of Soviet philosophy of 1920s–1930s. However one of his—and the only!—paper on logic, was published "in the order of discussion" in 1928 in the journal "Matematicheskii Sbornik"—with a purely mathematical, not philosophical content. The paper presents, so thought I.E. Orlov, the formal presentation of a logic which is strictly adequate to modern science. From a mathematical point of view, in his attempt to tie intuitionistic logic with a modal operator by introducing the appropriate system S4, I.E. Orlov claimed to have overcome the paradox of material implication.

I.E. Orlov published his first scientific work in 1916 (it concerned the nature of the inductive method and inductive proof), but in the years of revolution and civil war (1917–1920), he was silent. Since the beginning of the 1920s, in a completely different intellectual and ideological situation, he criticized, from the dialectic position G. Cantor's theory of sets, the logical machine of Shchukarev, and, in the philosophy of physics, he defended the truth of a mechanistic worldview; he also discussed the problems of psychology and even music.

His logical work of 1928 for the first time attracted the attention of A.A. Zinoviev in 1962 (being a logician he is more known as a Soviet dissident and writer). In 1978, V.M. Popov [15] declared I.E. Orlov to be the conceptual predecessor of relevant logic. This claim was later justified by A.V. Chagrov (1990) and R. Routley (1991). K. Došen noticed that I.E. Orlov can be considered the ancestor of substructural logics [16]. S.N. Artemov interpreted ideas of I.E. Orlov in the context of the logic of provability [17]. E. Alves in 1992 [18], and N.C.A. da Costa, J.-Y. Beziau, and O. Bueno in 1995 [19] judged that I.E. Orlov's ideas enable us to add him to the forerunners of paraconsistent logic.

Acknowledgment The author is grateful to M. Fuller for valuable suggestions, which enabled him to improve this paper.

References

1. C. Eisele, The New Elements of Mathematics by Charles S. Peirce, in *Men and Institutions in American Mathematics*, ed. by J.D. Tarwater, J.T. White, J.D. Miller (Lubbock, Texas Tech University Press, 1976), pp. 111–122
2. Carus, Paul (ed.), The nature of logical and mathematical thought. Monist **XX**(1), 33–75 (1910)
3. C.S. Peirce (P.C.), Non-Aristotelian logic. Monist **XX**(1), 158–159 (1910)
4. E.Y. Bakhmutskaya, About early works related to the foundations of mathematics of S.O. Shatunovsky. Historico-Math. Stud. **16**, 207–218 (1965) (in Russian)
5. E.A. Sidorenko, Ideas of non-monotonic and paraconsistent logic in the works of P. Florensky. Log. Investig. **4**, 290–303 (1997) (in Russian)
6. B.V. Birukov, B.M. Shuranov, Russian neo-Kantians: on the verge of paraconsistency idea, in *Modern Logic: Problems of Theory, History, and Applications* (St.-Petersburg University press, 1996), pp. 125–127 (in Russian)
7. V.A. Bazhanov, Nicolai A. Vasiliev (Nauka., Moscow, 1988) (in Russian)
8. V.A. Bazhanov, *N.A. Vasiliev and his Imaginary Logic: The Revival of One Forgotten Idea* (Kanon, Moscow, 2009) (in Russian)
9. V.A. Bazhanov, The Origins and Emergence of Non-Classical Logic in Russia (nineteenth century until the turn of the twentieth century), in *Zwischen traditioneller und moderner Logik. Nichtklassiche Ansatze* (Mentis-Verlag, Paderborn, 2001), pp. 205–217
10. A.I. Mal'tsev, *Selected works*, vol. 1 (Nauka, Moscow, 1976) (in Russian)
11. A. Fraenkel, Y. Bar–Hillel, *Foundations of Set Theory* (North-Holland Publ.Co., Amsterdam, 1958)
12. R. Feys, *Modal Logics* (E. Nauwelaerts, Louvain, 1965)
13. A.I. Arruda, in *A Survey of Paraconsistent Logic*, ed. by A.I. Arruda, R. Chuaquai, N.C.A. Da Costa. Mathematical logic in Latin America (Amsterdam, N.Y., Oxford, North-Holland, 1980), pp. 1–41
14. V.A. Bazhanov, The scholar and the "wolfhound era": The fate of Ivan E. Orlov's ideas in logic, philosophy, and science. Sci. Context **16**, 535–550 (2003)
15. V.M. Popov, The Decidability of Relevant Logic System RAO, in *Modal and Intensional Logics* (Institute of Philosophy, Moscow, 1978), pp. 115–119 (in Russian)
16. K. Došen, The first axiomatization of relevant logic. J. Philos. Log. **21**, 339–356 (1992)
17. S.N. Artemov, Personal letter to author (2003)
18. E. Alves, The first axiomatization of paraconsistent logic. Bull. Section Log. **21**, 19–20 (1992)
19. N.C.A. Da Costa, J.-Y. Beziau, O.S. Bueno, Aspects of paraconsistent logic. Bull. IGPL **3**(4), 597–614 (1995)

V.A. Bazhanov (✉)
Ulyanovsk State University, Transportnaya str., 2, apt. 5, 432048 Ulyanovsk, Russia
e-mail: vbazhanov@yandex.ru; vbazhanov@gmail.com

Constructive Mathematics in St. Petersburg, Russia: A (Somewhat Subjective) View from Within

Vladik Kreinovich

Abstract In the 1970 and 1980s, logic and constructive mathematics were an important part of my life; it's what I defended in my Master's thesis, it was an important part of my PhD dissertation. I was privileged to work with the giants. I visited them in their homes. They were who I went to for advice. And this is my story.

Keywords Constructive mathematics · History of constructive mathematics · Russian mathematics

Mathematics Subject Classification (2010). Primary 01A60 · Secondary 01A72 · 03F60

Why Constructive Mathematics Is One of the Most Important Activities in the World: As Well As Physics and Game Theory

What do we humans want?

Why Science and Physics Are Important We want to *understand the world*, we want to predict what will happen—including what will happen if we do nothing and what will happen if we perform certain actions. This is what *physics*—and science in general—is about. Physicists come up with equations describing how the state of the world changes with time, and we would like to use these equations to come up with the actual predictions.

Why Is Constructive Mathematics Important? How do we go from equations to predictions? At first glance, this is what mathematicians (especially specialists in numerical methods) are doing—and sometimes they are doing it—but in general, mathematics is about proving theorems, not generating numbers.

In Russia, many of us heard a story (possibly a legend) that once a famous mathematician, a colleague of the Nobelist physicist Lev Landau, asked Landau what he was working on. Landau wrote down a complex system of partial differential equations describing the physical phenomena that interested him at that time. After a few months, a happy mathematician came back to Landau with a thick manuscript: "I have solved your problem! It was not easy, but I have proven that your system of equations has a solution!" :-)

© Springer International Publishing Switzerland 2016
F.F. Abeles, M.E. Fuller (eds.), *Modern Logic 1850-1950, East and West*, Studies in Universal Logic, DOI 10.1007/978-3-319-24756-4_11

This may be an exaggeration, definitely Kolmogorov and other prominent applied mathematicians helped efficiently solve many complex practical problems—but this story shows that there is a need to formally distinguish between proving theorems and actually producing solutions.

This distinction is what *constructive mathematics* is about: crudely speaking, constructive mathematics is about algorithms—in constructive mathematics, existence means that we can already produce the corresponding description—and not simply that we have proven its existence.

We Also Want to Change the Word: Another Reason why Constructive Mathematics Is Important In addition to understanding, we also want to *change* the world, we want to find the appropriate actions and designs that will lead to the best possible outcomes. This is what engineering is about:

- We want to design a bridge that would withstand the prevailing winds and possible hurricanes and earthquakes;
- We want to design an efficient and safe airplane;
- We want to come up with a control strategy for a vehicle which would, for example,

    – lead an emergency vehicle to its destination in the shortest possible time or
    – make a bus spend as little fuel as possible while following the prescribed route.

In all these problems, we want to actually produce a solution. Here, it is even more important to actually produce the corresponding design or control algorithm.

Yes, numerical methods aim to do just that, they even use the word "algorithm", but often, what they call an algorithm is not exactly what computer scientists would call an algorithm. Rather it is a blueprint for an algorithm. For example, Newton's method for finding a root is a potentially infinite iterative process.

- We are not given any specific recommendation on when to stop[1]; and
- We are not sure that this method will always work—usually, we know that in many cases, it does not work.

We need a way to clearly distinguish between such heuristic "algorithms" and algorithms in the computer science sense: when the sequence of steps is pre-determined and always leads to a correct solution. Constructive mathematics provides such distinction.

Why Game Theory Is Important Finally, when selecting an appropriate solution, we need to take into account the preferences and opinions of different people who are (or may be) potentially affected by the solution. The discipline that takes these preferences into account is well established, it goes under the somewhat misleading name of *game theory*.

People in political science and humanities in general, political leaders, spiritual leaders, business leaders, may think that they should solve these problems—and at present, in most cases, they are solving these problems now. But the goal of game theory has always been to resolve many of these problems by applying appropriate mathematical methods—and in solving such problems, specialists in game theory and decision making have succeeded a lot.

---

[1]To be more precise, we are shown several possible recommendations, and told that none of them is perfect.

This Is Another Reason Why We Need Constructive Mathematics And again, in game theory and decision making, we do not just need existence proofs, we need algorithms, we need explicit solutions.

Summarizing: three things are most important:

- Physics—understood in the general sense, as a description of the physical world—which enables us to describe how the world changes;
- Constructive mathematics, which enables us to describe how to best affect the world;
- Game theory, which enables us to take into account preferences of different people.

From the Mathematical Viewpoint, These Three Research Areas Have Much in Common: They Are All About Important Partial Pre-Orders

Physics: Causality In physics, some things change by themselves, other things change because some objects affect other objects. Before we start studying *how* objects affect each other, it is very important to first understand *which* pairs of objects, which pairs of events can causally affect each other.

In other words, we need to understand the notion of *causality*, which, according to many physicists, is one of the most important notions of physics; see, e.g., [21].

The study of the causality relation is more important than it may seem at first glance. For example, in special relativity, even the linear structure on space-time can be determined based only on the causality relation; this result was first proven by the Russian geometer A.D. Alexandrov in 1949 [3, 5] and became widely known after a somewhat stronger result was proven by E.C. Zeeman (later of catastrophe theory fame) in 1964 [123].

From the mathematical viewpoint, causality is a *partial order*. To be more precise, it is a partial order only in relativistic physics. In Newtonian physics, with the possibility of instantaneous effect, simultaneous events can affect each other, i.e., we have $a \leq b$ and $b \leq a$, but $a \neq b$. So, causality is a *pre-order*.

Constructive Mathematics: Derivability Relation In constructive mathematics, there is also a natural ordering relation.

Namely, in some cases, we derive the corresponding algorithmic result "from scratch"—similarly to the fact that in mathematics, we sometimes prove results directly from the axioms. However, in most cases, both in traditional and in constructive mathematics, we use previous results.

Ideally, we should know how exactly we use the previous results—i.e., we need to know the actual proofs. However, in many cases, it is sufficient to know which results can be derived from other results.

The study of such a "derivability" relation is known as *logic*; a derivability relation corresponding to constructive mathematics is known as *constructive logic*. Logic is indeed often helpful in proving results in both traditional and constructive mathematics. From the mathematical viewpoint, derivability is also a partial order—to be more precise, it is a *pre-order*, since for two different statements $a \neq b$, we can have a implying b and b implying a.[2]

Game Theory: Preference Relation Finally, in game theory, there is also a natural pre-order.

Indeed, to make a decision that takes into account individual human preferences, we need to know these human preferences. Again, ideally, we should know *why* a person prefers one alternative to another and how strong the corresponding preference is. But first we need to know which alternatives are preferable and which are not—i.e., first we need to know each person's preference relation—yet another partial pre-order.

Moreover, in decision making theory, we can restore the numerical characteristics of human behavior—so-called *utility values*—based on the corresponding preference order; see, e.g., [23, 84, 97, 105].

My Personal Story: How I Came to Constructive Mathematics

I Was Interested in Mathematics and Physics I have always been fascinated by mathematics and physics. I participated in Olympiads in math and physics, I went to a math circle led by university students. When the time came for me to enter high school, I went to a special high school with an emphasis on math and physics.

Enter Game Theory When I was in high school, Igor Frenkel, then a student at a similar math high school (and a winner of city math Olympiads; he is now a professor at Yale) gave me, for my birthday, the best birthday present I ever got—an exciting book on game theory. I was awed by the fact that many real-life problems can potentially be solved by reasonably convincing mathematics.

I also saw that while this theoretically is possible, the available algorithms would require an unrealistic computation time to solve complex real-life conflict situations. This was one of the first cases when I realized that many open problems are not about answering purely mathematical questions (although there are many such questions in game theory as well), but rather about coming up with efficient algorithms which would implement the known ideas and techniques.

Game Theory: There Is Room for Optimism While the overall optimization may not be achievable, it is clear that algorithms have helped practical decision-makers. It is also clear that there is a strong need for new algorithms, algorithms which can produce optimal decisions, decisions which are better than heuristic suboptimal decisions people that use now based on their intuition and expertise.

Mathematics and Physics Beyond Game Theory: My High School Experience A game theory book further increased my interest in mathematics and physics. I wanted to read more. However, new books on mathematics and physics were difficult to buy. So to find a good book, one had to regularly go to one of the academic old books stores, where

---

[2]Moreover, many important mathematical theorems establish exactly such equivalences: when we know necessary and sufficient conditions for some property, this brings a sense of completion and satisfaction.

we would sometimes find monographs, edited books, journal issues. (This is, by the way, why I so much appreciated Igor Frenkel's gift.)

I often went to an academic old book store on Liteiny Prospect with my classmate Nikolay "Kolya" Vavilov. Kolya's father was a professor, so he knew in person—or heard about—many of the city's mathematicians, and the corresponding interesting personal stories added to my fascination.

Space-Time Geometry and Physics For example, when we came across a book on space-time geometry and space-time physics by Pimenov [102], Kolya explained to me that Pimenov spent some time in jail for his political activities.

This was not that surprising: in Stalin's times, many families had someone arrested—including my own grandfather. Many scientists and engineers were jailed, including:

- Tupolev (of the airplane fame),
- Korolev (later the leader of the successful Soviet space program),
- Lev Landau,

and many others (and there were lucky ones, who got out alive).

There was a known story that after Tupolev was arrested, the KGB told him that he could atone for his political "sins" and get released by forming a jail-based team and designing a good plane for the Motherland. They asked him to make a list of possible helpers. Tupolev was understandably afraid that the KGB would be tempted to arrest innocent people—just to make his jail team stronger—so he made a list of all the numerous specialists he knew—thinking that the KGB would not arrest everyone. It turned out that most people on his list had already been arrested.

But this was during Stalin's time, and, as Kolya explained, the unusual thing about Pimenov is that he was in jail not in Stalin's time, but under Khruschev, the Communist leader who denounced Stalin's crimes and freed people from jails and concentration camps. Kolya also mentioned that Pimenov was a student of A.D. Alexandrov—a geometer who used to be President ("Rector") of St. Petersburg University in the 1950s and 1960s (until he moved to Siberia to promote science there).

According to Kolya, Pimenov was probably the most beloved of Alexandrov's students—for his great scientific ideas and results—and probably the most hated—since Pimenov publicly accused Alexandrov of complicity with Stalin's crimes and of praising Stalin's outrageous behavior in his official speeches and articles (I think this was an unfair accusation: millions had to do that, those who refused were usually jailed themselves.)

Logic and Constructive Mathematics Kolya attracted my attention to many articles in Zapiski Seminarov LOMI, a local mathematical journal, written by Yuri Matiyasevich and Vladimir Lifschitz, two young talented mathematicians who, according to Kolya, were driven not only by their love of science, but also by their competition with each other.

I later knew both, I think the competition part was, to put it mildly, exaggerated, but the papers were interesting, and their talents clear.

I Joined the Mathematics Department I was fascinated by game theory, by algorithms, by physics. I was especially fascinated by the foundations of physics—so I wanted to major either in physics or in the philosophy of physics. Fate—in the avatar of our Communist dictators—decided otherwise.

It is was well known that Jews were not allowed to become students of philosophy or physics at St. Petersburg University. So, I joined the Mathematics department.

Seminars Talk about a kid in a toy store. I immediately found three seminars which satisfied all three of my needs, and I started actively attending all three of them.

First, I attended a seminar on space-time geometry and physics led by Revolt Pimenov himself. A few years before that, Pimenov started a deep analysis of space-time and physics in general based on the causality relation.[3]

I also started going to a seminar on game theory led by Nikolay N. Vorobiev, the leader of Russian game theory researchers [116, 117].

And finally, I started going to seminars on logic and constructive mathematics. In contrast to space-time physics and game theory, there were actually three different seminars:

- A city-wide official seminar, where completed results would be presented to a very general audience, including people from different schools;
- A working seminar, in which preliminary results and open problems were presented, as well as interesting papers published by others (the seminar leaders regularly assigned to seminar participants to review and present);
- An informal seminar "on systems", led by Sergey Maslov, where raw ideas were welcome, and where, in addition to logicians, interested (and interesting) people from humanities would often give presentations.

My purpose is to describe what happened at the seminars on constructive mathematics. To get a better understanding of this, let us first briefly recall what happened earlier, before the Fall 1969 when I started attending their seminars.

A Brief History of Constructive Mathematics up to the 1960s

Brouwer's Ideas: Intuitionism The need to have efficiency in mathematics started with Brouwer's *intuitionism* [11].

Brouwer was not happy with the fact that in classical logic and in classical mathematics, a statement $A \vee \neg A$ is always true. This seemed to conflict with a reasonable intuitive understanding of "or", according to which knowing $A \vee B$ would means that we either know A or we know B. Indeed, for many open mathematical statements A, we do not know whether these statements are true or false. Brouwer therefore decided to change mathematics in such a way that it would be in better accordance with this reasonable intuition.

To capture this intent, he called this new mathematics *intuitionistic mathematics*—and he called the corresponding logic *intuitionistic logic*.

Can Intuitionism Ideas Be Described in Formal Terms? Brouwer's use of the term "intuitionism" was even more appropriate since he believed that the problem with the

---

[3]It looks like this ideas was up in the intellectual air, since at that same time, in addition to Pimenov, similar ideas were proposed by the famous geometer Busemann [12] and by physicists Kronheimer and Penrose [77].

law of excluded middle $A \vee \neg A$ comes from over-emphasizing formalisms—which are inevitably imperfect and thus, lead us astray. He believed that we should always use our intuition as an ultimate test—and he doubted that a formalism would be able to capture, for example, his ideas about the law of the excluded middle $A \vee \neg A$.

These doubts were dispelled by Heyting [34], who showed, in 1930, that a large portion of then intuitionistic mathematics can actually be formalized; see also [35].

Intuitionistic Mathematics and Logic Promote Effectiveness In intuitionistic logic:

- The knowledge of $A \vee B$ means that we know either A or B,
- The knowledge of $\exists x\, Ax$ means that we can effectively produce x for which $A(x)$ is true,
- The knowledge of $\forall x\, \exists y\, A(x, y)$ means that, given x, we can effectively produce y for which $A(x, y)$ holds.

How Can We Describe Effectiveness? Effectiveness could not be formally described at that time since in the early 1930s, there was no formal notion of an effective procedure (what we now call an *algorithm*). This formal notion came later, with the pioneering papers by Turing [114] and Church [16].

Enter Constructive Mathematics By the late 1940s, the notion of an algorithm was universally accepted:

- Different versions of this definition were proven to be equivalent,
- Most procedures recognized as algorithms were shown to be covered by these definitions.

This enabled researchers to formulate the main ideas of constructive mathematics in precise terms: that $\forall x\, \exists y\, A(x, y)$ means that there exists an algorithm that, given x, returns y for which $A(x, y)$ is true.

The first idea of constructive mathematics came from Andrei A. Markov—and, as usual in the history of mathematics (and in history in general), his path to constructive mathematics was not as straightforward as it may seem now.

Andrei Andreevich Markov Jr.[4] at first chose topology as his area of mathematical interests, and he got interested in the problem of checking whether two given compact manifolds are homeomorphic. The traditional definition of a manifold is not very constructive, but it is known to be equivalent to a very constructive definition: like an assembly-required toy, each compact manifold can be represented as a finite collection of polyhedra, with faces marked so that faces marked with the same mark are glued together. (From the topological viewpoint, we can always assume that all the vertices of all the polyhedra have rational coordinates.)

In the 2-D case, there is a known algorithm for checking when two such manifolds are equivalent. Markov decided to analyze how to extend this algorithm to a 3-D case. If he succeeded in producing an algorithm, then he would just have described it as an efficient procedure, and there would have been no need for him to go into any details into what an algorithm means in the general case—all he would have needed was to show that his particular algorithm is efficient. Luckily for foundations of mathematics, Markov was proving a negative result—that no such algorithm is possible.

---

[4]the son of A. A. Markov Sr., of the Markov processes and the Markov chains fame.

However, there was no well-established notion of an algorithm operating on manifolds—and without a precise mathematical notion, it is impossible to prove that no algorithm can check homeomorphism.

So, to transform his intuition into a precise proof, Markov started looking into how to formalize the notion of an algorithm operating on manifolds. To do this, he started by describing algorithms operating on real numbers.

Constructive Mathematics: A General Idea Intuitively, a constructible object has a description in terms of a finite sequence of symbols. As we all know, inside a computer, every symbol is represented as a sequence of 0 s and 1 s, so every sequence of symbols is also represented by a binary sequence. Therefore, every constructible object can be represented as a sequence of 0 s and 1 s.

The simplest mathematical objects are natural numbers. So, from the mathematical viewpoint, it is natural to interpret each code of a constructible object as a natural number. A seemingly natural is to identify each binary sequence with the corresponding number. For example, a binary sequence 11 corresponds to a natural number 3, since 11_2 is a binary code for the decimal number 3_{10}.

However, this idea needs a modification. For example, two different binary sequences 0011 and 11 would then be described by the same code 3. We can avoid this problem if we first add an extra 1 in front of the original binary sequence and then convert the resulting binary sequence into a decimal code. In this case, the sequence 0011 will be transformed into a sequence 10011 and thus, will be represented by a number $10011_2 = 19_{10}$, while a sequence 11 is transformed into 111 and is thus represented by a different code $111_2 = 7_{10}$.[5]

Real Numbers in Constructive Mathematics In Markov's constructive mathematics, e.g., a constructive real number is simply an algorithm that transform a natural number k into a rational number r_k in such a way that $|r_k - r_\ell| \leq 2^{-k} + 2^{-\ell}$. The meaning of r_k is that r_k is a 2^{-k}-approximation to the desired real number.

Each is a code in some programming language. So, we can also represent this algorithm r as a sequence of 0 s and 1 s—hence, as an integer code.

Real-Valued Functions in Constructive Mathematics A constructive function f from real numbers to real numbers is a function that inputs the code of a real number x and returns the code of the real number $f(x)$.[6]

Logic of Constructive Mathematics Logical statements related to constructive mathematics are interpreted in accordance with a general idea.

---

[5]The fact that we can represent sequences of symbols by natural numbers was first discovered by Gödel and is therefore called *Gödelization*. This idea was new in the 1930s, but with the computers, it is so trivial that we feel that over-using this term to describe an otherwise clear idea may only confuse readers. Besides, the original Gödelization algorithm involved exponentiation $2^a \cdot 3^b \cdot \ldots$; in the 1930s, this was a reasonable idea but now, with the clear distinction between feasible (polynomial-time) and exponential-time (non-feasible) algorithms, it does not make sense to introduce an unnecessary exponential time into something as trivial as representing strings in a computer.

[6]It should be mentioned that constructive functions can only be applied to mathematically constructible real numbers—moreover, to compute the value $f(x)$, we must know the exact code of the program that generates the original number x.

For example, the implication $\exists x\, P(x) \to \exists y\, Q(y)$ means that there exists a constructive function f from reals to reals that is always applicable and for which $P(x)$ implies $Q(f(x))$. In other words, the above implication is interpreted as $(\exists f \in Con)(\forall x(P(x) \to Q(f(x))))$, where $f \in Con$ means that a natural number f is a code of a constructive function.

Similar interpretations can be made for more complex logical formulas as well; see, e.g., [87, 88]. As a result, we arrive at an algorithm that transforms an arbitrary formula into a form $\exists x\, A$, where A is an *almost negative* formula (in the sense that only decidable formulas can occur after $\exists$, $\vee$.) The corresponding algorithm was first explicitly described by Nikolay Alexandrovich Shanin, one of the first converts from topology to constructive mathematics and the future leader of the St. Peterburg School of Constructive Mathematics, in [107]; see also [89].[7]

Based on this idea, Markov, Shanin, and other researchers analyzed different mathematical results to see which results are constructive and which are not; see, e.g., [13, 78, 87, 88, 90, 108].

The Markov Principle One important tool in their analyses was Markov's *principle of constructive selection*—which now is known as the *Markov Principle*. The intuitive meaning behind this principle is related to the fact that, as it is well known, there is no algorithmic way to check whether a given algorithm will stop on given data. The Markov Principle says, in effect, that if it is *not* true that the algorithm never stops, this means that this algorithm *will* stop. In more precise terms, if we have a decidable property $P(x)$ (i.e., a property for which $\forall x\, (P(x) \vee \neg P(x)))$, then $\neg\neg\exists x P(x)$ implies $\exists x\, P(x)$.[8]

Negative Reaction to Constructive Mathematics: Why

The First Reaction of the Mathematical Community to Constructive Mathematics Was Rather Negative The way we have just described it, the activity of constructive mathematics is reasonable and useful both for understanding mathematics and for applications of mathematics.

However, originally, the first reaction of most mathematicians to constructive mathematics was negative. There were at least five reasons for this negative reaction.

First Reason: Methodological In their papers and talks, researchers in constructive mathematics did not just propose new ideas and results, they argued that, in effect, all the previous mathematical results and theories made no sense and should be replaced by their constructive versions. For example, Shanin liked to emphasize that when a property is proven to be true only almost everywhere, this result is practically useless, since we still do not have a single example of a point at which this property holds: "pochti vezde znachit neizvestno gde". I think many mathematicians would agree with this statement—but not

---

[7]It is worth mentioning that the algorithm SH is known to be equivalent (under a suitable coding in Heyting's formalized intuitionistic arithmetic) to recursive realizability introduced by Kleene [42].

[8]From the classical viewpoint, the constructive logic of Markov's school can be completely described using the three above-described basic principles: recursive realizability, the Markov principle, and classical logic for sentences containing no constructive problems, i.e., $\exists$, $\vee$-free sentences [94, 113].

with Shanin's conclusion that the result about the property being true almost everywhere makes no sense and should not be published.

Other Reasons There were other reasons, of course, why the initial reception of constructive mathematics was negative.

Some of these reasons were related to the abundance of negative results and counterexamples in constructive mathematics. In the beginning, the idea of looking for a constructive proof sounded reasonable: e.g., we have a theorem that proves the existence of a solution to a differential equation, but we do not know how to actually find this solution, so let us come up with such an algorithm. In these terms, the problem sounds like the need to find an algorithm. Somewhat surprisingly, it turned out that in many cases, such an algorithm does not exist.

- A. Turing proved, in effect, that no algorithm can detect whether two real numbers are equal or not.
- E. Specker was one of the first to move from general algorithmic impossibility to specific examples, by showing, in [111], that the maximum of a computable bounded increasing sequence can be non-computable.

Second Reason: Communication Problem Counterexamples were the second reason for mathematicians' negative reaction to constructive mathematics. For example, in traditional calculus, there is a theorem according to which a continuous function $f(x)$ on an interval $[a, b]$ always attains its supremum at some point x. In constructive mathematics, there is a counterexample to this classical theorem: there exists a constructive function $f(x)$ from reals to reals that does not attain its supremum value on a given interval in any constructive point. When presented in this form, it is an interesting negative result about algorithms: that we cannot algorithmically produce a point x_0 at which $f(x_0) = \sup_{x \in [a,b]} f(x)$.

However, most mathematicians understood this result—by literally interpreting the constructivists' existential quantifier—as claiming that no such point x_0 exists at all. Since their intuition of real numbers included non-constructive numbers (e.g., numbers coming from physical measurements), this non-existence could not be explained by just considering mathematically constructible real numbers.

Third Reason: Overemphasis on Negative Results The second reason is closely related to the third reason—originally, constructive mathematicians placed too much emphasis on counterexamples and negative results (showing that there is no universal algorithm for solving different general problems), while under-emphasizing the more useful part of constructive mathematics: providing positive algorithmic results.

If a general algorithm is impossible, then usually it is possible to have algorithms that work under certain conditions, and/or algorithms that solve a slightly weaker problem. For example, in the above problem, it is possible, for any given accuracy ε, to algorithmically produce a point x_0 for which $f(x_0) \geq \sup f(x) - \varepsilon$. From the viewpoint of practically solving optimization problems, this is quite enough.

Fourth Reason: Original Papers Are Difficult to Read The fourth reason was that the original papers were very difficult to read. Constructive mathematics tries to describe algorithms, algorithms that deal with higher-order objects—like $f(x)$ takes an algorithm as an input and returns an algorithm as an output. In the early 1950s, before the first programming languages appeared, there was no easy way to describe complex algorithms in a clear understandable way.

Even now, with multiple user-friendly programming languages, it is difficult to describe higher-order algorithms, with functions as inputs and functions as outputs, in an unambiguous and easily readable way. It is difficult to read these algorithms even now—even for computer scientists. Imagine how a mathematician would have felt about such code in the 1950s.

When I started learning constructive mathematics, we did not read Shanin's fundamental papers such as [108], since they were too difficult. Instead, we relied on an instructor's descriptions and later re-wordings.

Fifth Reason: Political There was a special political reason for this negativity. The main ideas of constructive mathematics arose in the late 1940s and early 1950s, when Stalin was still alive. That was a period when he purged the sciences which were considered to be ideologically impure:

- In 1948, genetics was condemned as a capitalist science, with researchers fired, jailed, and shot;
- Then came cybernetics and linguistics.

After these three campaigns, it looked like Stalin decided to go after physicists. A vicious media campaign was launched against "capitalist" relativity theory and quantum physics. Luckily, this campaign stopped—probably because physicists were considered to be useful in designing and improving atomic bombs. A few people who were denounced and arrested—among then, Vladimir Fock, known to physicists for Fock spaces—were soon released. Fock even had—a rarity in those days—all his belongings and manuscripts returned to him intact.

If not physics, then what? Everyone was afraid that their science was to be the next target.

And then, as A. D. Alexandrov described later, a "bomb" exploded on the ideological front: someone in the communist party noticed the philosophical differences between strict constructive mathematics—where only constructive objects exist—and traditional mathematics. He suggested that there be a "philosophical discussion"—similar to the one that preceded the bloody purge in genetics. Disaster was looming. So, A. D. Alexandrov (President of St. Petersburg University) and A. N. Kolmogorov (the most famous Soviet mathematician of that time) came up with a smart plan.

They convinced the party bosses that mathematics is too complex a science to start a discussion (at least a discussion without proper preparation). Instead, they proposed to first write a definitive book on the methodology and ideology of mathematics.

As A. D. Alexandrov explained, they were motivated by the known story about a legendary Molla Nasreddin. In this story, the Shah liked his pet donkey so much that he believed—as many pet owners do—that his pet donkey was more intelligent than most people. So, he asked Molla to teach his donkey. Molla was afraid to disobey the murderous Shah, so he agreed—but with a warning that he needed at least 15 years to do it. When his horrified wife asked how he was planning to do it, he cheerfully replied: "Do not worry. In 15 years, either I will be dead, or the Shah, or the donkey".

Alexandrov and Kolmogorov turned out to be right: while they were working on the book, Stalin died, and the book—a good book actually, re-published by Dover [4]—went out without the need to send anyone to jail.

The ending was happy, but this story left a bad taste in the mouths of many mathematicians. Somewhat understandably, since mathematicians could not do much about the communist dictatorship that nearly killed them, this negative feeling was often directed towards constructive mathematicians who allegedly provoked the government's attack.

Constructive Mathematics in the 1970s: A Boom

When I started going to the seminars, all four reasons were slowly being overcome, and constructive mathematics—and logic—were blossoming.

Matiyasevich's Solution of the 10th Hilbert Problem The big boost came from the 1970 result by Yuri Matiyasevich who solved [92, 93] the 10th of the Hilbert's 23 problems [36], challenges that nineteenth century mathematics presented to the twentieth century. The 10th problem was about finding an algorithm for solving Diophantine equations and systems of equations—i.e., polynomial equations in which all variables are natural numbers. Matiyasevich proved that no such general algorithm is possible.

Interestingly, what may have seemed, at first, like one of the many negative results turned out to be a very positive result. What Matiyasevich actually proved was that every set which can be eventually generated by some algorithm (such as, e.g., the set of all prime numbers or the set of all prime twin pairs for which both n and $n + 2$ are primes) can an be represented as the set of all possible non-negative values of a polynomial of (several) integer-valued variables.

How I Learned the Details of Matiyasevich's Result I myself learned the details of this result—my apologies to Yuri for the coming English-language metaphor—from the "horse's mouth", i.e., from Yuri himself.

Yuri was giving a talk at the general meeting of the St. Petersburg Mathematical Society, and I was late for his talk and missed the first half. I was very upset about this, since I thought I missed a unique opportunity to learn the details. However, my colleague, Evgeny "Zhenya" Dantsin, suggested that I simply approach Yuri after the lecture, that Yuri would be glad to repeat his descriptions to me.

On my own, as a freshman student, I would not have had the chutzpah to approach a famous mathematician with such a request, but after this advice, I did—and Yuri gladly did explain things to me.[9]

Matiyasevich's Result Brought Attention to the Logicians Matiyasevich's result focused everyone's attention in the logic group, in particular, to their results in

---

[9]While I truly appreciate what Yuri did, I want to add that this was an example of the attitude that was prevalent (and actively cultivated) in our department in general, and among logicians in particular: paraphrasing Rudyard Kipling's Mowgli, we all had a strong feeling that we are all "of one blood", that we are all brothers and sisters in mathematics and in science.

constructive mathematics—and the positive character of Matiyasevich's result conveyed that many results of constrictive mathematics have positive algorithmic aspects.

The Attitude of Constructive Mathematics Towards Non-Constructive Mathematics Became More Tolerant   The attitude of constructivists themselves somewhat mellowed. Once in a while, Shanin would repeat—parodying the official line about Marxism—that constructive mathematics is the only scientifically correct approach, but he became much more tolerant of other approaches.

When confronted with the difference between his new views and his more rigid view a few years back, he would always say, half-jokingly, that since all the atoms in the body change every 7 years, he is no longer his former physical self and has therefore the right to change his opinions.

Shanin was the only one to have such serious qualms about non-constructive objects. Everyone else in the group agreed that there is some meaning to non-constructive mathematics—moreover, that there is usually even some constructive meaning in seemingly non-constructive proofs and results, and the challenge is how to extract this meaning.

Constructivism Papers Became More Readable   The readability of papers in constructive mathematics had also greatly improved. A big push for this readability came with a book by Bishop [7], a renowned mathematician who became interested in effective constructions and ended up writing a ground-breaking book on constructive mathematics. Bishop did not use explicit algorithms and did not prove many negative results, his approach was more general, but most of his results could be easily interpreted in Markov-Shanin constructive terms.

Before that, there was a feeling that to learn constructive mathematics, one has to grind his/her teeth and go though barely comprehensible formulas. It turns out that there is a road to constructive mathematics—a road that a working mathematician can rather easily follow.

Logicians tried their best to make their papers clear and understandable. Each paper accepted for publication for *Zapiski* was assigned to another author for what we called "eating each other": thorough checking of every single formula and every single phrase. After that, Yuri Matiyasevich and Anatol Slissenko, fearless and tireless editors, would go over every word on their own, making many suggestions (and, to our embarrassment sometimes, corrections) along the way.

I remember how Anatol half-jokingly suggested that we erase his pencil marks before coming the next time, so that he would be able to make a different suggestion this time. This was somewhat painful but proudly painful: we all felt like Lev Tolstoy who re-wrote his *War and Peace*, I think, six times. The resulting text was not exactly of Tolstoy caliber, but still clearly improved.

Constructive Mathematics in the 1970s: Main Challenges

The Main Idea of Constructive Mathematics: A Reminder   What were the challenges that motivated our research? To understand these challenges, let us recall the general idea of constructive mathematics:

- We start with a general class of problems,

- We try to analyze whether a general algorithm is possible for solving all the problems from this class.

Challenges naturally emerged from all the aspects of this idea: objects, analysis, and algorithms.

First Challenge: The Need to Extend Constructive Mathematics to More Complex Mathematical Objects The first class of challenges came from the fact that most traditional results of constructive mathematics dealt with reasonably simple mathematical *objects*, such as numbers and functions. In modern mathematics and its applications, much more complex objects are used. We need to extend constructive mathematics to these more general objects.

Second Challenge: To Be Useful for Data Processing, Algorithms Must Be Able to Handle Possibly Non-Constructive Data Traditional constructive mathematics dealt only with computable objects—e.g., only with computable real numbers, computable functions, etc. In practice, we need to process data coming from measurements, and, according to modern physics, the corresponding data are not necessarily computable: e.g., the results of quantum measurements are inherently random.

We therefore need to extend the algorithms of constructive mathematics to algorithms for handling these not-necessarily-computable objects.

Third Challenge: The Need for General Ways of Analyzing Problems The *analysis* of a problem in constructive mathematics was too ad hoc. Crudely speaking, every new result was, in effect, worthy of a Master's thesis or a PhD dissertation.

If we wanted constructive methods to be widely used, we could not afford a situation in which so much effort is needed to analyze the constructiveness of a situation, we needed to develop general results which would make such an analysis easier.

Fourth Challenge: When An Algorithm Is Possible, Is It Feasible? On the algorithm stage, if an *algorithm* has been produced, how efficient is it? An algorithm whose running time exceeds the lifetime of the Universe is clearly not very feasible. If this algorithm is not feasible, is a feasible algorithm possible?

If it is not feasible on existing computers, can computers using some novel physical phenomena make these problems feasibly solvable? And if the problem is not feasibly solvable in general, when is it feasibly solvable?

Fifth Challenge: What If No General Algorithm Is Possible? On the other hand, if a general algorithm for solving all the instances of the original problem is proven to be not possible, then the natural questions are:

- How can we relax the problem to make it possible?
- Is it possible to find a reasonable subclass of problems for which the solution is algorithmically possible?
- Is it possible to relax the requirements of the problem and have an algorithm for solving a weaker problem?
- Can computers using some novel physical phenomena make these problems algorithmically solvable?

These were the challenges that we worked on. Let us now briefly enumerate the results of this work.

First Challenge: Dealing with More Complex Objects—Which Objects Do We Need?

We Need to Look Into Possible Application Areas Algorithms are most useful for applications. Thus, to understand which objects we should concentrate on, we need to look at possible applications: which mathematical objects are needed to describe the physical world?

Newton's Mechanics Let us start with traditional Newtonian physics (for details of the corresponding physics descriptions, see, e.g., [21]). In Newtonian physics:

- We have a 3-D Euclidean space $\mathbb{R}^3$ and a 1-D time $\mathbb{R}$.
- The world consists of particles.
- The state of the world at any moment of time t can be described by listing the spatial locations $x_i(t)$ of all these particles $i = 1, 2, \ldots$
- Newton's equations—a system of ordinary differential equations—describe how the coordinate $x_i(t)$ of each particle i changes with time.

This model perfectly describes, e.g., celestial mechanics.

This is a description which is well covered by traditional constructive mathematics.

Newton's Mechanics: Need for Approximate Descriptions and the Resulting Mathematical Objects *Theoretically*, Newton's equations are all we need to describe Newton's physical world. However, from the *practical* viewpoint, the corresponding number of particles is too large—e.g., we have 10^{23} atoms in each macro-volume. Even modern computers, no matter how fast they are, cannot handle that many computations. So, we need to simplify the above description.

First, to describe the dynamics of a single particle i, we cannot realistically use the positions of all the other particles to predict how the location $x_i(t)$ changes. Instead, we must use a simpler description that would capture the effect of all these particles. This description is known as a *field*. For example, the gravity field describes the joint effect of all the attracting particles—without us having to specify which part of the attractive force comes from which particle.

Second, since we have too many particles, we cannot describe the state of all of them, we can only describe their averages—e.g., the density of a body at a given location instead of the exact location of each particle.

Finally, since our description is inevitably approximate, we often cannot describe the exact dynamics, we can only make approximate predictions. In precise terms, instead of the exact value, we take into account that many different values are possible, and we can predict the probabilities of different values.

From this viewpoint:

- We need *functions* to describe densities and fields;
- We need *probability distributions* to describe uncertainty—probability distributions on numbers and on functions.

Resulting Challenge for Constructive Mathematics We need to describe functions, and we need to describe probability distributions.

Functions can be naturally describe in constructive mathematics, but probability distributions are not so easy to describe—even probability distributions corresponding to a single random variable. This difficulty is related to the fact that in constructive mathematics, every function is continuous—informally, if we have an algorithm that is applicable to all computable real numbers, then the resulting function can be proven to be continuous.[10]

This continuity creates a challenge when we try to describe probability distributions in constructive terms. For example, a natural way to describe a probability distribution is by describing its cumulative distribution function $F(x) = \text{Prob}(X \leq x)$. This function is continuous for, e.g., a normal distribution, but it is clearly discontinuous for a random variable X which takes the value 0 with probability 1. For this random variable, using the probability density function (pdf) will not help, since the corresponding pdf is not defined when $x = 0$.

The situation is even more complex for random *functions*—i.e., probability measures on the class of functions.

Relativity Theory Modern physics made the description of the physical world even more complex.

This complexity started with General Relativity, in which the space-time is a general *manifold*.

Already Markov showed how to describe manifolds in constructive terms, but manifolds with singularities are a challenge.

Quantum Physics Quantum physics leads to yet another class of objects. Specifically, in quantum mechanics, to describe a single particle, instead of a single 3-D vector x, we need a *wave function*, i.e., a complex-valued function $\psi(x)$ which assigns to each possible location x an "amplitude". We can then estimate the probability density of a particle at location x as $|\psi(x)|^2$. To handle quantum mechanics, we therefore need to extend the traditional constructive theory from real-valued to *complex-valued* functions.

The situation becomes even more complex in quantum field theory, where instead of a function $f(x)$, we need a *functional*, i.e., a mapping $\psi(f)$ which assigns a complex value to each function f. To describe the dynamics of such states, we need *operators* which map functions into functions, etc.

In relativistic gravity, the state of the world is a manifold M with functions defined on this manifold. So, in quantum gravity, we need a wave function $\psi(M)$ which assigns a value to each such manifold M.

Non-Separable Spaces: An Additional Problem Some of these constructions lead to *non-separable* spaces, i.e., spaces which do not have everywhere dense countable subsets. This a big problem for constructive mathematics, since usually, in constructive spaces, each object is approximated by objects represented by a finite number of symbols. There are countably many such objects, as a result of which all usual constructive spaces are separable.

---

[10]This result makes physical sense: in real life, if we process real values which are obtained with a higher and higher degree of accuracy by performing more and more accurate measurements, then we should be able to return the result at some point, before we know the detailed value of the inputs x—which is exactly what continuity is about.

Summarizing: we need to describe:

- Probability distributions,
- Manifolds with singularities,
- Functions of complex variables,
- Objects of higher order (functionals, operators, etc.), especially objects that form non-separable spaces.

Collaboration with Other Disciplines Was Encouraged

Collaboration Is Needed Complex objects come from disciplines such as physics. Thus, to generate an adequate constructive version of the corresponding notions, it is important to collaborate with researchers from other disciplines.

Such a collaboration, and, more generally, interest in other disciplines was welcomed and encouraged.

Students Were Encouraged to Take Classes Outside Their Discipline Once we started working on our Master's theses, there was no formal requirement to take any classes outside the discipline (this is an arrangement very typical for Master's programs in the academic world). However, Shanin always emphasized that while there was no *requirement* to take classes outside math, a student will be considered a true gentleman or a true lady if he or she takes a year-long class or two semester-long classes elsewhere. (I myself took General Relativity.)

Seminars Enhanced Collaboration At Sergey Maslov's seminar on systems, we would hear talks by linguists, historians, geoscientists, even writers and poets. We all loved it.

Conferences Provided Another Opportunity For example, at a school on computational complexity at a ski resort in Tsahkazdor, Armenia, during non-logical talks, many participants would quietly leave to enjoy the great skiing weather, while we—logicians from St. Petersburg—would stay and enjoy the good "intellectual weather".

Sergey Maslov often valued these non-logical talks even more than the more technical ones. In Tsahkadzor, he described his opinion with a rhyme: "Ia priehal v Tsakhadzor rasshiriat' svoy krugozor" ("I came to Tsahkadzor to broaden my horizon").

First Challenge: Dealing with Complex Objects in Constructive Mathematics—Main Results

As a result of collaboration with researchers from other disciplines, constructive mathematicians from St. Petersburg came up with constructive representations of the corresponding complex objects. Let us list the corresponding representations one by one.

Probability Distributions For random variables, a constructive description of *probability distributions* was proposed by Kossovsky [53, 54].

For random processes, the corresponding description was given in [56, 57], of the example of historically the first Wiener measure—a probability measure that describes Brownian motion.

Manifolds and, More Generally, Metric and Pseudo-Metric Spaces For *manifolds*, an important result was obtained by Zhenya Dantsin: he proved the constructive version of *Sard's Lemma*, according to which the critical values of a smooth function f from one manifold to another has Lebesgue measure 0.

Some results about constructive non-smooth metric and pseudo-metric spaces—presented at the seminar but not published at that time—later appeared in [18, 61, 70, 72, 74].

In particular, for our results about space-time models (later published in [72]) Dima Grigoriev and I received first prize at the department's best student paper competition.

An interesting aspect of studying general metric spaces is estimating their size. A natural way to estimate the size of a metric space S is to use the characteristic like ε-*entropy*, which is defined as the smallest number of points such that every point from S is ε-close to one of the these points. This characteristic takes only integer values and thus it is a discontinuous (hence, not computable) function of ε. A constructive way to describe ε-entropy and other similar characteristics is given in [59] (see also [63]).

Functions of Complex Variables Several problems related to functions of *complex variables* were handled in Bishop's book [7].

Significant further progress was made by Vladimir Orevkov; see, e.g., [100].

Objects of Higher Type A general constructive description of *objects of higher type*—functional, operators, etc.—was proposed by Victor Chernov in [14] (see also [15]).[11]

Non-Separable Spaces A constructive approach to *non-separable spaces* was developed, with Victor Chernov's guidance, by our French research visitor Maurice Margenstern [85], based on the example of the space of almost periodic functions.[12]

General Set-Theoretic Objects An even more general scheme—including constructive versions of all objects of the *set-theoretic* hierarchy—was described in an unpublished paper by Michael Gelfond and Vladimir Lifschitz. Their constructive version of set theory was based on the standard ZF.[13]

---

[11] It is worth mentioning that the resulting approach turned out be similar to the approach proposed in a somewhat different context by Yuri Ershov (see, e.g., [20]).

[12] Almost periodic functions were invented by Harald Bohr, a mathematician brother of the Nobelist physicist Niels Bohr.

[13] Shortly after that, another version of constructive set theory—this time based on type theory—was proposed by Per Martin-Löf [91] (see also [29, 99, 112]). Since Martin-Löf did not need to deal with the more complex axioms of ZF, his theory is much clearer and simpler than the Gelfond and Lifschitz's version—which is probably one of the reasons the reason why they never published their version.

Second Challenge: Algorithms Dealing with Not-Necessarily-Computable Objects

As we have mentioned, to process real-life data, we need algorithms which can process non-constructive objects as well.

Random Sequences For example, according to quantum physics, sequences of observations are not computable, they are *random* (with respect to some computable probability measure).[14]

If we simply allow random sequences (in the formal sense proposed by Kolmogorov and Martin-Löf; see, e.g., [81]), then we get a theory which is very similar to standard constructive mathematics; this was proven by Levin [80].[15]

Need to Go Beyond Random Sequences A restriction to random sequences makes sense if we believe quantum physics to be the ultimate theory of the universe. But since most physicists think that any theory may be later modified, a better idea may be *not* to impose such theory-specific restriction on possible inputs, and consider all possible real numbers as inputs.

General Inputs For an algorithm to be able to handle general inputs, a computable function $f(x)$ should use only approximate values of x, but *not*—as traditional constructive mathematics—the code of the algorithm which computes consecutive approximations to x.

- Some such "approximation-only" algorithms were presented in Bishop's book [7].
- A general description of such algorithms for objects of arbitrary type is given in above-cited Chernov's papers [14, 15].
- Vladimit Lifschitz, in [82], provided a formalism in which such generic number can be described in constructive terms—as "fillings".

Later, this field of research crystallized as *computable analysis*; see, e.g., [103, 118].

Third Challenge: Need for General Ways for Analyzing Problems, Towards General Constructivity Proofs

Another challenge was to find general proofs of constructivity—which would replace previous time-consuming case-by-case proofs.

Almost Negative Statements This activity started with statements that *do not contain "or" or existential quantifiers*—statements which should, intuitively, be equally valid in

---

[14]In [47], it is shown that such non-algorithmic sequences are intuitively justified. Without them, discrete transition processes (e.g., radioactive decay) would potentially lead to devices checking whether a given Turing machine halts or not.

[15]It is worth mentioning that when he presented this work in St. Petersburg, he drew a target on his flyer—expecting that in this center of constructive mathematics, he would be attacked for suggesting that non-constructive sequences are possible.

the traditional and in constructive mathematics. However, the actual proof turned out not to be easy; this was done by Gelfond [26–28].

This class includes integral equalities and inequalities, inequalities and equalities involving max and min, and many other useful mathematical statements.

Statements Containing Strict Inequalities It turned out (see, e.g., [64]) that this class can be easily extended to statements which contain existence.

Terms T describing such statements can be obtained from variables (ranging over a given interval $[0, 1]$) and variable functions by using:

- Addition, subtraction, multiplication, max, min,
- Substitution of a computable constant instead of a variable,
- An operation $f(x_1, \ldots, x_n) \to \min_t f(t, x_2, \ldots, x_n)$,
- An operation $f(x_1, \ldots, x_n) \to \max_t f(t, x_2, \ldots, x_n)$,
- An operation $f(x_1, \ldots, x_n) \to a(f(x_1, \ldots, x_n))$ for a computable function $a(t)$ satisfying the Lipschitz condition,
- An integration operation $f(x_1, \ldots, x_n) \to \int_0^{x_1} f(t, x_2, \ldots, x_n) \, dt$.

Conditions are obtained from inequalities of the type $T > 0$ by using $\vee$, &, and quantifiers over real numbers.

It turns out that if such a condition is classically true, then it is true for some rational values of the variables and piecewise-linear functions with rational coefficients—and is, thus, constructively true.

Uniqueness Implies Computability A more non-trivial class of classical statements which are automatically constructively true are statements about the existence of roots. In general, the fact that a computable function can be proven to have a root does not make this root algorithmically computable, but if this root is *unique*, then it is computable.

This result was first proven by Lacombe [79] for functions of one or several real variables defined on a bounded set. It was extended to general constructive compact spaces by Vladimir Lifschitz in [82]. Variations and applications of this result can be found in [60–62, 66, 75].

This approach was later developed by Ulrich Kohlenbach (see, e.g., [44–46]).

Fourth Challenge: When an Algorithm Is Possible, Is It Feasible? From Constructive Mathematics to Feasible (Polynomial-Time) Mathematics

Some Algorithms of Constructive Mathematics Are Not Feasible An exhaustive-search algorithm that we outlined in the previous section is a typical example of algorithms generated by constructive mathematics.

Most of these algorithms take time which is exponential in terms of the input size (or even longer). Already for $n \approx 300$, the corresponding 2^n time becomes longer than the lifetime of the Universe—so these algorithms are not feasible even for reasonable-size inputs; see, e.g., [25, 65, 75, 101].

What Is Feasible? What happens if we only allow feasible algorithms? To answer this question, we need to have a formal definition of feasibility.

The current definition identifies feasible algorithms with algorithms that execute in polynomial time. It is well known that this definition is not perfect (but since no better one is known, researchers use it):

- For example, an algorithm that takes computation time $t(n) = 10^{300} \cdot n$ on inputs of size n is clearly not feasible, but it is a polynomial-time (even linear-time) algorithm.
- On the other hand, an algorithm which requires time $t(n) = \exp(10^{-9} \cdot n)$ is clearly feasible—at least for all inputs up to a dozen Gigabytes—but is not a polynomial-time algorithm.

Another problem with this definition is that the division into polynomial time and non-polynomial time is somewhat heuristic, motivated more by examples of feasible and non-feasible algorithms than by a deep theoretical analysis. This problem was somewhat eliminated by Vladimir Sazonov who showed that this division can be reformulated in less heuristic logical terms [106].

What If We Only Allow Feasible Algorithms? So what happens if we only allow feasible algorithms—i.e., using the modern formalization of feasibility, algorithms that require polynomial time? Several results along these lines have been developed in [55]; see also [63] and later comments by Gurevich [33].

It turns out that this feasible analysis is even more negative that of the usual constructive mathematics: while addition and multiplication of computable numbers are still feasible, almost everything else is NP-hard:

- Integration,
- Computing the maximum of a computable function,
- Even computing $\sin(x)$ or $\exp(x)$ of a value in a floating point format.[16]

Most of these results were later covered by a thorough analysis presented in a monograph by Ko [43]. In addition to negative results, this book contains many interesting efficient algorithms; for example, algorithms for analytical functions, integration (and many other operations) are feasible.

However, in the 1970s, feasible analysis was not welcomed too much. The seminar's opinion was that if the goal was to make constructive mathematics closer to computational practice, this goal failed.

Interval Computations As Applied Constructive Mathematics Much more successful was another approach to make constructive mathematics more realistic. Namely, Yuri Matiyasevich observed that while algorithms of constructive mathematics assume that we have inputs known with increasing accuracy, in practice, the accuracy is fixed. At any given moment of time, we only have a single measurement result $\tilde{x}$, corresponding to the currently available accuracy Δ; see, e.g., [104]. As a result, the only information that we have about the (unknown) actual value x of the measured quantity is that it belongs to the interval $\mathbf{x} = [\tilde{x} - \Delta, \tilde{x} + \Delta]$. Given a data processing algorithm $y = f(x_1, \ldots, x_n)$

---

[16]Fixed point and floating point formats have to be treated separately, since the transition from floating point to fixed point requires, in general, exponential time.

and intervals $\mathbf{x}_1, \ldots, \mathbf{x}_n$ corresponding to the inputs, we must therefore describe the corresponding range of possible values of y. This problem is called the problem of *interval computations*, or *interval analysis*.

In this problem, techniques borrowed from constructive mathematics work so well that many researchers—including Yu. V. Matiyasevich himself—consider interval analysis Applied Constructive Mathematics. Interval analysis has numerous practical applications ranging from robotics to planning spaceship trajectories to chemical engineering; see, e.g., [19, 37, 39–41, 75, 95].

The main idea of interval computations can be traced to Wiener [119, 120]. Its algorithms were developed by Ramon Moore in the late 1950s and early 1960s. Yuri Matiyasevich boosted this area by organizing conferences and by helping to launch a journal—then called *Interval Computations*—which remains, under the new, somewhat more general title *Reliable Computing*, the main journal of the interval computations community.

Can Other Physical Ideas Make Computations Feasible? If computations are not feasible on existing computers, maybe computers using some novel physical phenomena can make these problems feasibly solvable?

This indeed turned out to be true.

- For example, if causality-violating processes ("time machines") are possible, then we can solve many NP-hard problems in polynomial time [50].
- We can achieve a similar speed-up if in our space-time, the volume of a sphere grows exponentially with the radius—as it does, e.g., in Lobachevsky space—see [76, 86, 96].
- Other schemes of this type are described in [1, 51, 75].

Fifth Challenge: What If No General Algorithm Is Possible?

If a General Problem Is Not Computable, Can We Relax It to Make It Computable? For example, if—as in Specker's sequence—the limit is not computable in the usual sense, in what sense is it computable?

This ideas was pioneered already by N. A. Shanin, who developed several notions of constructive pseudo-numbers; the whole hierarchy of such notions was developed and analyzed by Kushner [78]—and we have already mentioned even more general Lifschitz's "fillings" [82].

With respect to this question, it is important to distinguish between:

- Problems which are "almost" computable and
- Problems which are strongly non-computable.

It turns out that in many cases, we can abstract from the specifics of a problem and describe this difference on the level of logic, by introducing an additional operation of *strong negation*. This idea was pioneered by Vorobiev in [115] and later developed by Bishop [7] and by Gurevich [32].

For example, while a negation to the statement $x > 0$ is the statement $x \leq 0$, a strong negation would mean $x < 0$. In this case, there is no algorithmic way to distinguish between $x \leq 0$ or $x > 0$, but we can easily distinguish between $x > 0$ and $x < 0$: it is sufficient to compute x with sufficient accuracy.

The idea of strong negation—in which, instead of a *single* property, we consider a *pair* of properties which are strong negations to each other—enables us to re-introduce the duality between "and" and "or" [7, 122], duality that is present in classical logic but which is missing in the traditional constructive logic.

If a General Algorithm Is Not Possible, Can We Find a Reasonable Subclass of Problems for Which the Solution Is Algorithmically Possible? Such classes are known. For example, for a general computable function that takes values of different signs at different sides of the interval, it is not possible to algorithmically find a root. However, if we restrict ourselves to computable analytical functions, the root can always be computed.

Interestingly. a restriction to functions described by analytical *expressions* does not help: most algorithmically unsolvable problems remain algorithmically unsolvable; see, e.g., [49].

Another idea is, instead of all *mathematically* possible inputs, to only allow inputs which are *physically* possible. As a reasonable formalization of physical possibility, we can take, e.g. physicists's belief that events with very small probabilities cannot occur. This may sound strange, but this is exactly the belief behind a much more intuitive conclusion that a cold kettle, when placed on a cold stove, will never start boiling by itself—in spite of all the molecular motion which can theoretically lead to such phenomena.

This idea was first analyzed in [22, 48] by appropriately modifying the Kolmogorov/ Martin-Löf's algorithmic definition of randomness [81]. This idea was further developed in [38, 67–69, 71, 73]. If turns out that under such a physics-motivated limitation, most negative results of constructive analysis disappear—and the corresponding problems become algorithmically solvable [68, 73].

What If We Use Novel Physical Phenomena? Maybe, if a problem is not computable, the use of some novel physical phenomenon can make this problems algorithmically solvable.

This is indeed possible. For example, as shown in [50], if we have access to flawless time machine, and either time or space are potentially infinite, then we can compute problems from the class Δ_1^1—way beyond the usual computability.

Another—probably more realistic—idea is to take into account that, according to physicists, no physical theory is perfect, every theory will eventually encounter situations when this theory will need to be modified; see, e.g., [21]. Interestingly, a natural formalization of this idea leads to the possibility of computing functions which are usually considered not to be computable [52, 73, 121]. In other words, using observations of the physical world (looking at the tea leaves?) can enhance our computational abilities.

An interesting aspect of this problem again goes back to logic:

- If—by virtue of some physical phenomena—we are able to algorithmically solve some class of problems,
- What other classes of problems will we then be able to solve?

In [58], we describe which classes of problems imply the ability to algorithmically solve all the problems from analysis.

Questions of this type were later described, in a very general way, by Harvey Friedman who pioneered the whole area of *reverse mathematics*; see, e.g., [24, 110].

This Was Really a Boom

This Was a Boom We had many interesting results, we had many great ideas. Gena Davydov once compared this period with Boldino Autumn, a most productive period in the life of the famous Russian poet Alexander Pushkin.[17]

We Were Optimistic Vladimir Lifchitz was very optimistic that in a few years, to most mathematicians,

- A natural question after proving an existence theorem would be—can we effectively produce the resulting object?
- A natural way to answer this question would be to use tools from constructive mathematics.

I am euphoric, Vladimir liked to say, and I am not afraid to use this word—and this is how most of us felt.

We Were Recognized Other departments felt that logic and constructive mathematics were booming. In addition to Matiyasevich's world-wide recognition, there were many other recognitions on a smaller scale:

- Vladimit Lifschitz got several research prizes,
- Dima Grigoriev and I shared a prize for the best student research paper, etc.

I remember how at a game theory seminar (which I, by the way, continued to attend), Nikolay Vorobiev encouraged the attendees to submit their paper for the best paper competition: OK, so Vladimir Lifschitz would get the first prize, but we could still aim for the second and third place prizes.

Researchers Approached Us Suggesting Collaboration Motivated by our successes, more and more researchers from other disciplines started discussing topics of possible collaboration with us, especially physicists. Let me give two examples.

First Example: Use of Global and Local Properties of Analytical Functions in Physics Leonid Khalfin, a physicist from St. Petersburg, had an interesting idea related to the use of complex numbers in quantum physics.

- Physicists gladly use the "global" effects of analyticity, such as the possibility to estimate complex integrals by using only the function's behavior over singularities.
- However, physicists rarely use the "local" properties of analyticity, for which there is often no physical meaning.

---

[17]Luckily, our reasons for boom were different from Pushkin's: he got stuck in the village of Boldino due to the quarantine caused by the deadly cholera epidemic.

In classical mathematics, global and local properties are provably equivalent. However, Khalfin conjectured that, since local properties do not seem to correspond to any meaningful (observable) properties, maybe a proper constructive version of the theory—which explicitly limits us to potentially observable quantities—will enable us to separate the global and local properties, and to enjoy the useful effects of global properties without having to assume the local ones.

The usual constructive mathematics does not help here, since in it, global and local properties are still equivalent. However, I still believe that if we limit ourselves to only feasible algorithms, maybe such a goal can be achieved.

Second Example: Attempts to Use Quantum Effects to Speed Up Computations Andrei A. Grib, another physicist from St. Petersburg, helped us explore the possible use of quantum effects in computations. In this research, we were inspired by a question formulated by George Kreisel: if we use quantum effects,

- Can we compute something that we could not compute before?
- Can we compute some things faster than what we could compute before?

Our analysis only lead to preliminary results, but we were proud that we were part of the general intellectual atmosphere that had led to the current boom in quantum computing algorithms (see, e.g., [98]), which has already generated famous results:

- Grover's algorithm that searches in an array of size n in time $O(\sqrt{n})$ [30, 31] and
- Shor's algorithm [109] that factors large integers in time polynomial in this integer's bit length and can, thus, potentially break most current codes—specially the RSA code underlying security on the web, the code which is based on the inherent difficulty of such factoring.

Rebels in Science, Rebels in Life: Not Everything Was Perfect

We Were Rebels Being in constructive mathematics in the community of mathematicians means going against the grain. Not surprisingly, folks who are rebels in their professional life were rebels in their politics as well.[18] Let me give a few examples.

Shanin Resigned from the University As a Protest When the University, in violation of all its rules, rejected Zhenya Danstin's candidacy for the PhD program—and it was very clear to everyone that his Jewish origin was the only reason—Shanin officially resigned from the university.

This was a usual tactic under the tsars, when one could gain private employment, but Shanin is the only professor I know who resigned from the Soviet University as a protest.

Contacts with "Enemies of the People" Were Encouraged In 1970, Revolt Pimenov, leader of the space-time seminar, was arrested for reading and distributing "illegal" books (Orwell, Solzhenitsyn, etc.), and for these "anti-Soviet activities", he was sentenced to

---

[18]It is not that everyone else willingly supported the Communist regime: when the first reasonably free elections where held in St. Petersburg in 1989, most communist candidates convincingly lost. However, many logicians went further than many others in their resistance.

exile to the Far North Republic of Komi. I—and many others—kept in touch with him. When the time came for my University-required practicum, I expressed my desired to work with Pimenov in Komi Republic.

Shanin, who was required to approve (or not) our practicum plans, asked only one question: "What will you practice there? Science or anti-Soviet activities?". He was happy with my honest answer "both", and to the Komi Republic I went—to the shock of local folks who were surprised to see a student of the prestigious St. Petersburg University officially sent to work with an exiled "enemy of the people".[19]

Comment It is not that everyone was against socialism as an idea—socialist Sweden was, to many of us then, a good example of how social equality can be established without shooting and jailing political opponents. Some logicians even kept a rosy image of Lenin as a true defender of the people. However, everyone was openly and fearlessly appalled by the violations of human rights that were ubiquitous during the communist dictatorship years.

Shanin Expressed His Disapproval of the Authorities When Solzhenitsyn was exiled, in violation of many international treaties signed by the Soviet Union, Shanin made a loud protest statement at the beginning of the seminar.

At that time, I considered such behavior normal, but later, when I moved to Novosibirsk (where such behavior was unheard of), and when I learned of cases when people were fired and jailed for such public protests, I realized how unusually brave St. Petersburg logicians were.

Maslov Fired, Probably Killed The endings were not always good. When in 1978, the communists staged a political "process" against the physicist and human right defender Yuri Orlov, Sergei Maslov wrote a letter to Brezhnev condemning the unfair closed trial as a violation of Soviet laws and many treaties signed by the Soviet Union, he was promptly fired from his teaching job.

Since he continued his political activities in spite of the continuous threats from the communists, it is quite possible that the KGB helped organize a suspicious car accident that killed him in 1982.

Why Were We Not As Successful As We Hoped? Maybe There Is Still Hope

What Went Wrong? We were so optimistic, we were so successful, so what went wrong? Why is constructive mathematics still not exactly mainstream?

Political Reasons Of course, there were reasons beyond our control. We lived under a totalitarian dictatorship. Journals and conference proceedings were all regulated by the state—and just like sausages were often difficult to buy, paper was scarce too. As a result,

---

[19]Pimenov, by the way, taught me to not be afraid of the KGB-installed electronic bugs in our homes: they already know, he said, that we are mostly against them, so they do not gain anything by hearing us say it one more time.

most published papers were short and thus, inevitably, not easy to read—which did not help their understandability.

Travel to conferences abroad was strictly limited—I was never allowed to go to a conference abroad until 1988, when Gorbachev's perestroika was in full swing and I was allowed to attend a conference in Bulgaria—only to be not allowed to go to a conference in (still communist-controlled) Poland.

A special censorship permission was needed to send a paper abroad, even to send a letter on abstract mathematics abroad—and the permission was often denied. Mathematical letters sent to me from colleagues abroad were opened and stamped before they were delivered to me, and I was summoned to the KGB and threatened with jail because I sent a few letters with my own formulas abroad—they showed me Xerox copies of my own letters.[20]

When a Western mathematician visited from abroad, he or she was under constant open surveillance. When Kip Thorne, a famous astrophysicist, visited Moscow and asked me to meet him in front of his hotel (local citizens were not allowed inside hotels for foreigners), a guy in a typical KGB "uniform" (coat, tie, white shirt) followed us wherever we went, his hand over his ear so that we would know that he was listening attentively.

Maybe We Were Too Picky This is all true. But, I think, there were also our own reasons. Yes, publication space was limited, but I think we were too picky in selecting what to publish, trying to be more saintly than the Pope. Too often, after a reasonable paper was presented, its reception was negative.

I remember that, at one of the seminars, when the chair desperately asked the audience for any positive remark or suggestion, someone replied that the author may consider, as a positive suggestion, a suggestion to grow upon oneself.

Many things that we considered to be not worth publishing—at least not worth publishing in detail—later turned out to be useful, and many of us later published some of it—but alas, still only a small portion of it (since everyone prefers to publish their most recent results). A lot of results and details were simply lost.

Maybe It Is Because Our Algorithms Were Not Feasible? Maybe the problem was that the abstract algorithms that we analyzed and developed—inspired by the practical need for such algorithms—turned out be not exactly practical.

But we *were* working to make them more practical, so why did we not succeed?

Maybe We Had Problems Communicating with People From Other Disciplines? Sometimes, especially when we tried to handle algorithms of interest to other disciplines such as physics, we suffered from a lack of understanding—but as Grisha Mints mentioned recently, even when understanding was there, for some mysterious reasons, the results were not as spectacular and as ground-breaking as we hoped.

Constructive Mathematics Is Alive and Well Why we did not succeed is still a mystery.

I still feel that there is a need for constructive mathematics—and there are constructive mathematicians around who are still producing interesting results (see, e.g., [2, 6, 8–10]), publishing books and papers, and organizing conferences.

---

[20]This was even more appalling to me, since Xerox services were highly rationed, I could rarely get a copy of needed papers, but the KGB seemed to have an unlimited ability to copy everything we sent.

Let Us Hope So maybe there will be a second coming of constructivism.

Let us hope, and—more importantly—let us work together to make it happen.

Acknowledgements   This work was supported in part by the National Science Foundation grants HRD-0734825 and HRD-1242122 (Cyber-ShARE Center of Excellence) and DUE-0926721.

The author is greatly thankful to all his colleagues for valuable discussions. My special thanks to Irving Anellis, who tirelessly kept alive interest in foundations and history of logic and foundations of mathematics, especially history of login in Eastern Europe. My sincere thanks to Francine F. Abeles and Mark E. Fuller for their great idea to have a book published in Irving's memory, and for their support, encouragement, and editing help. Thank you all.

References

1. S. Aaronson, NP-complete problems and physical reality. ACM SIGACT News **36**, 30–52 (2005)
2. O. Aberth, *Precise Numerical Analysis Using C++* (Academic, New York, 1998)
3. A.D. Alexandrov, On Lorentz transformations. Uspekhi Math. Nauk **5**(1), 187 (1950) (in Russian)
4. A.D. Aleksandrov, A.N. Kolmogorov, M.A. Lavrentev, *Mathematics: Its Content, Methods and Meaning* (Dover, New York, 1999)
5. A.D. Alexandrov, V.V. Ovchinnikova, *Remarks on the foundations of Special Relativity*. Leningrad Univ. Vestn. **11**, 94–110 (1953) (in Russian)
6. M.J. Beeson, *Foundations of Constructive Mathematics* (Springer, New York, 1985)
7. E. Bishop, *Foundations of Constructive Analysis* (McGraw-Hill, New York, 1967)
8. E. Bishop, D.S. Bridges, *Constructive Analysis* (Springer, New York, 1985)
9. D.S. Bridges, *Constructive Functional Analysis* (Pitman, London, 1979)
10. D.S. Bridges, S.L. Vita, *Techniques of Constructive Analysis* (Springer, New York, 2006)
11. L.E.J. Brouwer, On the Foundations of Mathematics, Ph.D Dissertation (in Dutch), Amsterdam. English translation in: L. E. J. Brouwer: Collected Works, Philosophy and Foundations of Mathematics, vol. 1, ed. by A. Heyting (Elsevier, Amsterdam/New York, 1975), pp. 11–101
12. H. Busemann, *Timelike Spaces* (PWN, Warszawa, 1967)
13. G.S. Ceitin, I.D. Zaslavskii, N.A. Ŝanin, Peculiarities of constructive mathematical analysis, in *Proceedings of the 1966 International Congress of Mathematicians* (Mir Publications, Moscow, 1968), pp. 253–261
14. V.P. Chernov, On constructive operators of finite types. Zap. Nauchnth Semin. LOMI **32**, 140–147 (1972)
15. V.P. Chernov, Point-less and continuous mappoings. Izv. Vuzov, Mat. (8), 51–59 (1985) (in Russian)
16. A. Church, An unsolvable problem of elementary number theory. Am. J. Math. **58**(2), 345–363 (1936); reprinted in [17], pp. 88–107
17. M. Davis (ed.), *The Undecidable: Basic Papers on Undecidable Propositions, Unsolvable Problems and Computable Functions* (Dover, New York, 2004)
18. G.P. Dimuro, A.C.R. Costa, V. Kreinovich, Modelling measurement processes as timed information processes in simplex domains, in *Proceedings of the 10th IMEKO TC7 International Symposium on Advances of Measurement Science*, vol. 1 (Saint-Petersburg, Russia, June 30–July 2, 2004), pp. 71–76
19. V.G. Dmitriev, N.A. Zheludeva, V. Kreinovich, Applications of interval analysis methods to estimate algorithms errors in measuring systems. Meas. Control Autom. **1**(53), 31–40 (1985) (in Russian)
20. Y.V. Ershov, *Theory of Numberings* (Nauka, Moscow, 1977)
21. R. Feynman, R. Leighton, M. Sands, *Feynman Lectures on Physics* (Basic Books, New York, 2005)
22. A.M. Finkelstein, V. Kreinovich, Impossibility of hardly possible events: physical consequences, in *Abstracts of the 8th International Congress on Logic, Methodology, and Philosophy of Science*, vol. 5, chap. 2 (Moscow, 1987), pp. 25–27
23. P.C. Fishburn, *Nonlinear Preference and Utility Theory* (John Hopkins, Baltimore, 1988)

24. H. Friedman, Some systems of second order arithmetic and their use, in *Proceedings of the International Congress of Mathematicians, Vancouver, B.C., 1974*, vol. 1 (Canadian Mathematical Congress, Montreal, 1975), pp. 235–242

25. M.E. Garey, D.S. Johnson, *Computers and Intractability: A Guide to the Theory of NP-Completeness* (Freeman, San Francisco,1979)

26. M.G. Gel'fond, On constructive pseudofunctions. Proc. Leningrad Math. Inst. Acad. Sci. **16**, 20–27 (1969) (in Russian). English translation: Seminars in Mathematics **16**, 7–10 (1971), Published by Consultants Bureau (New York-London)

27. M.G. Gel'fond, Relationship between the classical and constructive developments of mathematical analysis. Proc. Leningrad Math. Inst. Acad. Sci.**32**, 5–11 (1972) (in Russian). English translation: Journal of Soviet Mathematics **6**(4), 347–352 (1976)

28. M.G. Gelfond, Classes of formulas of classical analysis which are consistent with the constructive interpretation. Ph.D Dissertation, Leningrad Mathematical Institute of the Academy of Sciences, 1975 (in Russian)

29. J.G. Granström, *Treatise on Intuitionistic Type Theory* (Springer, New York, 2011)

30. L.K. Grover, A fast quantum mechanical algorithm for database search, in *Proceedings of the 28th ACM Symposium on Theory of Computing* (1996), pp. 212–219

31. L.K. Grover, Quantum mechanics helps in searching for a needle in a haystack. Phys. Rev. Lett. **79**(2), 325–328 (1997)

32. Y. Gurevich, Intuitionistic logic with strong negation. Stud. Logica. **36**, 49–59 (1977)

33. Y. Gurevich, Platonism, constructivism, and computer proofs vs. proofs by hand. Bull. EATCS (Eur. Assoc.Theor. Comput. Sci.) **57**, 145–166 (1995)

34. A. Heyting, Die formalen Regeln der intuitionistischen Logik. *Sitzungsberichte der Preussischen Akademie der Wissenschaften*, Physik-Math. Kl., (1930), pp. 42–71 and 158–169. English translation of Part I in P. Mancosu, *From Brouwer to Hilbert: The Debate on the Foundations of Mathematics in the 1920s* (Oxford University Press, New York/Oxford, 1998), pp. 311–327

35. A. Heyting, *Intuitionism: An Introduction*, 3rd revised edn. (North-Holland, Amsterdam, 1971)

36. D. Hilbert, Mathematical Problems, lecture delivered before the Int'l Congress of Mathematics in Paris in 1900. Translated in Bull. Am. Math. Soc. **8**, 437–479 (1902). Reproduced in: F.E. Browder (ed.), *Mathematical Developments Arising from Hilbert's Problems* (American Mathematical Society, Providence, 1976)

37. Interval Computations website, http://www.cs.utep.edu/interval-comp

38. A. Jalal-Kamali, O. Nebesky, M.H. Durcholz, V. Kreinovich, L. Longpré, Towards a 'generic' notion of genericity: from typical and random to meager, shy, etc. J. Uncertain Syst. **6**(2), 104–113 (2012)

39. L. Jaulin, M. Kieffer, O. Didrit, E. Walter, *Applied Interval Analysis, with Examples in Parameter and State Estimation, Robust Control and Robotics* (Springer, London, 2001)

40. R.B. Kearfott, *Rigorous Global Search: Continuous Problems* (Kluwer, Dordrecht, 1996)

41. R.B. Kearfott, V. Kreinovich (eds.), *Applications of Interval Computations* (Kluwer, Dordrecht, 1996)

42. S.C. Kleene, On the interpretation of intuitionistic number theory. J. Symb. Log. **10**, 109–124 (1945)

43. K. Ko, *Computational Complexity of Real Functions* (Birkhauser, Boston, 1991)

44. U. Kohlenbach. Theorie der majorisierbaren und stetigen Funktionale und ihre Anwendung bei der Extraktion von Schranken aus inkonstruktiven Beweisen: Effektive Eindeutigkeitsmodule bei besten Approximationen aus ineffektiven Eindeutigkeitsbeweisen. Ph.D. Dissertation, Frankfurt am Main, 1990

45. U. Kohlenbach, Effective moduli from ineffective uniqueness proofs. An unwinding of de La Vallée Poussin's proof for Chebycheff approximation. Ann. Pure Appl. Log. **64**(1), 27–94 (1993)

46. U. Kohlenbach, *Applied Proof Theory: Proof Interpretations and their Use in Mathematics* (Springer, Berlin/Heidelberg, 2008)

47. O.M. Kosheleva, V. Kreinovich, Derivation of the probabilistic character of physics from fundamental assumptions. Research Notes in Philosophy of Physics, University of British Columbia, no. 4, 1978, 5 p.

48. O.M. Kosheleva, V. Kreinovich, On the Algorithmic Problems of a Measurement Process. Research Reports in Philosophy of Physics, University of Toronto, Ontario, Canada, Department of Philosophy, no. 5, 1978, 63 p.

49. O.M. Kosheleva, V. Kreinovich, Analytical representation of constructivistic counterexamples. Recursive Functions Theory: Newsletter No. 21 (1979); 9 (publ. No. 230)
50. O.M. Kosheleva, V. Kreinovich, What can physics give to constructive mathematics. *Mathematical Logic and Mathematical Linguistics* (Kalinin, 1981), pp. 117–128 (in Russian)
51. O.M. Kosheleva, V. Kreinovich, Space-time assumptions behind NP-hardness of propositional satisfiability. Math. Struct. Model. **29**, 13–30 (2014)
52. O.M. Kosheleva, S. Soloviev, On the logic of using observable events in decision making, in *Proceedings of the IX National USSR Symposium on Cybernetics* (Moscow, 1981), pp. 49–51
53. N.K. Kossovsky, Some questions in the constructive theory of normed boolean algebras. Trudy Mat. Inst. Steklov **113**, 3–38 (1970). English translation Proc. Steklov Inst. Math. **113**(1), 1–41 (1970)
54. N.K. Kossovsky, Some Problems in the Constructive Probability Theory, in *Logic, Language, and Probability* (Springer, Netherlands, 1973), pp. 83–99
55. V. Kreinovich, Remarks on the margins of Kushner's book: on the constructive mathematical analysis restricted to polynomial time algorithms. Manuscript (Leningrad, 1973) (in Russian)
56. V. Kreinovich, Constructive theory of Wiener measure, in *Proceedings of the 3rd USSR National Conference on Mathematical Logic* (Novosibirsk, USSR, 1974), pp. 116–118 (in Russian)
57. V. Kreinovich, Applying Wiener measure to calculations, in *Proceedings of the 3rd Leningrad Conference of Young Mathematicians* (Leningrad University, 1974), pp. 6–9 (in Russian)
58. V. Kreinovich, What does the law of the excluded middle follow from?. Proc. Leningrad Math. Inst. Acad. Sci. **40**, 37–40 (1974) (in Russian). English translation: J. Sov. Math. **8**(1), 266–271 (1974)
59. V. Kreinovich, Constructivization of the notions of epsilon- entropy and epsilon-capacity. Proc. Leningrad Math. Inst. Acad. Sci. **40**, 38–44 (1974) (in Russian). English translation: J. Sov. Math. **8**(3), 271–276 (1977)
60. V. Kreinovich, Uniqueness implies algorithmic computability, in *Proceedings of the 4th Student Mathematical Conference* (Leningrad University, Leningrad, 1975), pp. 19–21 (in Russian)
61. V. Kreinovich, Categories of Space-Time Models, Ph.D. Dissertation, Soviet Academy of Sciences, Siberian Branch, Institute of Mathematics, Novosibirsk 1979 (in Russian)
62. V. Kreinovich, Review of bridges, constructive functional analysis. Z. Math. **401**, 03027 (1979)
63. V. Kreinovich, Review of 'Constructivization of the notions of $\varepsilon-$entropy and $\varepsilon-$capacity' by V. Kreinovich. Math. Rev. **58**(1), 16 (1979); Review No. 78
64. V. Kreinovich, A reviewer's remark in a review of, 'Some relations between classical and constructive mathematics' by M. Beeson. Math. Rev. **58**(2), 755–756 (1979); Review No. 5094
65. V. Kreinovich, Unsolvability of several algorithmically solvable analytical problems. Abstr. Am. Math. Soc. **1**(1), 174 (1980)
66. V. Kreinovich, Physics-motivated ideas for extracting efficient bounds (and algorithms) from classical proofs: beyond local compactness, beyond uniqueness, in *Abstracts of the Conference on the Methods of Proof Theory in Mathematics* (Max-Planck Institut für Mathematik, Bonn, 2007), p. 8
67. V. Kreinovich, Toward formalizing non-monotonic reasoning in physics: the use of Kolmogorov complexity. Rev. Iberoam. Intel. Artif. **49**, 4–20 (2009)
68. V. Kreinovich, Negative results of computable analysis disappear if we restrict ourselves to random (or, more generally, typical) inputs. Math. Struct. Model. **25**, 100–113 (2012)
69. V. Kreinovich, Towards formalizing non-monotonic reasoning in physics: logical approach based on physical induction and its relation to Kolmogorov complexity, in *Correct Reasoning: Essays on Logic-Based AI in Honor of Vladimir Lifschitz*, ed. by E. Erdem, J. Lee, Y. Lierler, D. Pearce. Lecture Notes in Computer Science vol. 7265 (Springer, Berlin/Heidelberg, 2012), pp. 390–404
70. V. Kreinovich, G.P. Dimuro, A.C. da Rocha Costa, From Intervals to? Towards a General Description of Validated Uncertainty. Technical Report, Catholic University of Pelotas, Brazil, 2004
71. V. Kreinovich, A.M. Finkelstein, Towards applying computational complexity to foundations of physics. Notes Math. Semin. St. Petersburg Dep. Steklov Inst. Math. **316**, 63–110 (2014). Reprinted in J. Math. Sci.**134**, 2358–2382 (2006)
72. V. Kreinovich, O. Kosheleva, Computational complexity of determining which statements about causality hold in different space-time models. Theor. Comput. Sci. **405**(1–2), 50–63 (2008)
73. V. Kreinovich, O. Kosheleva, Logic of scientific discovery: how physical induction affects what is computable, in *Proceedings of the International Interdisciplinary Conference Philosophy,*

Mathematics, Linguistics: Aspects of Interaction 2014 PhML'2014 (St. Petersburg, Russia, 2014), pp. 116–127

74. V. Kreinovich, O. Kosheleva, S.A. Starks, K. Tupelly, G.P. Dimuro, A.C. da Rocha Costa, K. Villaverde, From intervals to domains: towards a general description of validated uncertainty, with potential applications to geospatial and meteorological data. J. Comput. Appl. Math. **199**(2), 411–417 (2007)

75. V. Kreinovich, A. Lakeyev, J. Rohn, P. Kahl, *Computational Complexity and Feasibility of Data Processing and Interval Computations* (Kluwer, Dordrecht, 1998)

76. V. Kreinovich, M. Margenstern, In some curved spaces, one can solve NP-hard problems in polynomial time. Notes Math. Semin. St. Petersburg Dep. Steklov Inst. Math. **358**, 224–250 (2008). Reprinted in J. Math. Sci. **158**(5), 727–740 (2009)

77. E.H. Kronheimer, R. Penrose, On the structure of causal spaces. Proc. Camb. Philos. Soc. **63**(2), 481–501 (1967)

78. B.A. Kushner, *Lectures on Constructive Mathematical Analysis* (American Mathematical Society, Providence 1984)

79. D. Lacombe, Les ensembles récursivement ouvert ou fermés, et leurs applications à l'analyse récurslve. C. R. Acad. Sci. **245**(13), 1040–1043 (1957)

80. L.A. Levin, Randomness Conservation Inequalities. Inf. Control **61**(1), 15–37 (1984)

81. M. Li, P.M.B. Vitányi, *An Introduction to Kolmogorov Complexity* (Springer, Berlin/Heidelberg/New York, 2008)

82. V.A. Lifschitz, Investigation of constructive functions by the method of fillings. J. Sov. Math. **1**, 41–47 (1973)

83. V. Lifschitz, Constructive assertions in an extension of classical mathematics. J. Symb. Log. **47**(2), 359–387 (1982)

84. R.D. Luce, R. Raiffa, *Games and Decisions: Introduction and Critical Survey* (Dover, New York, 1989)

85. M. Margenstern, On constructive functionals on the space of almost periodic functions. Zap. Nauchnykh Semin. LOMI **40**, 45–62 (1974) (in Russian)

86. M. Margenstern, K. Morita, NP problems are tractable in the space of cellular automata in the hyperbolic plane. Theor. Comput. Sci. **259**(1–2), 99–128 (2001)

87. A.A. Markov, On constructive mathematics. Trudy Mat. Inst. Steklov **67**, 8–14 (1962). English translation in Am. Math. Soc. Translat. **98**(2), 1–9 (1971)

88. A.A. Markov, On the logic of constructive mathematics. Vestnik Moskov. Univ., Ser. I Mat. Mekh. **25**, 7–29 (1970) (in Russian)

89. A.A. Markov, An attempt of construction of the logic of constructive mathematics, in *Theory of Algorithms and Mathematical Logic*, ed. by B. Kushner, N. Nagorny (Computing Centre of Russian Academy of Sciences, Moscow, 1976), pp. 3–31 (in Russian)

90. A.A. Markov, N.M. Nagorny, *The Theory of Algorithms* (Reidel Publ., 1988)

91. P. Martin-Löf, *Intuitionistic Type Theory* (Bibliopolis, Naples, 1984)

92. Y.V. Matiyasevich, Diophantine character of recursively enumerable sets. Dokl. Akad. Nauk SSSR **191**(2), 278–282 (1970) (in Rusian). English translation in Sov. Math. Doklady **11**(2), 354–358 (1970)

93. Y.V. Matiyasevich, *Hilbert's Tenth Problem* (MIT, Cambridge, 1993)

94. G.E. Mints, Stepwise semantics of A. A. Markov. A supplement to the Russian translation of J. Barwise (ed.), *Handbook of Mathematical Logic (North Holland, 1982)* (Nauka Publications, Moscow, 1983), pp. 348–357 (in Russian)

95. R.E. Moore, R.B. Kearfott, M.J. Cloud, *Introduction to Interval Analysis* (SIAM, Philadelphia, 2009)

96. D. Morgenstein, V. Kreinovich, Which algorithms are feasible and which are not depends on the geometry of space-time. Geombinatorics **4**(3), 80–97 (1995)

97. H.T. Nguyen, V. Kreinovich, B. Wu, G. Xiang, *Computing Statistics under Interval and Fuzzy Uncertainty* (Springer, Berlin/Heidelberg, 2012)

98. M.A. Nielsen, I.L. Chuang, *Quantum Computation and Quantum Information* (Cambridge University Press, Cambridge, 2000)

99. B. Nordström, K. Petersson, J.M. Smith, *Programming in Martin-Löf's Type Theory* (Oxford University Press, Oxford, 1990)

100. V.P. Orevkov, A new proof of the uniqueness theorem for constructive differentiable functions of a complex variable. Zap. Naucn. Sem. Leningrad. Otdel. Inst. Steklov. **40**, 119–126 (1974) (in Russian)

101. C.H. Papadimitriou, *Computational Complexity* (Addison Wesley, San Diego, 1994)

102. R.I. Pimenov, *Kinematic spaces: Mathematical Theory of Space-Time* (Consultants Bureau, New York, 1970)

103. M.B. Pour-El, J.I. Richards, *Computability in Analysis and Physics* (Springer, Berlin, 1989)

104. S.G. Rabinovich, *Measurement Errors and Uncertainty: Theory and Practice* (Springer, Berlin, 2005)

105. H. Raiffa, *Decision Analysis* (McGraw-Hill, Columbus, 1997)

106. V.Y. Sazonov, A logical approach to the problem P=NP. Springer Lect. Notes Comput. Sci **88**, 562–575 (1980)

107. N.A. Shanin, On constructive understanding of mathematical statements. Proc. Steklov Inst. Math. **52**, 226–231 (1958) (in Russian). English translation in Am. Math. Soc. Translat. **23**(2) (1971), 109–189.

108. N.A. Shanin, *Constructive real numbers and constructive function spaces*, Trudy Mat. Inst. Steklov **67**, 15–294 (1962) (in Russian). Translated into English as a monograph by the Am. Math. Soc. (1968)

109. P. Shor, Algorithms for quantum computation: Discrete logarithms and factoring, in *Proceedings of the 35th Annual IEEE Symposium on Foundations of Computer Science* (1994), pp. 124–134

110. S.G. Simspon (ed.), *Reverse Mathematics 2001* (Association for Symbolic Logic, Wellsley, 2001)

111. E. Specker, Nicht konstruktiv beweisbare Sätze der analysis. J. Symb. Log. **14**, 145–158 (1949)

112. S. Thompson, *Type Theory and Functional Programming* (Addison-Wesley, 1991)

113. A. Troelstra, D. van Dalen, *Constructivism in Mathematics. An Introduction*, vols. I, II (North-Holland, Amsterdam, 1988)

114. A.M. Turing, On computable numbers, with an application to the Entscheidungsproblem. Proc. Lond. Math. Soc. Ser. 2 **42**, 230–265 (1936/1937). Corrections ibid **43**, 544–546 (1937); reprinted in [17], pp. 115–153

115. N.N. Vorob'ev, A constructive calculus of statements with strong negation. Problems of the Constructive Direction in Mathematics. Part 3, Trudy Mat. Inst. Steklov **72**, 195–227 (1964) (in Russian)

116. N.N. Vorobiev, *Game Theory* (Springer, New York, 1977)

117. N.N. Vorobiev, *Foundations of Games: Non-Cooperative Games* (Nauka Publications, Moscow, 1984) (in Russian)

118. K. Weihrauch, *Computable Analysis* (Springer, Berlin/Heidelberg, 2000)

119. N. Wiener, A contribution to the theory of relative position. Proc. Camb. Philos. Soc. **17**, 441–449 (1914)

120. N. Wiener, A new theory of measurement: a study in the logic of mathematics. Proc. Lond. Math. Soc. **19**, 181–205 (1921)

121. M. Zakharevich, O. Kosheleva, If many physicists are right and no physical theory is perfect, then the use of physical observations can enhance computations. J. Uncertain Syst. **8**(3), 227–232 (2014)

122. I.D. Zaslavsky, *Symmetric Constructive Logic* (Armenian Academy of Sciences, Erevan, 1978)

123. E.C. Zeeman, Causality implies the Lorentz group. J. Math. Phys. **5**(4), 490–493 (1964)

V. Kreinovich (✉)

Department of Computer Science, University of Texas at El Paso, 500 W. University, El Paso, TX 79968, USA

e-mail: vladik@utep.edu

On Normalizing Disjunctive Intermediate Logics

Jonathan P. Seldin

Abstract In this paper it is shown that every intermediate logic obtained from intuitionistic logic by adding a disjunction can be normalized. However, the normalization procedure is not as complete as that for intuitionistic and minimal logic because some results which usually follow from normalization fail, including the separation property and the subformula property. However, in a few special cases, we can extend the normalization process to obtain new consistency proofs.

Keywords Intermediate logic · Normalization

Mathematics Subject Classification (2000) 03B55 · 03F05

By a *disjunctive intermediate logic* I mean a system of propositional or first order predicate logic obtained by adding to intuitionistic logic an axiom scheme of the form

$$C_1(A_1, A_2, \ldots, A_m) \vee C_2(A_1, A_2, \ldots, A_m) \vee \ldots \vee C_n(A_1, A_2, \ldots, A_m),$$

where $C_i(A_1, A_2, \ldots, A_m)$ is a formula scheme in which some or all of $A_1, A_2, \ldots, A_m$ occur as parameters.

Some examples of disjunctive intermediate logics are given by Umezawa [9] as follows:

(M) $\neg\neg A \vee \neg A$.

(P_n) $(A_1 \supset A_2) \vee (A_1 \supset A_3) \vee \ldots \vee (A_1 \supset A_n) \vee (A_2 \supset A_1) \vee (A_2 \supset A_3) \vee \ldots \vee (A_2 \supset A_n) \vee \ldots \vee (A_n \supset A_1) \vee (A_n \supset A_2) \vee \ldots \vee (A_n \supset A_{n-1})$, where $n \geq 2$ and, for any $(A_i \supset A_j)$, $i \neq j$. A special case is (P_2), when the axiom is $(A_1 \supset A_2) \vee (A_2 \supset A_1)$, and the logic is also known as (LC).

(R_n) $A_1 \vee (A_1 \supset A_2) \vee (A_2 \supset A_3) \vee \ldots \vee (A_{n-1} \supset A_n) \vee \neg A_n$, where $n \geq 2$.

(ME) $(\forall x)\neg\neg A(x) \vee (\exists x)\neg A(x)$.

(MEK°) $\neg\neg(\forall x)A(x) \vee (\exists x)\neg A(x)$.

(DP_2) $(\forall x)(A \supset B(x)) \vee (\exists x)(B(x) \supset A)$.

(FP_2) $(\exists x)(\forall y)(A(x) \supset B(y)) \vee (\exists y)(\forall x)(B(y) \supset A(x))$.

(GP_2) $(\exists y)(\forall x)(A(x) \supset B(y)) \vee (\exists x)(\forall y)(B(y) \supset A(x))$.

(FGP_2) $(\exists x)(\exists u)(\forall y)(\forall v)(A(x, v) \supset B(y, u)) \vee (\exists y)(\exists v)(\forall x)(\forall u)(B(y, u) \supset A(x, v))$.

(ER_n) $(\forall x)A_1(x) \vee (\exists x)(\forall y)(A_1(x) \supset A_2(y)) \vee \ldots \vee (\exists x)(\forall y)(A_{n-1}(x) \supset A_n(y))) \vee (\exists x)\neg A_n(x)$.

© Springer International Publishing Switzerland 2016
F.F. Abeles, M.E. Fuller (eds.), *Modern Logic 1850-1950, East and West*, Studies in Universal Logic, DOI 10.1007/978-3-319-24756-4_12

In addition, classical logic can be formulated in this form as follows:

(K) $\neg A \vee A$.

López-Escobar [7] gives another example:

(LIC_n) $(A_1 \supset A_2) \vee (A_2 \supset A_3) \supset \ldots \supset (A_{n-1} \supset A_n) \vee (A_n \supset A_1)$, where $n \geq 2$.

He also points out that classical logic can be axiomatized by adding to intuitionistic logic the axiom

(C) $(A \supset B) \vee A$.

(López-Escobar is considering only the implication fragments of these logics, but his definitions apply to full predicate logics.)

Note that (LC) is also (LIC_2).

The purpose of this paper is to look at natural deduction versions of these systems and to prove a normalization result for them. The normalization result obtained does not imply all of the results of normalization for classical or intuitionistic logic. In particular, the subformula property fails. Avron [1] has used the method of *hypersequents* to give a formulation that does satisfy the subformula property for one of the logics considered here, the logic (LC).

Since the intuitionistic rule

$$\frac{\bot}{A} \; \bot\text{I}$$

is not used in any essential way in obtaining these results, they also hold if the axioms in question are added to minimal logic instead of intuitionistic logic. If (K) is added to minimal logic, the result is the system which Curry [3] called (in a sequent L-version) LD. If (C) is added instead, the system obtained is the system called LE in [4], which is due to Bernays [2] and first extensively studied by Kripke [6].[1] The systems TD and TE are natural deduction formulations of the systems LD and LE respectively.

I would like to thank Richard Zach and Roger Hindley for their helpful comments and suggestions. I would also like to thank Fairouz Kamareddine for arranging for a grant from the Scottish Informatics and Computer Science Alliance (SICSA) for a Distinguished Visiting Fellowship for me to visit Scotland, which made possible the presentation at St. Andrews University. The preparation for that presentation began the process which led to in section "Some Special Cases" of this paper.

---

[1] See [4, pp. 260, 306]. The system was also studied by Kanger [5] before Kripke's extensive study.

The Natural Deduction Formulation

First, let us recall the system TM of minimal predicate logic.[2] It has no axioms, and its rules are as follows:

$$\wedge I \quad \frac{A_1 \qquad A_2}{A_1 \wedge A_2} \qquad\qquad \wedge E \quad \frac{A_1 \wedge A_2}{A_i}$$

$$\vee I \quad \frac{A_i}{A_1 \vee A_2} \qquad\qquad \vee E \quad \frac{A \vee B \quad \overset{[A]}{C} \quad \overset{[B]}{C}}{C}$$

$$\supset I \quad \frac{\overset{[A]}{B}}{A \supset B} \qquad\qquad \supset E \quad \frac{A \supset B \quad A}{B}$$

$$\forall I \quad \frac{A(c)}{(\forall x)A(x)} \qquad\qquad \forall E \quad \frac{(\forall x)A(x)}{A(t)}$$

$$\exists I \quad \frac{A(t)}{(\exists x)A(x)} \qquad\qquad \exists E \quad \frac{(\exists x)A(x) \quad \overset{[A(c)]}{C}}{C}$$

where in rules $\wedge E$ and $\vee I$, $i = 1$ or $i = 2$; in rules $\forall E$ and $\exists I$, t is any term; and in rules $\forall I$ and $\exists E$, c is an *eigenvariable*, which is a variable which does not occur free in any undischarged assumption (or, in the case of $\exists E$, in C). Also, $\neg A$ is defined to be $A \supset \bot$, so that negation satisfies the derived rules

$$\neg I \quad \frac{\overset{[A]}{\bot}}{\neg A} \qquad\qquad \neg E \quad \frac{\neg A \quad A}{\bot}$$

Recall also that the system TJ of intuitionistic logic is obtained from TM by adding the rule

$$\bot_I \frac{\bot}{A}$$

Now, let us write $C_i(\vec{A})$ for $C_i(A_1, A_2, \ldots, A_m)$. Then the axiom of a disjunctive intermediate logic has the form

$$C_1(\vec{A}) \vee C_2(\vec{A}) \vee \ldots \vee C_n(\vec{A}). \tag{1}$$

---

[2]This name is due to Curry; see [4, p. 280]. Actually, Curry reserved the name TM for propositional logic only, using the name TM* for the predicate logic, but since I am not considering the propositional case separately, I will drop the superscript asterisk.

For the systems he considers, López-Escobar [7] proposes replacing axiom (1) by the rule

$$\frac{[C_1(\vec{A})] \quad [C_2(\vec{A})] \qquad \quad [C_n(\vec{A})]}{\begin{array}{ccccc} C & C & \cdots & C \end{array}}{C} \; \mathrm{Ds}$$

Theorem 1 *Adding axiom (1) to TM (respectively TJ) is equivalent to adding rule* Ds *to TM (respectively TJ).*

Proof Suppose that we have added rule Ds to TM (or, for that matter TJ), and suppose we are given deductions

$$\begin{array}{ccccc} C_1(\vec{A}) & C_2(\vec{A}) & & C_n(\vec{A}) \\ \mathcal{D}_1 & \mathcal{D}_2 & & \mathcal{D}_n \\ C & , & C & , & \ldots, & C \end{array}.$$

Then, writing C_i for $C_i(\vec{A})$, E_2 for $C_2 \vee E_3$, E_3 for $C_3 \vee E_4$, ..., E_{n-1} for $C_{n-1} \vee C_n$, we can proceed as follows:

$$\cfrac{C_1 \vee E_2 \quad \cfrac{\overset{1}{[C_1]}}{\mathcal{D}_1} \quad \overset{2}{[C_2 \vee E_3]} \quad \cfrac{\overset{3}{[C_2]} \; \mathcal{D}_2}{C} \cdots \cfrac{\cfrac{\overset{2n-2}{[C_{n-1} \vee C_n]} \quad \cfrac{\overset{2n-1}{[C_{n-1}]} \; \overset{2n}{[C_n]}}{\mathcal{D}_{n-1} \; \mathcal{D}_n}{C \quad C}}{C} \; \vee \mathrm{E} - (2n-1), 2n}{\vdots}{C}}{C} \; \vee \mathrm{E} - 3, 4}{C}}{C} \; \vee \mathrm{E} - 1, 2$$

Conversely, suppose we have added rule Ds to TM. Then we can deduce axiom (1) as follows:

$$\cfrac{\cfrac{\overset{1}{[C_1]}}{C_1 \vee C_2 \vee \ldots \vee C_n} \; \text{repeated} \vee \mathrm{I} \quad \cdots \quad \cfrac{\overset{n}{[C_n]}}{C_1 \vee C_2 \vee \ldots \vee C_n} \; \text{repeated} \vee \mathrm{I}}{C_1 \vee C_2 \vee \ldots \vee C_n} \; \mathrm{Ds} - 1, 2, \ldots n$$

□

For the rest of the paper, we will assume that our disjunctive intermediate logic is obtained from TJ by adding rule Ds.

Definition 1 The *system* TI will be obtained from TJ by adding the rule Ds.

The Normalization Result

Recall that the deduction reduction steps for TJ (and also TM) are as follows: $\wedge$-reduction steps:

$$
\begin{array}{cc}
\mathcal{D}_1 & \mathcal{D}_2 \\
A_1 & A_2 \\
\hline
A_1 \wedge A_2 & \wedge\text{I} \\
\hline
A_i & \wedge\text{E} \\
\mathcal{D}_3 &
\end{array}
\qquad\qquad \text{reduces to} \qquad
\begin{array}{c}
\mathcal{D}_i \\
A_i \\
\mathcal{D}_3
\end{array}
$$

$\vee$-reduction steps:

$$
\begin{array}{c}
\begin{array}{ccc}
 & 1 & 2 \\
\mathcal{D}_0 & [A_1] & [A_2] \\
A_i & \mathcal{D}_1 & \mathcal{D}_2 \\
\hline
A_1 \vee A_2 \;\; \vee\text{I} & C & C
\end{array} \\
\hline
\qquad\qquad\qquad\qquad\quad \vee\text{E} - 1,2 \\
C \\
\mathcal{D}_3
\end{array}
\qquad \text{reduces to} \qquad
\begin{array}{c}
\mathcal{D}_0 \\
A_i \\
\mathcal{D}_i \\
C \\
\mathcal{D}_3
\end{array}
$$

$\supset$-reduction steps:

$$
\begin{array}{c}
\begin{array}{cc}
1 & \\
[A] & \\
\mathcal{D}_1 & \\
B & \mathcal{D}_2 \\
\hline
A \supset B \;\; \supset\text{I} - 1 & A
\end{array} \\
\hline
\qquad\qquad\qquad \supset\text{E} \\
B \\
\mathcal{D}_3
\end{array}
\qquad \text{reduces to} \qquad
\begin{array}{c}
\mathcal{D}_2 \\
A \\
\mathcal{D}_1 \\
B \\
\mathcal{D}_3
\end{array}
$$

$\forall$-reduction steps:

$$
\begin{array}{c}
\mathcal{D}_1(c) \\
A(c) \\
\hline
(\forall x)A(x) \;\; \forall\text{I} \\
\hline
A(t) \;\; \forall\text{E} \\
\mathcal{D}_3
\end{array}
\qquad\qquad \text{reduces to} \qquad
\begin{array}{c}
\mathcal{D}_1(t) \\
A(t) \\
\mathcal{D}_3
\end{array}
$$

$\exists$-reduction steps:

$$
\begin{array}{c}
\begin{array}{cc}
 & 1 \\
\mathcal{D}_1 & [A(c)] \\
A(t) & \mathcal{D}_2(c) \\
\hline
(\exists x)A(x) \;\; \exists\text{I} & C
\end{array} \\
\hline
\qquad\qquad\qquad \exists\text{E} - 1 \\
C \\
\mathcal{D}_3
\end{array}
\qquad \text{reduces to} \qquad
\begin{array}{c}
\mathcal{D}_1 \\
A(t) \\
\mathcal{D}_2(t) \\
C \\
\mathcal{D}_3
\end{array}
$$

$\vee R$-reduction steps: If R is an E-rule with C as its major (left) premise and $(\mathcal{D}_3)$ as the deduction(s) of its minor premises, if any, then

$$\cfrac{\cfrac{\begin{array}{cc} 1 & 2 \\ [A_1] & [A_2] \end{array}}{\mathcal{D}_0 \quad \begin{array}{cc} \mathcal{D}_1 & \mathcal{D}_2 \\ A_1 \vee A_2 & C & C \end{array}} \vee \mathrm{E}-1,2 }{\cfrac{C \qquad\qquad\qquad (\mathcal{D}_3)}{\begin{array}{c} E \\ \mathcal{D}_4 \end{array}} R}$$

reduces to

$$\cfrac{\mathcal{D}_0 \qquad \cfrac{\cfrac{\begin{array}{c} 1 \\ [A_1] \\ \mathcal{D}_1 \\ C \end{array} (\mathcal{D}_3)}{E} R \qquad \cfrac{\cfrac{\begin{array}{c} 2 \\ [A_2] \\ \mathcal{D}_2 \\ C \end{array} (\mathcal{D}_3)}{E} R}{}}{A_1 \vee A_2 \qquad\qquad E \qquad\qquad\qquad\qquad} \vee \mathrm{E}-1,2}{\begin{array}{c} E \\ \mathcal{D}_4 \end{array}}$$

$\exists R$-reduction steps: If R is an E-rule with C as its major (left) premise and $(\mathcal{D}_3)$ as the deduction(s) of its minor premises, if any, then

$$\cfrac{\cfrac{\begin{array}{c} \quad\quad 1 \\ \quad\quad [A(c)] \\ \mathcal{D}_1 \quad \mathcal{D}_2(c) \\ (\exists x)A(x) \quad C \end{array}}{C} \exists \mathrm{E}-1 \qquad (\mathcal{D}_3)}{\begin{array}{c} E \\ \mathcal{D}_4 \end{array} R}$$

reduces to

$$\cfrac{\mathcal{D}_1 \qquad \cfrac{\cfrac{\begin{array}{c} 1 \\ [A(c)] \\ \mathcal{D}_2(c) \\ C \end{array} (\mathcal{D}_3)}{E} R}{} }{\begin{array}{c} (\exists x)A(x) \qquad\qquad\qquad E \\ \qquad\qquad E \\ \qquad\qquad \mathcal{D}_4 \end{array} \exists \mathrm{E}-1}$$

To these reduction steps, we add one for rule Ds:

DsR-reduction steps: If R is an E-rule with C as its major (left) premise and $(\mathcal{D}_0)$ as the deduction(s) of its minor premises, if any, then

$$
\cfrac{
\cfrac{
\begin{array}{cccc}
\overset{1}{[C_1]} & \overset{2}{[C_2]} & & \overset{n}{[C_n]} \\
\mathcal{D}_1 & \mathcal{D}_2 & & \mathcal{D}_n \\
C & C & \cdots & C
\end{array}
}{C} \text{ Ds} - 1,2,\ldots,n \qquad (\mathcal{D}_0)
}{
\begin{array}{c} E \\ \mathcal{D}_{n+1} \end{array}
} R
$$

reduces to

$$
\cfrac{
\begin{array}{ccccc}
\cfrac{
\begin{array}{c}\overset{1}{[C_1]} \\ \mathcal{D}_1 \\ C \quad (\mathcal{D}_0)\end{array}
}{E} R
&
\cfrac{
\begin{array}{c}\overset{2}{[C_2]} \\ \mathcal{D}_2 \\ C \quad (\mathcal{D}_0)\end{array}
}{E} R
& \cdots &
\cfrac{
\begin{array}{c}\overset{n}{[C_n]} \\ \mathcal{D}_n \\ C \quad (\mathcal{D}_0)\end{array}
}{E} R \text{ Ds} - 1,2,\ldots,n
\end{array}
}{
\begin{array}{c} E \\ \mathcal{D}_{n+1} \end{array}
}
$$

Theorem 2 *Every deduction in this disjunctive intermediate logic can be reduced to a normalized deduction.*

Proof Similar to the proof by Prawitz [8, Chap. IV Theorem 1] of the normalization of minimal and intuitionistic predicate logic. Extend the definition of segment so that if a premise of rule Ds is in a segment, then so is the conclusion. Then Prawitz's procedure for removing maximum segments works for this system.   □

Remark 1 I conjecture that it is possible to prove strong normalization (that every sequence of reduction steps terminates), but I have not tried to find a proof.

A Gentzen L-formulation

Let us recall that the Gentzen formulation LM of minimal logic is defined as follows: the axiom scheme is

Ax $A \vdash A$.

The structural rules are as follows:

$$
*C \quad \frac{\Gamma_1, B, A, \Gamma_2 \vdash C}{\Gamma_1, A, B, \Gamma_2 \vdash C}
$$

$$
*K \quad \frac{\Gamma \vdash C}{\Gamma, A \vdash C}
$$

$$*W \quad \frac{\Gamma, A, A \;\vdash\; C}{\Gamma, A \;\vdash\; C}$$

$$\text{Cut} \quad \frac{\Gamma, A \;\vdash\; C \quad \Gamma \vdash A}{\Gamma \;\vdash\; C}$$

The operational rules are:

$$*\wedge \quad \frac{\Gamma, A_i \;\vdash\; C}{\Gamma, A_1 \wedge A_2 \;\vdash\; C} \qquad\qquad \wedge* \quad \frac{\Gamma \;\vdash\; A_1 \quad \Gamma \;\vdash\; A_2}{\Gamma \;\vdash\; A_1 \wedge A_2}$$

$$*\vee \quad \frac{\Gamma, A_1 \;\vdash\; C \quad \Gamma, A_2 \;\vdash\; C}{\Gamma, A_1 \vee A_2 \;\vdash\; C} \qquad \vee* \quad \frac{\Gamma \;\vdash\; A_i}{\Gamma \;\vdash\; A_1 \vee A_2}$$

$$*\supset \quad \frac{\Gamma \;\vdash\; A \quad \Gamma, B \;\vdash\; C}{\Gamma, A \supset B \;\vdash\; C} \qquad \supset* \quad \frac{\Gamma, A \;\vdash\; B}{\Gamma \;\vdash\; A \supset B}$$

$$*\forall \quad \frac{\Gamma, A(t) \;\vdash\; C}{\Gamma, (\forall x)A(x) \;\vdash\; C} \qquad \forall* \quad \frac{\Gamma \;\vdash\; A(c)}{\Gamma \;\vdash\; (\forall x)A(x)}$$

$$*\exists \quad \frac{\Gamma, A(c) \;\vdash\; C}{\Gamma, (\exists x)A(x) \;\vdash\; C} \qquad \exists* \quad \frac{\Gamma \;\vdash\; A(t)}{\Gamma \;\vdash\; (\exists x)A(x)}$$

Here, in rules $*\wedge$ and $\vee*$, $i = 1$ or $i = 2$; in rules $*\forall$ and $\exists*$, t is any term; and in rules $\forall*$ and $*\exists$, c is a variable which does not occur free in Γ (or C).

The system LJ is obtained from LM by adding the rule

$$\perp J \quad \frac{\Gamma \;\vdash\; \perp}{\Gamma \;\vdash\; A}$$

A *cut-free derivation* is a derivation in which rule Cut does not occur.
The Gentzen-style L-rule corresponding to Ds is

$$Dx \quad \frac{\Gamma, C_1(\vec{A}) \;\vdash\; C \quad \Gamma, C_2(\vec{A}) \;\vdash\; C \quad \dots \quad \Gamma, C_n(\vec{A}) \;\vdash\; C}{\Gamma \;\vdash\; C}$$

Definition 2 The *system LI* is obtained from the system LJ by adding rule Dx.

Theorem 3 *If*

$$\Gamma \;\vdash\; A$$

is derivable in system LI, then there is a deduction of it in TI.

Proof An easy induction on the derivation of $\Gamma \;\vdash\; A$. □

Theorem 4 *If there is a normal deduction of*

$$\Gamma \vdash A$$

in TI, then there is a cut-free derivation of it in LI.

Proof Similar to the proof of Prawitz [8, Appendix A, Theorem]. An extra case for Ds is needed, where Ds is neither an I-rule nor an E-rule; this case is easy using Dx. □

Theorem 5 *If*

$$\Gamma \vdash A$$

can be derived in LI, then there is a cut-free derivation of it.

Proof Theorems 2, 3, and 4. □

Remark 2 Curry [4, p. 262] uses as his rule Nx a special case of Dx. However, he does not use the corresponding rule for his rule Px, which he takes instead in the form

$$\frac{\Gamma, A \supset B \vdash A}{\Gamma \vdash A}$$

Some Special Cases

There are properties that follow from normalization for minimal and intuitionistic logic which do not hold for TI or LI. These include the separation property (which says that only if a connective or quantifier appears in an undischarged assumption or in the conclusion are its rules used in the deduction) and the subformula property (which says that every formula occurring in a deduction occurs in one of the undischarged assumptions or in the conclusion). It is easy to see that rule Ds (or Dx) may result in the discharge (disappearance from the cut-free deduction) of a connective or quantifier or of a subformula that does not occur in another undischarged assumption or in the conclusion. Furthermore, it is easy to see that rule Dx is a special case of Cut. This suggests that the normalization procedure given so far in this paper is not complete.

However, there are some special cases in which we can get more complete results by adding new reduction steps. These are cases in which we can at least get the usual consistency result as a consequence of normalization. Now since all intermediate logics are subsystems of classical logic, we already know that they are consistent, but finding a way to obtain consistency directly from the normalization does tell us something about the normalization procedure. And we can get this result for some systems.

The usual way normalization is used to prove consistency is to note that in a completely normalized deduction, it is impossible to have a proof with no undischarged premises of an atomic formula. This is because, if the *main branch* of the deduction is defined as in [8, Chap. III, Sect. 2], then the formula at the top of the main branch cannot be discharged, because the only rules that can discharge a premise at the top of the main branch are ¬I and ⊃I. (Rules ∨E and ∃E discharge assumptions, but none of those assumptions can sit atop the main branch. Recall that in a normalized deduction, there can be no occurrence of an I-rule whose conclusion is the major premise of an E-rule. I have been writing the main branch to the left.) Since rule Ds can discharge premises above either premise, this argument will not work for TI. Indeed, the notion of a main branch has not yet been defined for deductions that include rule Ds. But if we can find a suitable way to extend the definition of main branch for some systems TI and if we can add reduction steps that remove the possibility that a discharged assumption atop that branch is the major premise for an inference by an E-rule, then consistency will follow. We can do this for some of the systems given here.

For example, consider the situation in which the first disjunct is the major premise of rule ¬E or ⊃E and the second disjunct is the minor premise. This is the case for the examples (M), (K), and (C).

Let us use (M) as an example. Let us extend the definition of main branch to include a branch headed by a premise $\neg\neg A$ that is discharged by rule Ds. Then, if such an occurrence of $\neg\neg A$ is the major premise for an inference by ¬E, the deduction must have the form

$$
\cfrac{
\cfrac{
\overset{1}{[\neg\neg A]} \quad \overset{\mathcal{D}_1}{\neg A}
}{\bot,} \neg E
\qquad
\begin{array}{c}
\overset{1}{[\neg\neg A]} \\ \mathcal{D}_2 \\ C
\end{array}
\mathcal{D}_3
\qquad
\begin{array}{c}
\overset{2}{[\neg A]} \\ \mathcal{D}_4 \\ C
\end{array}
}{\underset{\mathcal{D}_5}{C}} \; Ds - 1 - 2
$$

Let us extend the definition of *maximum formula* of [8, Chap. II, Sect. 1] to include the indicated occurrence of $\neg\neg A$. This deduction can be reduced to

$$
\begin{array}{c}
\mathcal{D}_1 \\ \neg A \\ \mathcal{D}_4 \\ C \\ \mathcal{D}_5
\end{array}
$$

Let us call this kind of reduction step a *Ds-main-reduction step*.

Theorem 6 *The system TI with (M) with Ds-main-reduction steps can be normalized.*

Proof Follow the order of reduction steps specified by Prawitz [8, Chap. III, Sect. 1 Theorem 2] with the extended definition of maximum formula indicated above.   □

Corollary 1 *The consistency of TI with (M) follows from the normalization theorem.*

Proof Suppose there were a deduction of $\vdash \perp$ with no undischarged assumptions. Since the conclusion is atomic, there is no I-rule at the bottom of the main branch. If the formula at the top of the main branch were discharged by rule Ds anywhere in that branch, there would be a Ds-main-reduction step that could still be carried out. Hence, the formula at the top of the main branch cannot be discharged in the deduction, contrary to hypothesis.   □

The proofs for TI with (K) and (C) are similar.

If we are adding Ds to TJ, we get a new consistency proof for logic (M) or classical logic. If we add it to TM, we get a new consistency proof for TD and TE.

But there are other systems we can treat similarly, for nowhere in the proof of Theorem 6 or Corollary 1 have we used the fact that there are only two disjuncts. The proof will work if there are other premises for rule Ds, as long as the discharged assumption for one of them is the major premise for ¬E or ⊃E and at least one of the other discharged assumptions is the minor premise for the same inference. This is the case for (R_n) for $n \geq 2$. The rule Ds here takes the form

$$\frac{[A_1]\ [A_1 \supset A_2]\ [A_2 \supset A_3] \qquad [A_{n-1} \supset A_n]\ [\neg A_n]}{C \qquad C \qquad C \quad \ldots \quad C \qquad C}{C}$$

Here we designate the main branch to lead to the discharged assumption $[A_1 \supset A_2]$. Then the discharged assumption $[A_1]$ is the minor premise if $[A_1 \supset A_2]$ is the major premise for an inference by ⊃E, and the above argument will lead to a similar proof of consistency.

Acknowledgements   This work was supported in part by a grant from the Natural Sciences and Engineering Research Council of Canada. The first three sections of the paper were presented at the 39th Annual Meeting of the Western Canadian Philosophical Association, Calgary, 25–27 October, 2002. The paper with the first part of the fourth section (the part without the proofs of Theorem 6 and its corollary) was presented to a seminar at St. Andrews University in Scotland on 3 September, 2013, hosted by Dr. Roy Dyckhoff.

References

1. A. Avron, The method of hypersequents in the proof theory of propositional non-classical logics, in *Logic: Foundations to Applications*, eds. by W. Hodges, M. Hyland, C. Steinhorn, J. Truss (Oxford Scientific, Oxford, 1996), pp. 1–32
2. P. Bernays, Review of H.B. Curry. The system LD. J. Symb. Log. **17**(1), 35–42 (1952); On the definition of negation by a fixed proposition in the inferential calculus. J. Symb. Log. **17**, 98–104 (1952); J. Symb. Log. **18**, 266–268 (1953)
3. H.B. Curry, The system LD. J. Symb. Log. **17**(1), 35–42 (1952)
4. H.B. Curry, *Foundations of Mathematical Logic* (McGraw-Hill, New York/San Francisco/Toronto/London, 1963). Reprinted by Dover, 1977 and 1984
5. S. Kanger, A note on partial postulate sets for propositional logic. Theoria (Sweden) **21**, 99–104 (1955)
6. S.A. Kripke, The system LE. Westinghouse Talent Search, February 1958. Quoted from [4, p. 381] (submitted, 1958)
7. E.G.K. López-Escobar. Implicational logics in natural deduction systems. J. Symb. Log. **47**(1), 184–186 (1982)

8. D. Prawitz, *Natural Deduction* (Almqvist & Wiksell, Stockholm/Göteborg/Uppsala, 1965). Reprinted by Dover, 2006
9. T. Umezawa. On logics intermediate between intuitionistic and classical predicate logic. J. Symb. Log. **24**(2), 141–153 (1959)

J.P. Seldin (✉)

Department of Mathematics and Computer Science, University of Lethbridge, Lethbridge, AB, Canada

e-mail: jonathan.seldin@uleth.ca; http://www.cs.uleth.ca/~seldin

A Natural Axiom System for Boolean Algebras with Applications

R.E. Hodel

Abstract We use an equivalent form of the Boolean Prime Ideal Theorem to give a proof of the Stone Representation Theorem for Boolean algebras. This proof gives rise to a natural list of axioms for Boolean algebras and also for propositional logic. Applications of the axiom system are also given.

Keywords Boolean algebras · Boolean rings · Lindenbaum algebras · Propositional logic · Regular open algebras · Stone's representation theorem

AMS Classifications: 06E05 · 03B05

Introduction

Let $\langle B, \wedge, ', 0 \rangle$ be an algebraic structure; in detail, B is a set with more than one element, $\wedge$ is a 2-operation on B, $'$ is a 1-ary operation on B, and $0 \in B$. Our goal is to isolate a list of axioms for B that suffice to prove Stone's Representation Theorem for Boolean algebras. This has already been done by Frink (see [2, 8]); however, our axioms are somewhat different and seem to have certain advantages. The difference is due to the fact that his proof uses AC, but it is well known (see [1, 3]) that a weaker axiom suffices, namely the Boolean Prime Ideal Theorem. Our axioms reflect the use of this weaker axiom. In more detail: AC is equivalent to the Tukey-Teichmüller Theorem (TT); in a recent paper [4] the author proved that there is a restricted version of (TT), denoted (rTT), that is equivalent to the Boolean Prime Ideal Theorem. Our proof uses (rTT), which can be stated as follows:

(rTT) Let X be a set, let $'$ be a 1-ary operation on X, and let $\mathcal{A}$ be a non-empty collection of subsets of X such that for all $A \subseteq X$,

1. $A \in \mathcal{A}$ if and only if every finite subset of A is in $\mathcal{A}$ ($\mathcal{A}$ has *finite character*);
2. if $A \in \mathcal{A}$ and $x \in X$, then $A \cup \{x\} \in \mathcal{A}$ or $A \cup \{x'\} \in \mathcal{A}$ ($\mathcal{A}$ has the *extension property*).

Then for all $A \in \mathcal{A}$, there exists $U \in \mathcal{A}$ such that $A \subseteq U$ and for all $x \in X$, $x \in U$ or $x' \in U$.

We will also give applications of our axiom system for Boolean algebras. In particular, we obtain a natural list of axioms and rules of inference for propositional logic. We also show that the axioms are tailor-made for easy proofs of important classical results, namely

© Springer International Publishing Switzerland 2016   249
F.F. Abeles, M.E. Fuller (eds.), *Modern Logic 1850-1950, East and West*, Studies in Universal Logic, DOI 10.1007/978-3-319-24756-4_13

that each of the following is a Boolean algebra: every Boolean ring, propositional logic (modulo an equivalence relation), the collection of regular open sets of a topological space.

Proof of Stone's Theorem

Stone's Theorem asserts that every Boolean algebra B is isomorphic to a field of sets. More precisely, there is a set X and a non-empty collection $\mathcal{A}$ of subsets of X such that B is isomorphic to $\mathcal{A}$, where $\mathcal{A}$ satisfies the following two conditions: for all $A, B \in \mathcal{A}$, $A^c \in \mathcal{A}$ ($A^c = \{x : x \in X \text{ and } x \notin A\}$) and $A \cap B \in \mathcal{A}$ (the axioms for a field of sets).

We take the following point of view: Given the algebraic structure $\langle B, \wedge, ', 0 \rangle$, what axioms are required to show that B is isomorphic to a field of sets? Any such choice of axioms will automatically make B a Boolean algebra, where $a \vee b$ is defined by $(a' \wedge b')'$ and 1 is defined by $0'$. Here are the required axioms (all $a, b \in B$):

(B1) $\wedge$ is commutative and associative;
(B2) $a \wedge a' = 0$ and $a \wedge 0 = 0$;
(B3) if $a \wedge x = 0$ and $a \wedge x' = 0$ for some $x \in B$, then $a = 0$;
(B4) if $a \wedge b' = 0$ and $a' \wedge b = 0$, then $a = b$.

Throughout this section we assume that $\langle B, \wedge, ', 0 \rangle$ is an algebraic structure that satisfies (B1)–(B4). Axioms (B1) and (B2) are well-known properties of a Boolean algebra, and it is easy to check that every Boolean algebra satisfies (B3) and (B4). Note that (B3) and (B4) give us a strategy for showing $a = 0$ or $a = b$. To illustrate:

Lemma 1 *The following hold for all $a \in B$:*

(a) $a \wedge a = a$ ($\wedge$ is idempotent);
(b) $a \neq a'$.

Proof For (a), we use (B4) and show that $(a \wedge a) \wedge a' = 0$ and $(a \wedge a)' \wedge a = 0$. For the first equation, use (B2). For the second equation, use (B3) with $x = a$. To prove (b), suppose there exists $a \in B$ such that $a = a'$. Then $a = 0$ as follows: $a = a \wedge a = a \wedge a' = 0$. Now let $b \in B$ be arbitrary; we use (B3) with $x = a$ to show that $b = 0$, a contradiction of $|B| > 1$. We have: $b \wedge a = b \wedge 0 = 0$ and $b \wedge a' = b \wedge 0 = 0$. $\quad\square$

Definition 1 Let U be a non-empty subset of B. Then

(1) U has the *finite meet property* if $a_1 \wedge \cdots \wedge a_n \neq 0$ for every finite subset $\{a_1, \cdots, a_n\}$ of U;
(2) U is an *ultrafilter* if U has the finite meet property and for all $a \in B$, $a \in U$ or $a' \in U$ (but not both).

Lemma 2 *Let U be an ultrafilter on B. For all $a, b \in B$, $a \wedge b \in U$ if and only if $a \in U$ and $b \in U$.*

Proof First assume $a \wedge b \in U$, but suppose $a \notin U$. By property (2) of an ultrafilter, $a' \in U$ and we have: $(a \wedge b) \wedge a' = a \wedge a' \wedge b = 0$, a contradiction of the finite meet property. For the other direction, assume $a \in U$ and $b \in U$ but $(a \wedge b) \notin U$. Then $(a \wedge b)' \in U$ and we have $(a \wedge b)' \wedge (a \wedge b) = 0$, again a contradiction. □

The finite meet property obviously has finite character, but in addition it also has the extension property (here is where axiom (B3) is critical).

Lemma 3 (Extension Property) *Let A be a subset of B that has the finite meet property and let $x \in B$. Then $A \cup \{x\}$ or $A \cup \{x'\}$ has the finite meet property.*

Proof We consider the case $A \neq \varnothing$ (see Lemma 1(b) for the case $A = \varnothing$). If not, there exist $a_1, \cdots, a_n, b_1, \cdots, b_k \in A$ such that $a_1 \wedge \cdots \wedge a_n \wedge x = 0$ and $b_1 \wedge \cdots \wedge b_k \wedge x' = 0$. It follows that $z \wedge x = 0$ and $z \wedge x' = 0$, where

$$z = a_1 \wedge \cdots \wedge a_n \wedge b_1 \wedge \cdots \wedge b_k.$$

Since $z \wedge x = 0$ and $z \wedge x' = 0$, (B3) applies and we obtain $z = 0$, a contradiction of the assumption that A has the finite meet property. □

Corollary 1 *Let A be a subset of B that has the finite meet property. Then there is an ultrafilter U such that $A \subseteq U$.*

Proof Let $\mathcal{A} = \{C: C \subseteq B$ and C has the finite meet property$\}$. Clearly $A \in \mathcal{A}$ and $\mathcal{A}$ has finite character; by Lemma 3, $\mathcal{A}$ also has the extension property. Now apply (rTT) to obtain an ultrafilter U such that $A \subseteq U$. □

We now turn to the proof of Stone's Theorem. The following sets and notation are used. Given $\langle B, \wedge, ', 0 \rangle$, let

$$\mathbf{ULT} = \{U: U \text{ is an ultrafilter on } B\};$$

$$\mathcal{A} = \{[a]: a \in B\}, \text{ where } [a] = \{U: U \in \mathbf{ULT} \text{ and } a \in U\}.$$

Theorem 1 (Stone Representation) *Let $\langle B, \wedge, ', 0 \rangle$ be an algebraic structure that satisfies (B1)–(B4). Then B is isomorphic to a field of sets and therefore is a Boolean algebra, where $a \vee b = (a' \wedge b')'$ and $1 = 0'$.*

Proof First note that $[a] \subseteq \mathbf{ULT}$ for each $a \in B$ and thus $\mathcal{A}$ is a collection of subsets of $\mathbf{ULT}$. Moreover, for all $a, b \in B$,

(1) $[a] \cap [b] = [a \wedge b]$ (Lemma 2);
(2) $\mathbf{ULT} - [a] = [a']$ (property (2) of an ultrafilter);
(3) $[0] = \varnothing$ (property (1) of an ultrafilter).

By (1)–(3), $\mathcal{A}$ is a field of sets and moreover the function $\Phi : B \rightarrow \mathcal{A}$, defined by $\Phi(a) = [a]$, satisfies the following properties and thus is a homomorphism from B onto $\mathcal{A}$:

(1) $\Phi(a \wedge b) = \Phi(a) \cap \Phi(b)$;
(2) $\Phi(a') = \Phi(a)^c$;
(3) $\Phi(0) = \varnothing$.

It remains to prove that Φ is one-to-one. Let $a \neq b$; by (B4), $a \wedge b' \neq 0$ or $a' \wedge b \neq 0$, say $a \wedge b' \neq 0$. Then $\{a, b'\}$ has the finite meet property and therefore by Corollary 1 there is an ultrafilter U such that $\{a, b'\} \subseteq U$. We now have $a \in U$ and $b \notin U$, therefore $[a] \neq [b]$ and so $\Phi(a) \neq \Phi(b)$ as required. $\qquad\qquad\qquad\qquad\qquad\qquad\qquad\qquad\qquad\square$

Constructive Proof

Let $\langle B, \wedge, {}', 0\rangle$ be an algebraic structure that satisfies the axioms (B1)–(B4). We now give a constructive proof that B is a Boolean algebra, where $\vee$ and 1 are defined by $a \vee b = (a' \wedge b')'$ and $1 = 0'$ respectively.

Lemma 4 *The following hold for all $a, b \in B$:*

(a) $a'' = a$;
(b) $a \vee a' = 1$;

Proof For (a), use (B4) and prove that $a'' \wedge a' = 0$ and $a''' \wedge a = 0$; for the second equation, use (B3) with $x = a'$. For (b): $a \vee a' = (a' \wedge a'')' = 0' = 1$. $\qquad\square$

Lemma 5 (Associative, Commutative, and Idempotent Laws for $\vee$) *The following hold for all $a, b, c \in B$:*

(a) $a \vee (b \vee c) = (a \vee b) \vee c$;
(b) $a \vee b = b \vee a$;
(c) $a \vee a = a$.

Proof For (a): use the definition $a \vee b = (a' \wedge b')'$ and the associativity of $\wedge$. $\qquad\square$

Lemma 6 (Absorption Laws) *The following hold for all $a, b \in B$:*

(a) $a \wedge (a \vee b) = a$;
(b) $a \vee (a \wedge b) = a$.

Proof For (a): use (B4) and show $[a \wedge (a \vee b)] \wedge a' = 0$ and $[a \wedge (a \vee b)]' \wedge a = 0$. For the second equation, use (B3) with $x = a \vee b$. For (b):
$$a \vee (a \wedge b) = [a' \wedge (a \wedge b)']' = [a' \wedge (a' \vee b')]' = a'' = a.\qquad\square$$

Lemma 7 *If $a \wedge x = 0$ and $a \wedge y = 0$, then $a \wedge (x \vee y) = 0$.*

Proof By (B3), it suffices to prove $[a \wedge (x \vee y)] \wedge x = 0$ and $[a \wedge (x \vee y)] \wedge x' = 0$. Use the hypothesis $a \wedge x = 0$ for the first and use (B3) with y for the second.   □

Lemma 8 (Distributive Laws) *The following hold for all $a, b, c \in B$:*

(a) $a \wedge (b \vee c) = (a \wedge b) \vee (a \wedge c)$;
(b) $a \vee (b \wedge c) = (a \vee b) \wedge (a \vee c)$.

Proof It suffices to prove (a). To do this, use (B4) and show that

(1) $[(a \wedge b) \vee (a \wedge c)]' \wedge [a \wedge (b \vee c)] = 0$;
(2) $[(a \wedge b) \vee (a \wedge c)] \wedge [a \wedge (b \vee c)]' = 0$.
   Now (1) can be written as $[(a \wedge b)' \wedge (a \wedge c)' \wedge a] \wedge (b \vee c) = 0$, and this follows easily from Lemma 7. To prove (2), it suffices by Lemma 7 to prove
(3) $[a \wedge (b \vee c)]' \wedge (a \wedge b) = 0$;
(4) $[a \wedge (b \vee c)]' \wedge (a \wedge c) = 0$.
   To prove both (3) and (4), use (B3) with $x = b \vee c$.   □

As noted earlier, O. Frink used Zorn's Lemma and a proof of Stone's Theorem to find axioms for $\langle B, \wedge, ', 0 \rangle$. They are: $\wedge$ is associative, commutative, and idempotent; $a \wedge b' = 0$ if and only if $a \wedge b = a$. For a discussion of Frink's axioms and a constructive proof, see [8]. It is an interesting exercise to show that $\langle B, \vee, ', 1 \rangle$ (with the dual versions of (B1)–(B4)) is equivalent to axiomizations due to Lewis-Langford and to Malliah (see p. 87, 89 of [8]).

Applications

As a first application, we construct a deductive system for propositional logic based on negation and disjunction. A nice feature of this axiomization is an easy proof of the Deduction Theorem, a fundamental result of proof theory. We will work with the *dual* algebraic structure $\langle B, \vee, ', 1 \rangle$ with axioms

(B1) $\vee$ is commutative and associative;
(B2) $a \vee a' = 1$ and $a \vee 1 = 1$;
(B3) if $a \vee x = 1$ and $a \vee x' = 1$ for some $x \in B$, then $a = 1$;
(B4) if $a \vee b' = 1$ and $a' \vee b = 1$, then $a = b$.

A deductive system (or formal system) **F** has three main components: the set FOR of formulas of the system; a subset of FOR that serves as the axioms of the system; a finite list of rules of inference, each of which states that under certain circumstances, conclusion B follows from the finite set of hypotheses $A_1, \cdots, A_n$. Within this framework, the notion of a formal proof can be defined: for $A \in$ FOR and $\Gamma \subseteq$ FOR, a *proof in* **F** *of A using* Γ is a finite sequence $A_1, \cdots, A_n$ of formulas with $A_n = A$ and such that for $1 \leq k \leq n$, one of the following holds: A_k is an axiom of the formal system; $A_k \in \Gamma$; $k > 1$ and A_k is the conclusion of a rule of inference of **F** whose hypotheses are among $A_1, \cdots, A_{k-1}$. In this case we also say that A is a *theorem* of Γ and denote this by $\Gamma \vdash_{\mathbf{F}} A$.

We now construct a formal system **P** for propositional logic. Let PROP be an infinite set of propositional variables; the set FOR of formulas is then defined inductively as follows:

each $p \in$ PROP is a formula; if A and B are formulas, so are $\neg A$ and $(A \vee B)$; every formula is obtained by a finite number of applications of these two rules. To improve readability, we omit the outermost parentheses and occasionally use brackets [ ] instead of parentheses. We use A, B, C to denote formulas and Γ to denote a set of formulas. The axioms and rules of inference of **P** are as follows:

Axioms for P   There are three axiom schemes (A, B, C denote arbitrary formulas):

(**EM**) $\neg A \vee A$ (law of excluded middle);
(**ASSOC**) $\neg[(A \vee B) \vee C] \vee [A \vee (B \vee C)]$ (associative axiom);
(**CM**) $\neg(A \vee B) \vee (B \vee A)$ (commutative axiom).

Rules for P   There are two rules of inference:

$$(\textbf{EXP}) \qquad \frac{A}{A \vee B} \qquad\qquad (\textbf{CUT}) \qquad \frac{A \vee B, \neg A \vee B}{B}$$

We can summarize this choice of axioms and rules of inference as follows: the axioms (ASSOC) and (CM) come from (B1); the axiom (EM) and the rule (EXP) come from (B2) (where the axiom $1 \vee b = 1$ is interpreted as: if $a = 1$, then $a \vee b = 1$), and the rule (CUT) comes from (B3). The axiom (B4) has no direct role, but instead motivates the choice of an equivalence relation $\sim_\Gamma$ so that FOR/$\sim_\Gamma$ is a Boolean algebra, namely

$$A \sim_\Gamma B \text{ if and only if } \Gamma \vdash_\textbf{P} \neg A \vee B \text{ and } \Gamma \vdash_\textbf{P} \neg B \vee A.$$

We now turn to a proof of the Deduction Theorem. First of all, from these axioms and rules of inference we can immediately obtain three *derived* rules of inference (new rules that can be used as a step in a formal proof):

$$(\textbf{DS}) \frac{\neg A \vee B, A}{B} \quad \frac{A \vee B, \neg A}{B} \quad (\textbf{ASSOC}) \frac{(A \vee B) \vee C}{A \vee (B \vee C)} \quad (\textbf{CM}) \frac{A \vee B}{B \vee A}.$$

Here is a derivation of the first version of (DS) = disjunctive syllogism; the two rules (ASSOC) and (CM) follow immediately from the corresponding axiom and (DS).

(1) $\neg A \vee B$ Hypothesis
(2) A Hypothesis
(3) $A \vee B$ EXP RULE(2)
(4) B CUT RULE(1,3)

Deduction Theorem for P   *If* $\Gamma \cup \{A\} \vdash_\textbf{P} B$, *then* $\Gamma \vdash_\textbf{P} \neg A \vee B$.

Proof   We first prove a special case: if $B = A$, or B is an axiom of **P**, or $B \in \Gamma$, then $\Gamma \vdash_\textbf{P} \neg A \vee B$. In the case $B = A$, we are asked to show $\Gamma \vdash_\textbf{P} \neg A \vee A$, an axiom of **P**. Suppose B is an axiom or $B \in \Gamma$. The following verifies $\Gamma \vdash_\textbf{P} \neg A \vee B$:

(1) B Axiom of **P** or a formula in Γ
(2) $B \vee \neg A$ EXP RULE
(3) $\neg A \vee B$ CM RULE
    The general case is by induction on the number of steps in the proof of B using $\Gamma \cup \{A\}$ (B varies, A does not). For a proof of B using $\Gamma \cup \{A\}$ in one step, the special case applies.

Now consider a proof of B using $\Gamma \cup \{A\}$ in n+1 steps, and assume the result holds for proofs of formulas using $\Gamma \cup \{A\}$ in at most n steps. If $B = A$, or B is an axiom, or $B \in \Gamma$, the special case applies. So we assume that B is obtained by a rule of inference. There are two possibilities; we consider the case in which the rule is (CUT) and leave (EXP) to the reader. If (CUT) is used, the proof of B using $\Gamma \cup \{A\}$ in $n+1$ steps looks something like this:

$$A_1, \cdots, \boxed{C \vee B}, \cdots, \boxed{\neg C \vee B}, \cdots, \boxed{B}.$$

Apply the induction hypothesis to the two formulas $C \vee B$ and $\neg C \vee B$ to obtain $\Gamma \vdash_{\mathbf{P}} \neg A \vee (C \vee B)$ and $\Gamma \vdash_{\mathbf{P}} \neg A \vee (\neg C \vee B)$. Here is the required verification of $\Gamma \vdash_{\mathbf{P}} \neg A \vee B$ (a theorem of Γ can be used as a line in a proof using Γ):

(1) $\neg A \vee (C \vee B)$ Theorem of Γ
(2) $\neg A \vee (\neg C \vee B)$ Theorem of Γ
(3) $(C \vee B) \vee \neg A$ CM RULE(1)
(4) $(\neg C \vee B) \vee \neg A$ CM RULE(2)
(5) $C \vee (B \vee \neg A)$ ASSOC RULE(3)
(6) $\neg C \vee (B \vee \neg A)$ ASSOC RULE(4)
(7) $B \vee \neg A$ CUT RULE
(8) $\neg A \vee B$ CM RULE

$\square$

Lindenbaum-Tarski Algebras The goal here is to show that propositional logic (modulo an equivalence relation) is a Boolean algebra. We will start with the formal system **P** and construct a Boolean algebra $\langle$ FOR$/\sim_\Gamma$, $\vee$, $'$, $1 \rangle$ for each consistent $\Gamma \subseteq$ FOR. To begin, define a relation $\sim_\Gamma$ on FOR as follows: for $A, B \in$ FOR:

$$A \sim_\Gamma B \Leftrightarrow \Gamma \vdash_{\mathbf{P}} \neg A \vee B \text{ and } \Gamma \vdash_{\mathbf{P}} \neg B \vee A$$

$$\Leftrightarrow \Gamma \cup \{A\} \vdash_{\mathbf{P}} B \text{ and } \Gamma \cup \{B\} \vdash_{\mathbf{P}} A \text{ [Deduction Theorem]}.$$

Let FOR$/\sim_\Gamma = \{[A] : A \in$ FOR$\}$, where $[A] = \{B: B \in$ FOR and $A \sim_\Gamma B\}$. The operations $\vee$ and $'$ on FOR$/\sim_\Gamma$ and the constant 1 are defined by

$$[A] \vee [B] = [A \vee B], [A]' = [\neg A], 1 = [\neg p \vee p].$$

The proof that $\sim_\Gamma$ is an equivalence relation and that the definitions of $\vee$ and $'$ are independent of the choice of representative elements requires some work. However, the Deduction Theorem and several derived rules of inference help, for example a stronger version of (CUT):

(STRONGCUT) $\dfrac{A \vee B, \neg A \vee C}{B \vee C}$

The following equivalence holds for every formula A: $[A] = 1 \Leftrightarrow \Gamma \vdash A$. Thus, to verify that $\langle \text{FOR}/\!\sim_\Gamma, \vee, ', 1 \rangle$ satisfies (B3) and (B4), we must show:

(B3) if $\Gamma \vdash_P A \vee B$ and $\Gamma \vdash_P A \vee \neg B$, then $\Gamma \vdash_P A$ (use (CUT));
(B4) if $\Gamma \vdash_P A \vee \neg B$ and $\Gamma \vdash_P \neg A \vee B$, then $[A] = [B]$ (definition of $\sim_\Gamma$).

Construction of a Boolean Algebra from a Boolean Ring There is an interesting back-and-forth relationship between Boolean algebras and Boolean rings (see [10]). We sketch the part that constructs a Boolean algebra from a given Boolean ring. Let $\langle B, +, \times, 0, 1 \rangle$ be a Boolean ring; that is, a ring with multiplicative identity 1 such that $a \times a = a$ for all $a \in B$. It is a well-known result that the following hold in any Boolean ring: $a + a = 0$ and $a \times b = b \times a$ (all $a, b \in B$). Let $\langle B, \wedge, ', 0 \rangle$ be defined by

$$a \wedge b = a \times b, a' = 1 + a.$$

To check that $\langle B, \wedge, ', 0 \rangle$ satisfies (B2)–(B4), it suffices to show the following:

(B2) $a \times (1 + a) = 0$ and $a \times 0 = 0$;
(B3) if $a \times x = 0$ and $a \times (1 + x) = 0$, then $a = 0$;
(B4) if $a \times (1 + b) = 0$ and $(a + 1) \times b = 0$, then $a = b$.

Also note that $a \vee b = (a' \wedge b')' = [(a+1) \times (b+1)] + 1 = a + b + (a \times b)$.

Regular Open Algebra Recall the following notation: for a subset V of a topological space X, V^- denotes the topological closure of V and V° the topological interior of V. A subset U of X is *regular open* if $U^{-\circ} = U$; note that $U \subseteq U^{-\circ}$ automatically holds for any open set U. If A is any subset of X, then $A^{-\circ} = A^{-\circ-\circ}$ and therefore $A^{-\circ}$ is a regular open set. Let

$$RO(X) = \{U : U \text{ is a regular open subset of } X\}.$$

Our goal is to prove that $\langle RO(X), \cap, ', \varnothing \rangle$ is a Boolean algebra, where $'$ is defined by $V' = X - V^-$. For this we need two Lemmas.

Lemma 9 *Let V be an open set and let $V' = X - V^-$. Then*

(a) $V' = (X - V)^\circ$;
(b) V' is a regular open set.

Proof of (a): Clearly $(X - V^-) \subseteq (X - V)$, and since $(X - V^-)$ is an open set, $(X - V^-) \subseteq (X - V)^\circ$. For the opposite direction: $(X - V)^\circ \cap V = \varnothing$, and since $(X - V)^\circ$ is an open set, $(X - V)^\circ \cap V^- = \varnothing$. Thus $(X - V)^\circ \subseteq (X - V^-)$ as required.
Proof of (b): We need to prove that $(X - V^-)^{-\circ} \subseteq (X - V^-)$, and by (a), it suffices to show that $(X - V^-)^{-\circ} \subseteq (X - V)^\circ$. We have: $(X - V^-)^- \subseteq (X - V)^- = (X - V)$ and therefore $(X - V^-)^{-\circ} \subseteq (X - V)^\circ$. □

Lemma 10 *Let U and V be regular open sets. Then*

(a) $U \cap V$ is a regular open set;
(b) if $U \cap V' = \varnothing$, then $U \subseteq V$.

Proof For (a): $(U \cap V)^{-\circ} \subseteq U^{-\circ} = U$ and likewise $(U \cap V)^{-\circ} \subseteq V$. To prove (b): by hypothesis, $U \subseteq V^{-}$ and therefore $U \subseteq V^{-\circ} = V$.   □

Theorem 2 $\langle \mathrm{RO}(X), \cap, ', \varnothing \rangle$ *is a Boolean algebra.*

Proof To verify the axioms (B3) and (B4), it suffices to check the following:

(B3) if $U \cap V = \varnothing$ and $U \cap V' = \varnothing$, then $U = \varnothing$ [use part (b) of Lemma 10];
(B4) if $U \cap V' = \varnothing$ and $U' \cap V = \varnothing$, then $U = V$ [use part (b) of Lemma 10].   □

In summary, $\langle \mathrm{RO}(X), \cap, ', \varnothing \rangle$ is a Boolean algebra. Now let $U, V \in \mathrm{RO}(X)$, and let us describe $U \vee V$ as a regular open set. By definition, $U \vee V = (U' \cap V')'$ and we have:

$$U \vee V = (U' \cap V')' = [X - (U' \cap V')]^{\circ} = [(X - U') \cup (X - V')]^{\circ}$$

$$= (U^{-} \cup V^{-})^{\circ} = (U \cup V)^{-\circ}.$$

The Boolean Algebra P(ω)/fin Here fin is the collection of all finite subsets of ω. Define an equivalence relation $\sim$ on P(ω) by

$$A \sim B \Leftrightarrow A - B \text{ and } B - A \text{ are both finite.}$$

Let P(ω)/fin $= \{[A]: A \subseteq \omega\}$, the collection of all equivalence classes for $\sim$. Then $\langle$ P(ω)/fin, $\wedge, ', [\varnothing] \rangle$ is a Boolean algebra, where $\wedge, '$, and $[\varnothing]$ are defined by

$$[A] \wedge [B] = [A \cap B];$$

$$[A]' = [A^{c}];$$

$$[\varnothing] = \{A : A \subseteq \omega \text{ and } A \text{ is finite}\}.$$

The verification of the axioms (B3) and (B4) reduces to checking the following:

(B3) if $A \cap B$ and $A \cap B^{c}$ are both finite, then A is finite;
(B4) if $A \cap B^{c}$ and $A^{c} \cap B$ are both finite, then $A - B$ and $B - A$ are both finite.

References

1. S. Feferman, Some applications of the notion of forcing and generic sets. Fund. Math. **56**, 325–345 (1965)
2. O. Frink, Jr., Representations of Boolean algebras. Bull. Am. Math. Soc. **47**, 755–756 (1941)
3. J.D. Halpern, A. Lévy, The Boolean prime ideal theorem does not imply the axiom of choice, in *Axiomatic Set Theory*. Proceedings of Symposia in Pure Mathematics, vol. 13(part 1) (American Mathematical Society, Providence, 1971), pp. 83–134
4. R. Hodel, Restricted versions of the Tukey-Teichmüller Theorem that are equivalent to the Boolean prime ideal theorem. Arch. Math. Logic **44**, 459–472 (2005)
5. P. Howard, J. Rubin, *Consequences of the Axiom of Choice*. Mathematical Surveys and Monographs, vol. 59 (American Mathematical Society, Providence, 1998)
6. T. Jech, *The Axiom of Choice* (North Holland, Amsterdam, 1973)

7. S. Koppelberg, *Handbook of Boolean Algebras*, vol. 1, eds. by D. Monk, R. Bonnet (North Holland, Amsterdam, 1989)
8. R. Padmanabhan, S. Rudeanu, *Axioms for Lattices and Boolean Algebras* (World Scientific, Hackensack, 2008)
9. H. Rubin, J. Rubin, *Equivalents of the Axiom of Choice, II* (North-Holland, Amsterdam, 1985)
10. M. Stone, The theory of representations for Boolean algebras. Trans. Am. Math. Soc. **40**, 37–111 (1936)
11. A. Tarski, On some fundamental concepts of metamathematics, in *Logic, Semantics, Metamathematics* (Oxford University Press, Oxford, 1956)
12. O. Teichmüller, *Braucht der Algebraiker das Auswahlaxiom?* Deutsche Math. **4**(1939), 567–577
13. J. Tukey, *Convergence and Uniformity in Topology*. Annals of Mathematics Studies, vol. 2 (Princeton University Press, Princeton, 1940)

R.E. Hodel (✉)

Department of Mathematics, Duke University, Durham, NC 27708, USA

e-mail: hodel@math.duke.edu